EXPRESSION AND INTERPRETATION OF NEGATION

AN OT TYPOLOGY

Studies in Natural Language and Linguistic Theory

VOLUME 77

For other titles published in this series, go to
www.springer.com/series/6559

EXPRESSION AND INTERPRETATION OF NEGATION

AN OT TYPOLOGY

HENRIËTTE DE SWART

Utrecht University, The Netherlands

Springer

Henriëtte de Swart
Utrecht University
The Netherlands
h.deswart@uu.nl

ISBN 978-90-481-3161-7 e-ISBN 978-90-481-3162-4
DOI 10.1007/978-90-481-3162-4
Springer Dordrecht Heidelberg London New York

Library of Congress Control Number: 2009932669

Printed on acid-free paper

Springer is part of Springer Science+Business Media (www.springer.com)

Preface

This book applies recently developed tools in strong and weak bidirectional optimality theory (OT) as well as an evolutionary modeling of OT in a bidirectional setting to the empirical domain of negation across a wide range of languages. I have long been intrigued by the patterns of semantic variation we find in natural language, and negation has always been one of the topics I was fascinated by. In the past, I have proposed analyses of language-specific observations about *not...until* in English (de Swart 1996), Dutch negative polarity items (NPIs) occurring outside the c-command domain of the licensor (de Swart 1998b), the interaction of negation and aspect in French (de Swart and Molendijk 1999), scope ambiguities with negative quantifiers in Germanic (de Swart 2000), and negative concord in Romance (de Swart and Sag 2002).

Although I felt my proposals were contributing to a better understanding of the phenomena under consideration, they did not lead to an explanatory theory of cross-linguistic variation in the area of negation. Meanwhile, the discussion of semantic universals and cross-linguistic variation in meaning assumed more importance in the literature (cf. von Fintel and Matthewson 2008), which made it all the more urgent to develop such a theory. Other proposals came along in the literature, exploiting syntactic and lexical notions of variation, and making claims about universal grammar and typological generalizations. But I always took the distinction between negative concord and double negation to be semantic in nature, and I kept looking for the possibility to account for cross-linguistic variation in the grammar.

When I became acquainted with OT, I acquired a new set of tools for linguistic analysis. Furthermore, OT is embedded in a broader conception of language as part of our cognitive system and provides a new perspective on universal grammar and typological variation. It quickly occurred to me that this might be the appropriate framework to work out my ideas about semantic variation. For a while, I was struggling to make syntactic and semantic insights meet, but with the development of bidirectional OT in the project *Conflicts in Interpretation*, we obtained a new conception of the syntax-semantics interface (cf. Hendriks et al. 2009).

Many people contributed to the genesis of this book. I owe much to Frans Zwarts and Jack Hoeksema for raising my curiosity about negation while I was working in Groningen. I thank the members of the PICS working group on negation (Francis Corblin, Danièle Godard, Jacques Jayez, Lucia Tovena, and Viviane Déprez) for teaching me everything they knew about the subject in French and other Romance languages. It was a lot of fun to work out the bidirectional OT model with Petra Hendriks, Helen de Hoop, Joost Zwarts, Gerlof Bouma, and Irene Krämer, and their friendly help was crucial when I was developing the basic ideas behind this book.

Financial support for our research by the NWO-Cognition program is hereby gratefully acknowledged (grant 051-02-070 for the project *Conflicts in Interpretation*). I thank the audiences at workshops and conferences in Utrecht, Nijmegen, Georgetown, Hopkins, New York, and Berlin for helpful feedback on my presentations. My proposals were first published as de Swart (2006). The ideas I presented there are worked out in more empirical, typological, and theoretical detail in this book. Over the years, many people volunteered data and helped me make sense of them, and I would like to thank them all!

This book would never have been completed had Rudie Botha not invited me to join the Netherlands Institute for Advanced Study in the Humanities and Social Sciences (NIAS) research group "Restricted Linguistic Systems as Windows on Language Evolution" in 2005–2006. The NIAS created a wonderful environment and provided excellent support for the completion of the manuscript. I would also like to gratefully acknowledge the financial support provided by NWO grant 365-70-015 for my sabbatical year.

Two anonymous reviewers read the manuscript for Springer, and wrote extensive reports. I also got valuable feedback from the series editor Liliane Haegeman. I considered their comments and incorporated whatever I felt necessary. I hope this has led to improvements in the final version. Of course, all remaining errors are my own.

The reader is invited to discover the rich inventory of the expression and interpretation of negation in natural language throughout this book. I hope (s)he will see the range and limits of the typological variation, and appreciate how the interaction of a small number of functionally and cognitively motivated principles embedded in an optimization approach to language accounts for the observations made.

May 2009

Contents

List of Figures

List of Tables

List of Tableaux

Chapter 1
Negation in a Cross-Linguistic Perspective

Introduction and Overview Chapter 1 introduces the empirical scope of the study on the expression and interpretation of negation in natural language. Background notions on negation in logic and language are introduced, and a range of linguistic issues concerning negation at the syntax–semantics interface are discussed. Cross-linguistic variation is a major topic, in both synchronic (typology) and diachronic (language change) perspectives.

Besides expressions of propositional negation, this book analyzes the form and interpretation of indefinites in the scope of negation. This raises the issue of negative polarity and its relation to negative concord. The main facts, criteria, and proposals on this topic developed in the literature are presented. The chapter closes with an overview of the book.

Optimality theory is used in this book to account for the syntax and semantics of negation in a cross-linguistic perspective. This theoretical framework is introduced in Chapter 2.

1.1 Negation in Logic and Language

The main aim of this book is to provide an account of the patterns of negation found in natural language. The expression and interpretation of negation in natural language have long fascinated philosophers, logicians, and linguists. Horn's (1989) *A Natural History of Negation* opens with the following statement: "All human systems of communication contain a representation of negation. No animal communication system includes negative utterances, and consequently, none possesses a means for assigning truth value, for lying, for irony, or for coping with false or contradictory statements." A bit further on the first page, Horn states: "Despite the simplicity of the one-place connective of propositional logic ($\neg$p is true if and only if p is not true) and of the laws of inference in which it participates (e.g. the Law of Double Negation: from $\neg\neg$p infer p, and vice versa), the form and function of

H. de Swart, *Expression and Interpretation of Negation: An OT Typology*, Studies in Natural Language and Linguistic Theory,
DOI 10.1007/978-90-481-3162-4_1, © Springer Science+Business Media B.V. 2010

negative statements in ordinary language are far from simple and transparent. In particular, the absolute symmetry definable between affirmative and negative propositions in logic is not reflected by a comparable symmetry in language structure and language use."

The scope of this book is more modest than Horn's seminal study, but I will nevertheless attempt to work out some of the issues highlighted by Horn. The focus is on negation as a universal category of human language, with negation as the marked member of the pair <affirmation, negation>, and as the unmarked member of the pair <(single) negation, double negation>. Cross-linguistic variation in the marking and interpretation of propositional negation and negative indefinites is central to the investigation.

1.1.1 Markedness of Negation

The fact that all human languages establish a distinction between affirmative and negative statements is the starting point of my investigation. The relation with animal communication systems is investigated in de Swart (2009), where I draw implications for language genesis from my study of negation in L2 acquisition. Modern studies on animal cognition make it possible to assign a mental representation of (pre-logical) negation to certain primates. Under the view that language evolved from thought, I connect these findings to data from early L2 acquisition, and hypothesize a stepwise evolution of negation, leading up to the truth-functional operator familiar from first-order logic. These connections will not be discussed in this book, which assumes the semantics of negation as defined in first-order logic. Negation will thus be analyzed as a truth-functional operator represented by the connective $\neg$.

The fact that negation is a universal concept of human communication does not explain the asymmetry between affirmation and negation in natural language, as Horn observes. In first-order logic, the propositions p and $\neg$p have the same status, and we can go back and forth between $\neg\neg$p and p without any change in meaning. Dahl (1979: 80) states that "although the semantics of Neg is connected with quite a few intricate problems, it still seems possible to give a relatively uncontroversial characterization of Neg in semantic terms. It is thus a necessary condition for something to be called Neg that it be a means for converting a sentence S_1 into another sentence S_2 such that S_2 is true whenever S_1 is false, and vice versa."

Dahl's definition of negation as a linguistic operator operating on truth values introduces an asymmetry between affirmation and negation. His definition is inspired by the observation that in natural language, negative sentences (1b, c) typically involve expressions not present in affirmative sentences (1a). Double negation sentences multiply the markings, and have a more complex structure than plain affirmative sentences (1d).

(1) a. Colyn believes that Phil plays chess.
b. Colyn believes that Phil does not play chess.
c. Colyn does not believe that Phil plays chess.
d. Colyn does not believe that Phil does not play chess.

In first-order logic, sentences like (1a) and (1d) are expected to have the same truth conditions. Negation in (1d) is truth-functional, but comes with a special communicative effect not present in (1a). The double negation of (1d) is known as the rhetorical figure of litotes. Litotes is not particular to English. Xiao and McEnery (2008) point out that the continuations of the Chinese example (2) in (2a) and (2b) convey a different meaning.

(2) Shixiong bu xiang tomorrow leave but dad say-out-Asp
Shixiong SN want mingtian zou, keshi diedie shuo-chulai-le,
Shixiong did not want to leave the next day, but now that his dad had said so,
a. jiu gan ying
then dare agree
'he dared to agree.'
b. jiu bu gan bu ying
then SN dare SN agree
'he did not dare not to agree.'

Pragmatic accounts of litotes are found in Horn (1989, 2001), van der Wouden (1994, 1997), and Blutner (2004). Postal (2000, 2004) is also concerned with syntactic and prosodic features of double negation in English. This book focuses on the truth-functional effects of single and double negation. However, we should always be aware of the fact that special prosody and syntactic restrictions, coupled with non truth-functional aspects of meaning are an integrative part of the semantics of double negation readings like (1d) and (2b).

As far as the expression of single negation meanings is concerned, I accept Horn's generalization that all natural languages have an expression for propositional negation. In all languages, this leads to a formal contrast between affirmation (1a) and negation (1b, c). Dahl (1979) takes negation to be a universal category of natural language. Inspired by Saussure, the Prague linguistic school developed a notion of markedness to deal with such asymmetries (Jakobson 1932, 1939, 1962, 1971). In a binary opposition, the unmarked term tends to be formally less complex (often with zero realization). Greenberg (1966) has observed that negation typically receives an overt expression, while affirmation usually has zero expression. Givón (1979) argues that negative structures are syntactically more constrained than their affirmative counterparts. The question arises whether we are only dealing with a morphosyntactic asymmetry, or whether the formal asymmetry is mirrored in interpretation. A semantic asymmetry is not supported by the standard interpretation of negation in (two-valued) first-order logic. However, Horn (1989: 161 sqq) cites psycholinguistic evidence concerning the acquisition of negation in L1 acquisition, and processing difficulties with negation as suggestive evidence in favor of the semantic markedness of negation. Haspelmath (2006) takes frequency asymmetries (rarity of meanings) to be the source of structural asymmetries. In Chapter 3, I argue that the relative infrequency of negative statements as compared to their affirmative counterparts makes it possible to derive the formal markedness of negation in a bidirectional evolutionary OT model.

Markedness is a relative notion in the sense that we always talk about the marked and unmarked members of a pair. Negation is the marked member of the pair <affirmation, negation>, but the unmarked member of the pair <(single) negation, double negation>. This underlies the highly marked character of sentences like (1d) and (2b), which is further supported by the special prosody and syntactic restrictions associated with double negation (see above). The markedness of double negation plays a crucial role in the argumentation developed in Chapter 6.

1.1.2 Sentence Negation

There is little controversy about the characterization of sentences like those in (1b-d) and (2b) as negative. However, as Horn (1989: 31 sqq) reminds us, it is not always easy to draw the line between affirmative and negative sentences. Consider the pairs of examples in (3) and (4).

(3) a. Mary did not manage to secure her job.
    b. Mary failed to secure her job.
(4) a. Colyn is not happy.
    b. Colyn is unhappy.

The different forms in (3) and (4) can be truthful descriptions of the same situation with slightly different nuances of meaning. This highlights the impossibility of characterizing (extra-linguistic) situations as either positive or negative.

Even if the discussion is restricted to negative sentences (linguistic expressions) and negative meanings (semantic representations in terms of a particular formalism such as first-order logic), it is not easy to determine whether sentences like (3b) and (4b) are affirmative or negative in nature. Certain verbs contribute an inherently negative meaning. *Fail* in (3b) patterns with *deny*, *refuse*, *reject*, *dissuade*, *doubt* in this respect. Horn (1989: 522 sqq) treats inherent negation as pragmatically more complex, because it relies on propositions evoked in earlier discourse. The phenomenon of inherent negation, illustrated in (3b) is outside the scope of this study.

Klima (1964) provides some diagnostics that come in useful in the distinction between sentence negation and constituent negation relevant to (4). The (a) examples in (5) and (6) pass the test for sentential negation; the (b) sentences contain constituent negation.

(5) *either* vs. *too* tags:
    a. Mary isn't happy, and John isn't happy either.
    b. Mary is unhappy, and John is unhappy {*either/too}.
(6) positive vs. negative tag questions:
    a. It isn't possible to solve that problem, is it?
    b. It is impossible to solve that problem, {#is it/isn't it}?

Additional tests have been proposed in the literature. Horn (1989: 185) warns that the tests sometimes give conflicting results, so uncertainties remain. I will assume here that it is possible to draw the line between sentence negation (4a) and constituent negation (4b). Chapter 6 (Section 1) comes back to affixal negation like *un-* (4b), and shows that the special semantic and syntactic status of adjectives like *unhappy* explains their interaction with negation particles such as *not* and negative indefinites like *nobody* in double negation as well as negative concord languages.

Other than that, this book concentrates on sentence negation, as illustrated in (1b-d), (2b), (3a), (4a), (5a) and (6a).

1.1.3 Square of Oppositions

Since Aristotle, it is customary to distinguish types of oppositions, and Horn (1989: Chapter 1) discusses them extensively. Contrariety and contradiction both come into play in the study of negation. Contrariety is a relation between two opposites, e.g. *good* vs. *bad*. Contraries cannot both be true, but both can be false. For instance, nothing can be good and bad at the same time, along the same dimension, but something can be neither good nor bad. Contradiction is a relation between members of a pair such that it is necessary for one to be true and the other false. This phenomenon is known as the 'law of the excluded middle'. Negation and affirmation are contradictions in this sense.

The notions of contradiction and contrariety come into play in the square of oppositions for the first-order quantifiers exemplified in (7).

(7) a. All students are happy.
  b. No students are happy.
  c. Some student is happy.
  d. Not all students are happy.

The pairs $\forall/\neg\forall$ and $\exists/\neg\exists$ are contradictories, because in any state of affairs, one member of each must be true, and the other false. Propositions are opposed as contraries when both the affirmation and the denial are universal. $\forall$ and $\neg\exists$ are contraries, as indicated in Figure 1.

The contradiction between $\exists$ and $\neg\exists$ will be central to the discussion of the status of indefinites under negation (Sections 3–5), because there is no agreement on the lexical semantics of negative indefinites in the literature. In fact, all four corners of the square of oppositions in Figure 1 have been explored as the possible lexical semantic representation of negative indefinites in some analysis or other. Fortunately, there is no disagreement about the truth conditions at the sentence level. The literature agrees that propositions involving indefinites under negation are universal in nature (involving $\forall\neg$ or $\neg\exists$), as opposed to their affirmative, existential counterparts (involving $\exists$).

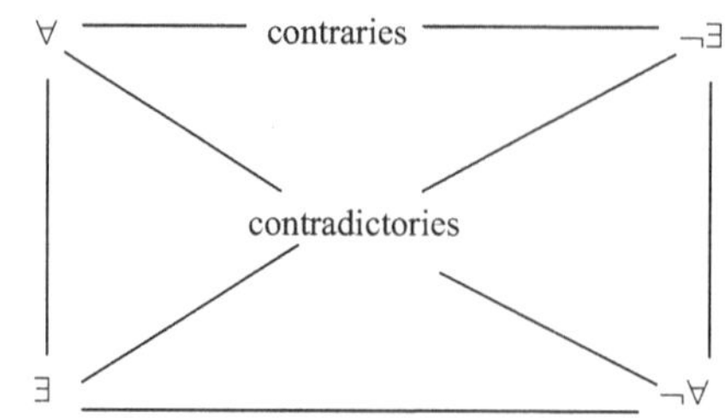

Figure 1 Square of oppositions for first-order quantifiers

1.2 Negation in Typology and Diachronic Linguistics

In English, sentence negation is realized by a negative particle (1b-d), (3a), (4a). In other languages, negative verbs express sentence negation. Payne (1985) provides examples of negative verbs (8a) and auxiliary negative verbs (8b).[1]

(8) a. Na'e 'ikai ke 'alu 'a Siale [Tongan]
ASP SN ASP go ABS Charlie
'Charlie did not go.'
b. Bi dukuwūn-ma ə-cə̄|-w duku-ra [Evenki]
I SN-PAST-1SG letter-OBJ write-PART

In (8a), the aspectual particle *na'e* bearing on the negative verb *'ikai* represents a complete and noncontinuing (simple past) action. The lexical verb *'alu* behaves like a complement clause verb. In (8b), the negative verb behaves like an auxiliary followed by the participle form of the main verb. The negative verb stem *ə-* inflects for tense and mood. Negative verbs have been understudied in linguistic theory, but see Mitchell (2006), Kaiser (2006) and Thomson (2006) for studies of negative verbs in Finno-Ugric languages, Finnish, and Bengali respectively.

Payne (1985) cites quite a few languages that use a negative verb. At the same time, he points out that the majority of natural languages use some kind of negative particle to express propositional negation. This book does not take negative verbs as in (8) into account, but focuses on negation particles and negative indefinites. Section 2 investigates the position of negation particles across languages. The study of negative indefinites is closely intertwined with the issue of negative polarity and negative concord, as worked out in Sections 3 and 4.

[1] Throughout this book, SN is used to gloss the marker of sentential negation, in order to avoid any confusion with Neg-expressions, used as the technical term to refer to negative indefinites (cf. Section 5 and Chapters 4 and 5 for more details).

1.2.1 Preverbal and Postverbal Negation

Syntacticians and typologists have extensively studied the position of the negation marker in the sentence. Greenberg (1966), Dahl (1979) and Dryer (1988, 2007) provide well-known examples of such studies. The main issue discussed in the literature concerns the position of negation with respect to the verb. The examples in (9) and (10) illustrate the preverbal and postverbal position of negation in a range of languages[2]:

(9) a. Maria *non* parla molto. [Italian]
Maria SN talks much.
'Maria doesn't talk much.'
b. *Nid* oedd Sioned yn gweithio. [formal Welsh]
SN be.IMPf.3SG Sioned PROG work
'Sioned was not working.'
c. ʔəli *ma*: ra:ħ lidda: ˀirə [Baghdad Arabic]
Ali SN went to the office
'Ali didn't go to the office.'
d. A vaga koŋ *ba* bɛnɛ [Koromfe]
ART dog.SG det.NONHUMAN.SG SN come.PAST
'The dog did not come.'
e. Mary does *not* talk much.

(10) a. Maria a parla *nen* tant. [Piedmontese]
Maria CL talks SN much.
'Maria doesn't talk much.'
b. Maria spricht *nicht* viel. [German]
Maria talks SN much.
'Maria doesn't talk much.'
c. Maria praat *niet* veel. [Dutch]
Maria talks SN much.
'Maria doesn't talk much.'
d. Mi-zɔk wi ndɔng *na* [Gbaya Kaka]
ISG-see person that SN
'I do not see those people.'

In most languages, negation systematically either precedes (9) or follows (10) the verb. English exemplifies a complex situation in which negation follows the auxiliary (4a), (5a), (6a), but precedes the main verb. This motivates the construction of *do*-support in sentences like (1b, c), (3a) and (9e) (cf. Chapter 3, Section 3 for an analysis).

---

[2] The Romance examples are from Zanuttini (1991, 1996). The Baghdad Arabic example is from Payne (1985). The Welsh example is from Borsley and Jones (2005). The Koromfe example and the Gbaya Kaka example are from Dryer (2007). Koromfe is a Niger-Congo language spoken in Burkina-Fasso and Mali; Gbaya Kaka is a Niger-Congo language spoken in Cameroon.

Dryer (1988) presents a systematic study of the placement of the marker of sentential negation in relation to the three main clausal elements of subject (S), object (O) and verb (V) in a worldwide sample of 345 languages. His results indicate that SOV languages are most commonly either SOVNeg or SONegV. NegSOV and SNegOV languages are infrequent. SVO languages are most commonly SNegVO, and V-initial languages are overwhelmingly NegV (i.e. NegVSO or NegVOS).

The patterns of negation in relation to the S, V and O system of the language are quite intriguing, but a full study of the placement of negation with respect to these three elements is outside the scope of this book. The position of the negative particle in relation to the verb will be the focus of this investigation, because this factor turns out to have important implications for the syntax–semantics interface.

There is an overall tendency for the negative marker to precede the verb. Out of 345 languages in the sample, Dryer (1988) finds that 227 (70%) place the negation marker before the verb. The patterns of preverbal (9) and postverbal negation (10) were first described by Jespersen (1917). Jespersen identifies a strong tendency "to place the negative first, or at any rate as soon as possible, very often immediately before the particular word to be negated (generally the verb)" (Jespersen 1924: 4). Horn (1989: 292–293) uses the term NegFirst for this tendency. NegFirst is motivated by communicative efficiency, i.e. to "put the negative word or element as early as possible, so as to leave no doubt in the mind of the hearer as to the purport of what is said" (Jespersen 1924: 297), quoted by Horn (1989: 293).

Although many languages have a preverbal marker of sentential negation, the examples in (10) indicate that NegFirst is not an absolute rule. In the OT system developed in Chapter 3, NegFirst is defined as a violable constraint that interacts with other constraints governing word order in the language. An opposing force coming from information structure favors a position of new or focused information late in the sentence (FocusLast). When this general tendency applies to negation, it favors a postverbal position of negation, so it is in conflict with NegFirst. The OT grammar of a language establishes a balance between these opposing tendencies in terms of constraint ranking.

1.2.2 Discontinuous Negation

The patterns in (9) and (10) represent cases in which a language expresses propositional negation by means of a single negative marker. A small number of languages use so-called discontinuous negation. In such languages, negation is expressed by two 'bits' of form, which appear in two different positions in the sentence, as illustrated in (11). In such cases, SN appears twice in the gloss.

(11) a. *Ne* bið he *na* geriht. [Old English]
SN is he SN righted
'He is not/never set right (=forgiven)'

b. Elle *ne* vient *pas*. [written French]
She SN comes SN.

c. *Ni* soniodd Sioned *ddim* am y digwyddiad. [formal Welsh]
SN mention.PAST.3SG Sioned SN about the event
'Sioned did not talk about the event.'

d. *Doedd* Gwyn *(*ddim*) yn cysgu. [informal Welsh]
NEG.be.IMPF.3SG Gwyn *(SN) PROG sleep
'Gwyn was not sleeping.'

e. baba wo-shìi nai tapa u. [Kanakuru]
father SN-he drink tobacco SN
'My father does not smoke tobacco.'

f. Haar suster het *nie* haar verjaarsdag vergeet *nie*. [Afrikaans]
Her sister has SN her birthday forgotten SN
'Her sister didn't forget her birthday.'

Even though there are two markers in the syntax, there is only one negation in the semantics, that is, all the sentences in (11) express a proposition of the form ¬p, with p an atomic proposition. However, negation is expressed by two 'bits' of form, one usually preceding the verb, the other following it, which is why I refer to it as discontinuous negation. The two markers are often (11a–e), though not always (11f), different lexical items.

The analysis of discontinuous negation raises important problems for the principle of compositionality of meaning. This foundational principle states that the meaning of a complex whole is a function of the meaning of its composing parts. If a sentence contains two expressions contributing negation, the question arises as to how to derive the single negation meaning of the sentences. The compositionality problem surfaces with negative indefinites as well. The compositionality problem, and possible solutions advanced in the literature, are discussed in more detail in Sections 4 and 5.

Example (11a) is from Mazzon (2004: 27), who indicates that discontinuous negation was a rather unstable phenomenon in the late Old English and Early Middle English period. The written French example in (11b) illustrates the bleaching of the preverbal *ne* to a co-negative, where the expressive force of negation is borne by the postverbal negator *pas* (cf. Godard 2004 and references therein). Formal Welsh reflects an older stage of the language in which the postverbal *ddim* is optional (11d) (Borsley and Jones 2005). In informal Welsh, the preverbal particle has disappeared, but it survives in an incorporated form in some verbs, such as *oedd-doedd* (11e). Although the verb appears in a negative form, it is unable to express semantic negation, and the presence of the postverbal adverb *ddim* is obligatory.

Discontinuous negation is not restricted to languages spoken in Europe. (11e) is cited as an example of discontinuous negation by Dryer (2007) in languages spoken on the African continent.

Typologically speaking, discontinuous negation does not occur in many languages, and when it does, it is usually not very stable in a diachronic sense (Haspelmath 1997). Modern English does not have a discontinuous negation anymore. In spoken French, preverbal *ne* is frequently dropped. In colloquial Welsh, the special negative form of the verb is limited to a small number of lexical verbs. This book argues that discontinuous negation is rare because it is uneconomical. Syntactically, discontinuous negation is of course rather costly: why use two markers to express a single negation, if one could do the job? Economy plays an important role in the analysis, but there are factors overruling economy in certain configurations.

Jespersen (1917) argues that discontinuous negation is a phase in a diachronic process in which preverbal negation is gradually replaced by postverbal negation. This process is commonly referred to as the 'Jespersen cycle'.

1.2.3 The Jespersen Cycle

Jespersen formulates the diachronic pattern of negation as follows: "The history of negative expressions in various languages makes us witness the following curious fluctuation: the original negative adverb is first weakened, then found insufficient and therefore strengthened, generally through some additional word, and this in turn may be felt as the negative proper and may then in course of time be subject to the same development as the original word' (Jespersen 1917: 4), quoted by Horn (1989: 452).

A few pages later, Jespersen adds: "Now, when the negative begins a sentence, it is on account of that very position more liable than elsewhere to fall out, by the phenomenon for which I venture to coin the term of *prosiopesis* (the opposite of what has been termed of old aposiopesis): the speaker begins to articulate, or thinks he begins to articulate, but produces no audible sound (either for want of expiration, or because he does not put his vocal chords in the proper position) till one or two syllables after the beginning of what he intended to say. (…) The interplay of these tendencies – weakening and strengthening and protraction – will be seen to lead to curiously similar, though in some respects different developments in Latin with its continuation in French, in Scandinavian and in English" (Jespersen 1917: 6).

The trajectory of the Jespersen cycle is well documented for English (Horn 1989, Mazzon 2004, Wallage 2005, 2008), French (Bréal 1897/1900, Clarke 1904, Tesnière 1959, Horn 1989, Godard 2004), Dutch/Flemish (Hoeksema 1997, Zeijlstra 2004, Breitbarth and Haegeman 2008) and German (Jäger 2008, Breitbarth, to appear). Although Borsley and Jones (2005) do not describe it in these terms, it is traceable for Welsh in their book.

Horn's (1989: 455) summary of the English and French development is given in Table 1. The preverbal negation *ne* in Old French is reinforced by the postverbal marker *pas*, which leads to the discontinuous negation *ne..pas* in modern written French. The discontinuous negation is currently giving way to a single postverbal negation in spoken French, even in the higher registers (Ashby 1981, 2001).

Table 1 Jespersen cycle in English and French

Old French	Jeo *ne* dis I SN say	Old English	Ic *ne* secge I SN say
Modern French (written/standard)	Je *ne* dis *pas* I SN say SN	Middle English	Ic *ne* seye *not* I SN say SN
Modern French (colloquial)	Je dis *pas* I say SN	Early Modern English	I say *not* I say SN
		Modern English	I do*n't* say I do SN say

In English, a similar development took place from the Old English preverbal negation *ne* via the discontinuous pattern *ne..not* in Middle English to the postverbal negation *not* in Early Modern English. Postverbal *not,* which originates from *nawiht/nogh/nahtet* 'nothing', has taken over the negative force in this phase. The *do*-support construction in Modern English signals a return to the preverbal position of negation, and supports Jespersen's view that the diachronic process is cyclic.

Chapter 3 provides an analysis of the Jespersen cycle in an optimality-theoretic model. This approach can explain why economy is overruled in certain grammars.

1.2.4 Negative Indefinites

In logic as well as linguistics, the analysis of sentence negation is closely intertwined with the treatment of quantifiers. If negation affects an indefinite in argument (12a) or adjunct position (12b, c), negation may be incorporated into the indefinite in languages like English.

(12) a. No one came.
$\neg\exists x\ Came(x)$
b. It never rains here.
$\neg\exists t\ Rain(t)$
c. The book was nowhere to be found.
$\neg\exists l\ Be\text{-}Found(b, l)$

Of course, the functional architecture of the clause is quite different from that of the nominal domain, so from a syntactic perspective, it may come as a surprise that propositional negation may be realized on a pronoun like *no one, nothing* or an adverb like *never, nowhere.* Semantically, sentences involving *not* and sentences involving *no one, never* are variants on the expression of truth-functional negation. Besides issues concerning the position and interpretation of the marker of sentential negation, the status of pronouns and adverbs such as English *no one, never, nowhere* in (12) is central to the syntax and semantics of negation. I borrow the terminology from Haspelmath (1997) and Penka (2006, 2007), and characterize these expressions as negative indefinites. I include temporal and spatial variables into the argument structure of lexical verbs in order to treat the cases in (12a-c) in the same way. The predicate-logical translations given in (12) reflect the enriched view of argument structure adopted.

In Chapter 4, I refer to negative indefinites as Neg-expressions, and give this term a precise theoretical status. The translations provided in (12) are straightforward, and it seems sensible to treat expressions like *no one* as quantifiers, and assign them the lexical semantics $\neg\exists x$. Further research reveals that the status of negative indefinites in natural language is much more complex than the examples in (12) might suggest. The lexical semantics one assigns to negative indefinites depends on one's views on negative polarity and negative concord. There is a wide range of proposals in the literature, which are spelled out in Sections 3–5.

1.3 Negative Polarity

Under the definition advanced by van der Wouden (1994: 1), negative polarity items are lexical elements with a restricted distribution: they occur in 'negative' contexts only (where 'negative' includes more than sentential negation, see below). This section discusses the status of negative polarity items as special indefinites occurring in the scope of negation, and the issues raised by the study of polarity items in natural language. The relation between negative polarity and negative concord will be addressed in Section 4.

1.3.1 Negative Polarity Items as Special Indefinites

Many languages use a special form of the indefinite if it occurs in the scope of negation. For propositional operators like negation or quantification, the semantic scope is defined as the proposition to which the operator is prefixed. English is a prime example of a language using so-called negative polarity items in negative contexts. Compare the sentences in (13) and (14).

(13) a. I did *not* buy *something.* $[\exists\neg, *\neg\exists]$
b. I did *not* buy *anything.* $[\neg\exists, *\exists\neg]$

(14) a. *Nobody* saw *something.* $[\exists\neg\exists, *\neg\exists\exists]$
b. *Nobody* saw *anything.* $[\neg\exists\exists, *\exists\neg\exists]$
c. *Nobody* said *anything* to *anyone.*

Examples (13a) and (14a) are grammatical if the indefinite takes wide scope over negation or the negative quantifier, but cannot be used to express the narrow scope of the indefinite. (13b) and (14b) mirror (13a) and (14a) in that *anything* obligatorily takes a narrow scope with respect to negation or the negative quantifier.

Support for the claim that a negative polarity item must be in the semantic scope of negation comes from pairs of sentences such as (15) (from de Swart 1998b).

(15) a. Sue did not read a book by Chomsky.
b. Sue did not read any book by Chomsky.

(15a) is ambiguous depending on the scope of the negation operator with respect to the existential quantifier introduced by the indefinite NP. The first-order representation of the two readings of (15a) in (16) makes this explicit.

(16) a. $\neg\exists x$ (Book-by-Chomsky(x) $\wedge$ Read(x)) Neg > $\exists$
b. $\exists x$ (Book-by-Chomsky(x) $\wedge$ $\neg$Read(x)) $\exists$ > Neg

Expressions like the English *anything* are called 'negative polarity items', because such items can only be felicitously used in contexts with a certain "negative" flavor, and they always take a narrow scope with respect to their licensor (Ladusaw 1979). Accordingly, (15b) only has the interpretation in (16a). Items like the English *something* are called 'positive polarity items', because they are allergic to negative contexts, and want to be interpreted outside the scope of negation (Baker 1970). Thus, (13a) only gets the reading akin to (16b). Not all indefinites are either positive or negative polarity items: plain indefinites like the English *a book* are neither, as illustrated by (15a).

Analyses of negative and positive polarity are offered by Ladusaw (1979, 1996), Zwarts (1986, 1995, 1998), van der Wouden (1994, 1997), Szabolcsi (2004), Giannakidou (1998, 1999, 2008) and others. This book does not address the phenomenon of positive polarity as such, but is restricted to negative polarity, and more particularly the relation between negative polarity items (NPIs) and negative indefinites (Neg-expressions).

Negative polarity items occur in a wider range of contexts than just negation, as emphasized by Ladusaw (1979, 1996).

(17) a. If you saw *anything*, please tell the police.
b. Did *anyone* notice *anything* unusual?
c. Few commuters *ever* take the train to work.

The examples in (17) illustrate that NPIs such as *anything* do not inherently carry a negative meaning. Rather they have existential force, with some additional meaning component characterized as 'widening' of a set of alternatives by Kadmon and Landman (1993), and Lahiri (1995, 1998), as indicating the bottom of a scale by Fauconnier (1975, 1979), Linebarger (1980, 1987), Krifka (1995), Israel (1996), and de Swart (1998b), as sensitive to scalar implicatures by Chierchia (2006), or to a non-deictic interpretation of the variable (Giannakidou 2008).

This meaning is particularly strong in so-called 'minimizers', i.e. indications of a small quantity that function as the bottom of the scale. The sentences in (18) have a strong idiomatic flavor. Their affirmative counterparts in (18a') and (18b') are not ungrammatical, but have a literal meaning only. The truth conditions in (16) only spell out the existential import of the negative polarity item.

(18) a. He didn't *lift a finger* to help me.
a'. #He lifted a finger to help me.
c. Nobody had *a red cent.*
b'. #Everybody had a red cent.
c. Every restaurant that charges so much as *a dime* for iceberg lettuce ought to be closed down.
c'. #Some restaurant that charges so much as a dime for iceberg lettuce ought to be closed down.

Negative polarity items are found in a wide range of languages. Zwarts (1986) studied negative polarity early on for Dutch, cf. also van der Wouden (1994, 1997), from whom the examples in (19) are taken. Haspelmath (1997: 193, 215) provides examples of negative polarity items from Basque (20) and Swedish (21) (cf. Laka 1990 for more on Basque). The Mandarin Chinese example in (22) is from Xiao and McEnery (2008).

(19) a. Geen monnik zal *ook maar* iets bereiken. [Dutch]
No monk will NPI something achieve
'No monk will achieve anything.'
b. Weinig monniken *kunnen* vader abt *uitstaan*
Few monks can father abbot stand
'Few monks can stand father abbot.'

(20) Ez dut *inor* ikusi. [Basque]
SN I:have:him anybody seen.
'I haven't seen anybody.'

(21) Ja har inte sett *någon.* [Swedish]
I have SN seen anybody.
I have not seen anybody.'

(22) zhe bing bu yewei-zhe women jiang jujue [Mandarin Chinese]
this actually SN mean-ASP we will refuse
xiang *renheren* chushou *renhe* dongxi
to anyone sell any thing
'This does not mean that we will refuse to sell anything to anyone.'

Section 3.2 provides additional examples from Hindi. Negative polarity is not restricted to the nominal domain, as the examples in (23) show.

(23) a. She doesn't have a car *yet.*
b. This is the cleverest idea I have seen *in years.*
c. I could *stand* it no more.
d. Hij *hoeft* zijn huis niet te verkopen. [Dutch]
He needs his house SN to sell
'He doesn't need to sell his house.'
e. Daniel n'a pas *du tout* aimé le concert. [French]
Daniel SN has SN of all liked the concert
'Daniel didn't like the concert at all.'

The examples in (23) indicate that negative polarity items also live in the adverbial and the verbal domain (cf. Hoeksema 1994, 2005, Tovena, Déprez and Jayez 2004).

1.3.2 *Issues in the Study of Negative Polarity Items*

For Ladusaw (1996), the study of negative polarity items raises three important issues: the question of the licensee, the question of the licensor, and the question of the licensing relation.[3] The term licensee refers to the lexical items used as NPIs. As illustrated in Section 3.1, a variety of expressions can behave like an NPI. A large class of NPIs involves minimizers such as *lift a finger* and *have a red cent*, the lexical semantics of which has been studied by Fauconnier (1975, 1979), Krifka (1995), Israel (1996), and others. Additive particles have been studied by Rullmann (2003) and Giannikidou (2008). Other categories of NPIs have been studied by Jack Hoeksema in a large ongoing corpus research of Dutch polarity items (cf. Hoeksema 2000, 2002, 2005, Rullmann and Hoeksema 2001 and references therein). In the remainder of this section, and in this book, I will be concerned only with pronominal and adverbial indefinites, such as the English *anything, anywhere*.

The question of the licensor involves the contexts in which NPIs are felicitous. The literature has shown that a wide range of expressions license NPIs, as exemplified in (13, 14, 17). Licensors generally create a downward entailing context (13, 14, 17a, c, 19) (Ladusaw 1979, Zwarts 1986, van der Wouden 1994, 1997) or a non-veridical context (17b) (Zwarts 1995, Giannakidou 1997, 1998, 1999, 2008). Non-veridical operators such as question operators block the inference from Op(p) to p, according to the definition in (24). Downward entailing operators such as *nobody, few students, at most five children* allow inferences to smaller sets, as observed in generalized quantifier theory (Barwise and Cooper 1981) (25).[4]

(24) An operator Op is veridical if and only if Op(p) $\rightarrow$ p.
An operator is non-veridical if and only if it is not veridical.
  a. It is possible that Jane is coming. –/→ Jane is coming.
  b. Jane is not coming. –/→ Jane is coming.
  c. Is Jane coming? –/→ Jane is coming.

(25) An operator Op is downward entailing if and only if Op(A) is true, and A' $\subseteq$ A, implies that Op(A') is true as well.
  a. Nobody read a book. → Nobody read a book by Chomsky.
  b. At most five children ate vegetables. → At most five children ate carrots.

[3] Ladusaw (1996), Krifka (1995), Chierchia (2006) and Giannakidou (2008) are also concerned with the status question: are sentences in which NPIs are not properly licensed ungrammatical, semantically ill-formed or pragmatically infelicitous? This issue is tangential to my concerns, so I refer the reader to the relevant literature for further discussion.

[4] De Swart (1998a: Chapter 7) offers an introduction to generalized quantifier theory.

A subset of the set of downward entailing operators has the property of anti-additivity. Anti-additivity is defined as in (26) (Barwise and Cooper 1981, Zwarts 1986, van der Wouden 1994, 1997).

(26) An operator Op is anti-additive if and only if Op(A) and Op(B) is equivalent to Op(A or B).
   a. Nobody danced and nobody sang. ↔ Nobody sang or danced.
   b. She never calls and she never writes ↔ She never calls or writes.
   c. Jane did not dance and Jane did not sing. ↔ Jane did not dance or sing.

van der Wouden (1994, 1997) defines 'medium' negative polarity items as expressions that require an anti-additive licensor. The Dutch NPI *ook maar* is an example (27a, b). 'Weak' NPIs such as *kunnen uitstaan* are also licensed by downward entailing operators (27c, d).

(27) a. *Weinig monniken zullen *ook maar* iets bereiken. [Dutch]
   Few monks will NPI something achieve
   'Few monks will achieve anything.'
   b. Geen monnik zal *ook maar* iets bereiken.
   No monk will NPI something achieve
   'No monk will achieve anything.'
   c. Weinig monniken *kunnen* vader abt *uitstaan.*
   Few monks can father abbot stand
   'Few monks can stand father abbot.'
   d. Niemand *kan* de schoolmeester *uitstaan.*
   Nobody can the schoolmaster stand
   'Nobody can stand the schoolmaster.'

van der Wouden (1994, 1997) distinguishes a third class of 'strong' NPIs, which are exclusively licensed by antimorphic operators. The class of antimorphic operators includes sentential negation (*not*), but not negative indefinites (*nobody*):

(28) An operator Op is antimorphic if and only if Op(A) and Op(B) is equivalent to Op(A or B) ànd Op(A) or Op(B) is equivalent to Op(A and B):
   a. Jane did not dance and Jane did not sing. ↔ Jane did not dance or sing.
   b. Jane did not dance or Jane did not sing. ↔ Jane did not (both) dance and sing.

Strong NPIs co occur with clausemate negation only, as illustrated by the following examples from Dutch (van der Wouden 1994).

(29) a. Het is niet *pluis* in Leeuwarden. [Dutch]
   'It is not safe in Leeuwarden.'
   b. *Het is nooit *pluis* in Leeuwarden.
   'It is never safe in Leeuwarden.'

The contrasts in (27) and (29) illustrate that negative polarity items are sensitive to different degrees of negativity, and that these degrees correspond with well-defined properties from generalized quantifier theory. Similar classes of polarity items are definable for German (Zwarts 1995) and Greek (Giannakidou 2008), so there is cross-linguistic support for the distinction between weak, medium and strong NPIs.

So far, it has been established that negative polarity items need to be licensed by an operator with particular semantic properties in a particular context.[5] However, NPIs and licensors cannot be related in just any syntactic configuration. An important syntactic constraint on the licensing relation is that negative polarity items generally have to occur in the c-command domain of their licensor.[6] The NPI is then in the direct scope of its licensor (Linebarger 1987). The definition of direct scope is in (30) (Szabolcsi 1997).

(30) An expression *a* has direct scope over an expression *b*, if and only if *b* is in the semantic scope of *a*, and *a* c-commands *b* in the syntactic structure.

The requirement on direct scope implies that syntax and semantics converge. The requirement on direct scope is visible in the contrasts in (31)–(34) (from de Swart 1998b, with observations going back to Klima 1964 and Ladusaw 1979).

(31) a. Phil did not say anything to me.
b. *Anyone did not talk to me.

(32) a. No one said anything to me.
b. *Anyone said nothing to me.

(33) a. Didn't anybody come?
b. *Anybody didn't come.

(34) a. Phil would not give me anything.
b. *Anything Phil would not give me.

---

[5]Potential counterexamples to this claim are expressions like English *less*, *squat* and German *einen Dreck*, which optionally allow the construction without a marker of negation, as illustrated in (i)–(iii):

(i) I could/couldn't care less.

(ii) Frida knows/doesn't know diddly squat about physics.

(iii) Das geht dir keinen/ einen Dreck an. [German]
That concerns you no/ a bit to.
'That doesn't concern you one bit.'

The status of these expressions and their relation to n-words in the Jespersen cycle is under scrutiny in the current literature. *Squat* is discussed by Postal (2000, 2004), *einen Dreck* by Richter and Sailer (2006). Thanks to an anonymous reviewer for drawing my attention to the relevance of these examples for the discussion at hand.

[6]A node *a* c-commands another node *b* in the syntactic tree if and only if every branching node dominating *a* also dominates *b*. Instead of imposing a configurational restriction on direct scope, it is also possible to define constraints on lists of argument structures in a lexalist theory such as HPSG (cf. Sag, Wasow and Bender 2003). The result is essentially the same. I use the configurational definition here, because tree-like representations are probably familiar to the reader. I don't adopt a formal theory of syntax in this book. What I mean with 'syntactic structure' is some version of surface-oriented syntax. The HPSG analysis advanced in Chapter 4 relies on argument structure and feature sharing. Crucially, movement, invisible syntactic structures (either 'deep structure' or 'logical form'), or empty categories are not assumed anywhere in the analysis (cf. Section 5 for discussion).

In English, the marker of sentential negation c-commands the direct object, but not the subject, so (31a) is fine, but (31b) is ungrammatical. The subject c-commands the direct object, but not vice versa, so (32a) is well-formed, but (32b) is ungrammatical. Question formation in English comes with a configuration in which negation c-commands the subject in the syntactic structure, so the grammaticality of (33a) contrasts with the infelicity of (33b). Object preposing brings the NPI outside of the c-command domain of negation, so (34b) is ill-formed, while (34a) is fine. Similar data have been discussed for Dutch (van der Wouden 1994) and French (Tovena, Déprez and Jayez 2004).

Exceptions to the direct scope constraint involve embedding of the NPI in a constituent that itself takes narrow scope with respect to negation, as in (35), cf. de Swart (1998b) for English and Dutch, going back to work by Uribe-Etxebarria (1994), Linebarger (1980, 1987). Tovena, Déprez and Jayez (2004) discuss this for French.

(35) a. That he had stolen anything was never proven.
b. A doctor who knew anything about acupuncture was not available.
c. Qu'il s'intéresse au moindre étudiant, ça me surprendrait. [French]
'That he cares for the least student, it would surprise me.'
d. Un méd ecin ayant la moindre connaissance de l'acupuncture se révéla impossible à trouver.
'A doctor with any knowledge of acupuncture was impossible to find.'

de Swart (1998b) offers an account of such exceptions through pragmatic reasoning involving scalar implicatures.

Apart from these special cases, the direct scope constraint is valid for English, and a wide range of other languages. However, it is not universal.[7] In Old English, indefinites could precede the preverbal negation *ne* without a problem, as illustrated by examples (36) from Mazzon (2004: 39).

(36) a. þæt hi æfre on *ænine* man curs *ne* settan [Old English]
that they ever on any man curse SN lay
'that they ever on any man curse not lay'
b. *Ængum ne* mæg se cræft losian.
anyone SN may his craft loose
'anyone not may the skill abandon'

Vasishth (2000, 2002) makes similar observations for Hindi (cf. also Lahiri 1995, 1998). Lahiri and Vasishth demonstrate that sentences like (37) exemplify negative polarity, not negative concord. Lahiri provides an account of NPI licensing in terms of implicatures, Vasishth exploits a multimodal categorial grammar framework.

[7] Modal verbs frequently escape the restriction to c-command, as already illustrated in (23c,d) above, but the discussion in this section focuses on indefinites, so I will not pursue this issue.

(37) a. *Koi-bhii nahĩĩ* aayaa [Hindi]
Anybody SN came
'Nobody came.'
b. *Koi-bhii nahĩĩ* khaat-aa th-aa sabzii
Anyone SN eat.IMP.MASC be.PAST.MASC vegetables
'No one used to eat vegetables.'

As we will see in Section 4, the direct scope requirement is used as a diagnostic to distinguish negative polarity items from n-words. The examples in (36), (37) show that this criterion is not infallible, but it works in many languages.

A full study of NPIs, their licensing conditions, and their cross-linguistic behavior is outside the scope of this book. However, the notion of negative polarity comes into play in the discussion of negative concord, as will become obvious in Section 4.

1.4 Negative Concord: Observations and Issues

Negative concord and negative polarity are two versions of the phenomenon of special indefinites interpreted in the scope of negation (cf. also Chapter 4). This section investigates similarities and differences between the two phenomena, and discusses analyses of negative concord that have been advanced in the literature.

Section 5 develops the compositional semantics of double negation and negative concord languages that constitutes the foundation of this book. Subclasses of negative concord languages are defined in Section 6 on the basis of their interaction with the marker of sentential negation.

1.4.1 Negative Polarity and Negative Concord

Negative polarity and negative concord are closely related phenomena. The Italian example (38a) (from Haegeman and Zanuttini 1996) is a direct counterpart of the English (38b).[8]

(38) a. *Non* ho visto *nessuno*. [Italian]
SN has seen nobody.
'I haven't seen anybody.'
b. I have*n't* seen *anybody*. [English]
c. $\neg\exists x$ See(I, x)

In the context of (38a), it is tempting to analyze *nessuno* as a negative polarity item on a par with English *anybody*. The identification with *anybody* would suggest that

[8] *Nessuno* is not glossed as 'anybody', but as 'nobody', in anticipation of the analysis to be developed in Section 5.

nessuno gets an interpretation in terms of existential quantification (∃). Function application would provide the desired truth conditions of both (38a) and (38b), spelled out in terms of the first-order logical formula (38c). However, other examples raise problems for this view.

Haegeman and Zanuttini (1996) show that *nessuno* can be the sole expression of negation in the sentence (39a). Example (39b) is ungrammatical, because the licensor of *anybody* is missing. Example (39c) is ungrammatical, because the licensor is not c-commanding the NPI. The appropriate translation of (39a) requires the use of *nobody* in (39d).

(39) a. *Nessuno* ha telefonato. [Italian]
Nobody has called
'Nobody has called.'
$\neg\exists x\ Call(x)$
b. *Anybody has called. [English]
c. *Anybody has not called.
d. Nobody has called.

The contrast between (38) and (39) indicates that *nessuno* seems to mean 'anybody' in some contexts, and 'nobody' in others. In sentences that combine two negative indefinites, the first one seems to behave like 'nobody', and the second one like 'anyone'. This is illustrated for the combination of *nessuno* ('nobody') and *niente* ('nothing') in (40).

(40) a. *Nessuno* ha detto *niente*. [Italian]
Nobody has said nothing.
'Nobody has said anything.'
$\neg\exists x\exists y\ Say(x,y)$
b. *Anybody has said anything.
c. Nobody has said anything.
d. #Nobody has said nothing.
$\neg\exists x\neg\exists y\ Say(x,y)$

Example (40a) expresses a single negation, even though the combination of *nessuno* and *niente* involves two formally negative expressions, which can have negative interpretations in contexts like (40a). The English translation (40c) involves the combination of a negative indefinite and a negative polarity item. The combination of two negative polarity items in (40b) is ungrammatical, because there is no licensor for the NPIs (cf. Section 3). The combination of two negative indefinites in (40d) is not ungrammatical, but the sentence does not have the same meaning as (40a): it conveys double, rather than single negation.

The pattern exemplified for Italian in (38)–(40) has been well described in the literature. Jespersen (1917) dubs the phenomenon double negation, Klima (1964) calls it neg-incorporation, and Labov (1972) proposes a negative attraction rule. Most current linguistic literature uses the term negative concord for cases where multiple occurrences of negation and indefinite pronouns that appear to be negative express a single negation, and I will follow this use. The indefinite pronouns participating in negative concord are termed n-words, following Laka (1990).

Negative concord is a widespread phenomenon in natural language, as Haspelmath (1997) observes. It is found in Romance, Slavic, Greek, Hungarian, nonstandard English, (West) Flemish, Afrikaans, Japanese and elsewhere. The literature concerning negative concord is quite extensive. A wide range of observations and proposals is presented in this chapter and in the rest of the book. For starters, Section 4.2 focuses on the distributional criteria that draw the line between negative polarity and negative concord.

1.4.2 Distributional Criteria

It is sometimes difficult to distinguish between NPIs and n-words in a language. Three distributional criteria have been advanced to separate the two classes at a descriptive level. Theoretical implications of these empirical observations are discussed in the following subsections.

One criterion used to distinguish between NPIs and n-words is based on the observation that n-words are strictly limited to anti-additive environments, whereas weak NPIs typically occur in a wider set of downward entailing or non-veridical contexts (cf. Section 3). In some cases, the n-word is infelicitous in environments where the NPI is licensed, as illustrated for Greek *too*-clauses and conditionals in (41) and (42) (from Giannakidou 1998, 2000) and for Japanese questions in (43) (from Watanabe 2004).

(41) I Ilektra ine poli kurasmeni ja na milisi se kanenan/*KANENAN.
the Electra be.3SG very tired for SUBJ talk.3SG to anyone/ no one.
'Electra is too tired to talk to anybody.' [Greek]

(42) a. An dhis tin Ilektra puthena/* PUTHENA, [Greek]
If see.2SG the Electra anywhere/nowhere
na tis pis na me perimeni.
SUBJ her say.2SG SUBJ me wait.3SG
'If you see Electra anywhere, tell her to wait for me.'

b. An eleje LEKSI, tha ton skotona.
if said.3SG word FUT him kill.1SG
If he had said a word, I would have killed him.

(43) *Nani-mo mi-mashi-ta ka? [Japanese]
what-mo see-polite-PAST Q
Intended meaning: did you see anything?

The occurrence of the emphatic n-word *PUTHENA* in the *if*-clause of the conditional in (42a) is labeled as ungrammatical, whereas the non-emphatic NPI *puthena* gets an existential interpretation. Note that the emphatic minimizer *LEKSI* in (42b) gets a non-negative interpretation on a par with the NPI in (42a) (translation provided by Giannakidou 1998).

In these cases, there are grammaticality contrasts. In other cases, the NPI and the n-word lead to different interpretations. The contrast between the French n-word *rien* in (44a) versus the NPI *quoi que ce soit* in (44b) in the antecedent of a conditional shows that the NPI leads to an existential interpretation, whereas the n-word is interpreted as negative (examples from Corblin et al. 2004).

(44) a. S'il ne dit *rien*, il doit soumettre ses devoirs par écrit.
If he SN says nothing, he must submit his homeworks in writing
'If he says nothing, he must submit his homework in writing.'
b. Si *quoi que ce soit* vous dérange, faites-le nous savoir.
If what that it be.SUBJ you disturbs, make it us know
'If anything at all bothers you, tell us.'

The antecedent of a conditional is a downward entailing environment (pace von Fintel 1999, Giannakidou 2007) in which the NPI *quoi que ce soit* is licensed, and is interpreted as an existential quantifier (44b) (cf. also 42 and Section 3). However, if the n-word *rien* occurs in this environment, it behaves like a negative quantifier similar to the English *nobody* rather than like an existentially quantified indefinite (44a).

This criterion is helpful to distinguish n-words from weak NPIs, but it does not apply to medium or strong NPIs, which require an anti-additive and an anti-morphic licensor respectively (cf. Section 3.2). In order to distinguish these NPIs from n-words, other criteria are called for.

The second criterion used to distinguish n-words and NPIs concerns differences in licensing configurations. Section 3 established the licensing condition for NPIs, implying that they need to appear in the context of a licensor with the appropriate semantic properties. N-words on the other hand can appear in the context of another n-word or the marker of sentential negation, but they do not have to. They are 'self-licensing' in the terminology of Ladusaw (1992).

The Italian data introduced in Section 4.1 illustrate the difference in licensing properties between n-words and NPIs. The n-word *nessuno* occurs in the c-command domain of the negation marker in (38a), but is felicitously used in the absence of a licensor in (39a). The infelicity of *anybody* in the English translation in (39b) shows that an NPI cannot be licensed in this configuration. In (39a), the n-word *niente* in object position is licensed by the n-word *nessuno* in subject position, but nothing licenses the n-word in subject position. The unacceptability of (40b) indicates that NPIs are not licensed in this configuration. Recall that it would not help to insert a negation marker (39c), for the NPI has to be in the direct scope of its licensor (cf. examples 31–34 in Section 3).

In languages such as Italian, we can see the 'self-licensing' nature of n-words at work in examples like (39a) and (40a), where no licensor is available for the n-word in subject position. The 'self-licensing' nature of n-words is harder to illustrate in languages in which a marker of sentential negation is obligatorily present in all sentences containing an n-word. Such languages are labeled 'strict' negative concord languages by Giannakidou (1998), and are opposed to the 'nonstrict' variety of negative concord found in Italian (see Section 6 for more discussion of the various negative concord systems).

Given that the negation marker is always present in sentences containing an n-word, configurations like (39a), (40a) are not found in strict negative concord languages. This leads Giannakidou (2008) to maintain the identification of n-words in these languages with 'strong' NPIs. Recall that strong NPIs are exclusively licensed by antimorphic operators such as sentential negation. However, even in these languages, NPIs and n-words display distributional differences. In particular, n-words are felicitous in preverbal subject position, or other positions that are outside the c-command domain of negation, but NPIs are not.

As Zeijlstra (2004: 220–222) points out, it is problematic to analyze n-words as NPIs, when they occur in positions at which other NPIs are banned. Thus the felicitous use of a French or Catalan n-word in subject position (46a, 47a) and the appearance of a Greek n-word in a topicalized preverbal object position (45a) support the view that n-words are not to be identified with NPIs.[9]

(45) a. KANENAN dhen idha. [Greek]
Nobody SN saw.1SG
'I saw nobody.'
b. *Kanenan dhen idha.
Anybody SN saw.1SG
(46) a. Personne n'est venu. [French]
Nobody SN has come
'Nobody came.'
b. *Qui que ce soit n'est (pas) venu.
Anybody SN has (SN) come
(47) a. No funcionen gaires coses. [Catalan]
SN 3PL-work many things
'There aren't many things working.'
b. *Gaires coses (no) funcionen.
c. Res (no) funciona.
nothing SN 3SG-work
'Nothing works.'

Kanenan, qui que ce soit and *gaires* are NPIs that are blocked from the preverbal position, because they are not in the direct scope of negation (45b, 46b, 47b).

In Greek, the distinction between NPIs and n-words is related to stress. The emphatic (capitalized) counterpart KANENAN functions as an n-word that appears felicitously in a topicalized preverbal position (45a). In French and Catalan, stress does not play a role, but NPIs and n-words belong to different classes of lexical items. The contrast between NPIs and n-words is repeated in (46) for the French n-word *personne* in preverbal subject position versus the NPI *qui que ce soit*, and in (47) for the Catalan n-word *res* as opposed to the NPI *gaires coses* (47c).

---

[9] The Greek data are from Giannakidou (1998); the Catalan examples are from Vallduví (1994).

As pointed out in Section 3, there are some exceptions to the constraint that NPIs have to be in the direct scope of their licensor, so we have to be careful. But in many languages (including Greek, French, Catalan), the presence of an item in preverbal subject position or in a topicalized preverbal position can be used to determine its status as an NPI or as an n-word, and the contrasts in (45)–(47) illustrate how this can be used as a criterion. An analysis that maintains the view of n-words as NPIs needs to provide a special account of the contrasts in (45)–(47). Giannakidou (2000, 2006) provides such an analysis, but the discussion of her ideas is postponed until Section 4.4.

The third and last criterion used to distinguish NPIs from n-words concerns fragment answers to questions (Ladusaw 1992, Vallduví 1994, Bernini and Ramat 1996, Haspelmath 1997). In languages like English, negative quantifiers (*nothing*) constitute fragment answers with a negative meaning, but NPIs (*anything*) do not (48).

(48) Q: What did he say? A: Nothing
*Anything
*(not) a word

N-words are 'self-licensing' in Ladusaw's terminology, because they constitute a well-formed fragment question to a question, and convey a negative meaning, just like negative quantifiers like *nobody* in English. NPIs are not felicitous in elided contexts, because the licensor is missing. In order to maintain the view that n-words are NPIs, a special analysis of fragment answers is called for, which distinguishes between *nothing* and *anything* in a different way (see Section 4.5 for a proposal, and critical discussion).

The examples in (49) through (55) illustrate the contrast between n-words and NPIs with examples from Giannakidou (1998) for Greek, Herburger (2001) for Spanish, Watanabe (2004) for Japanese, Przepiórkowski and Kupść (1999) for Polish, Progovac (1994) for Serbo-Croatian, and Haspelmath (1997) for Hindi.

(49)	Q: Qu'est-ce que tu as vu? What did you see?	A: Rien. Nothing	[French]
(50)	Q: Quién viste? Whom saw.2SG 'Who did you see?'	A: A nadie. Nobody A: *A un alma A soul	[Spanish]
(51)	Q: Pjon ihes? Who did you see?	A: KANENAN Nobody A: *kanenan Anybody	[Greek]
(52)	Q: Nani-o mita no? what-ACC saw Q 'What did you see?'	A: Nani-mo what-o 'Nothing'	[Japanese]

(53) Q: Kto pomógł Tomkowi? A: Nikt. [Polish]
who-NOM helped Tom-DAT nobody-NOM
'Who helped Tom?' 'Nobody.'

(54) Q: Koga Milan voli? A: Ni(t)koga [Serbo-Croatian]
who Milan loves no one.ACC
'Who does Milan love?' 'Nobody'

(55) Q: raam-ne kyaa khaayaa? A: *kuch bhii. [Hindi]
Ram-ERG what ate anything
'What did Ram eat?'

As Haspelmath (1997: 198) observes, this criterion is not always decisive. The felicity of an expression in fragment answers does not exclude the possibility that the n-word behaves like an NPI in other contexts. The examples in (59) below illustrate this for Italian.

Furthermore, according to Giannakidou (1998), Greek emphatic bare singulars that behave like minimizers are used as fragment answers with a negative meaning, as illustrated in (56).[10]

(56) Q: Ipe tipota i Ilectra olo to vradi? [Greek]
said.3SG anything the Electra all the evening
'Did Electra say anything all evening?'
A: LEKSI
word
'Not a word.'

In general however, fragment answers provide a clear distributional difference between NPIs and n-words, as illustrated by the systematic contrasts in (49)–(55). The implications of these data for the theory of negative polarity and negative concord are discussed in Section 4.5.

Two further differences between NPIs and n-words have been pointed out in the literature. First, unlike negative polarity licensing, negative concord is a clause-bound phenomenon. Second, the combination of an n-word with morphological negation (as in *unable, impossible*) is not an instance of negative concord, but leads to double negation readings. Both issues are treated in Chapter 6 (Sections 1 and 2). Chapter 4 (Section 5) exploits the contrast between clause-bound negative concord and long distance NPI licensing to account for the distribution of n-words in Serbo-Croation and Hungarian.

Linguists generally agree on the empirical differences between NPIs and n-words outlined in this section. There is much less agreement on the implications of these observations for the analysis of the range of expressions under consideration, in particular when it comes to the quantificational status of n-words. This debate is presented in Sections 4.3 and 4.4.

---

[10] Not all native speakers agree with Giannakidou's judgments, but further discussion of these data is postponed until Section 4.5.

1.4.3 *The Quantificational Status of N-words*

The data presented in Sections 4.1 and 4.2 highlight the difficult issue of the quantificational status of n-words. The problem of negative concord is usually defined at the syntax-semantics interface.[11]

Semantic theories are founded on the principle of compositionality of meaning. The principle of compositionality of meaning defines the meaning of a complex whole as a function of the meaning of its component parts and the way they are put together. The analysis of negative concord thus requires a lexical semantics of the n-word as well as an integration of the semantic contribution of the n-word into the meaning of the sentence as a whole. This requires any theory of negative concord to make choices in the lexical and compositional semantic toolkit as well as the syntactic set-up.

As far as the syntax is concerned, the pertinent question is how much 'underlying' or 'logical' structure the analysis appeals to. As far as the compositional semantic toolkit is concerned, the main decision to make is whether to remain strictly at a first-order level, or whether to allow second order operations from generalized quantifier theory. The remainder of this chapter will make it clear that theories mostly differ on these two points.

Suppose that first-order logic is to function as the tool to describe the meaning of a natural language sentence. First-order logic offers an inventory of predicates, individual arguments, connectives and quantifiers, and uses function application as the standard mode of composition. Function application implies that constructions of predication and quantification are built up by relating expressions as functors that apply to arguments. Regular indefinites are commonly translated in terms of the existential quantifier $\exists$ in first-order logic (57a). For negative polarity items, such as the English *anything*, a representation in terms of existential quantification is also in order (57b) (cf. also Section 3).

(57) a. Someone came in late.
$\exists x$ Came-Late(x)
b. Nobody said anything.
$\neg\exists x \exists y$ Say(x,y)

For n-words, a compositional interpretation in first-order logic is less straightforward. Consider the Italian patterns in (38)–(40) again, repeated here in (58).

(58) a. *Non* ho visto *nessuno*. [Italian]
SN has seen nobody
'I haven't seen anybody.'
$\neg\exists x$ See(I, x)

[11] Tubau (2008) is an exception: she focuses on syntax–morphology interface conditions on the expression of negation. Her study does not pay much attention to the interpretive issues which are central to this book, though.

b. *Nessuno* ha telefonato.
Nobody has called
'Nobody has called.'
¬∃x Call(x)

c. *Nessuno* ha detto *niente*.
Nobody has said nothing
'Nobody has said anything.'
¬∃x∃y Say(x,y)

Everyone agrees on the meaning of these sentences, and it is easy to spell out their truth conditions in first-order logic. However, it is hard to see what lexical semantics to assign to the n-word in order to compositionally arrive at the semantics of the sentence as a whole. As observed in Section 4.1, it seems that n-words should sometimes be translated in terms of the existential quantifier ∃ (*nessuno* in 58a, and *niente* in 58c), and sometimes in terms of ¬∃ (*nessuno* in 58b, 58c). As pointed out by Zeijlstra (2004), this is a highly problematic outcome because of the relation of contradiction between these two quantifiers (cf. Figure 1 in Section 1).

The question to be addressed is then the following. If (58a–c) is interpreted in terms of first-order logic with negation, universal/existential quantification, and function application serves as the mode of composition, what is the lexical semantics of n-words like *nessuno* and *niente* that needs to be adopted in order to derive the desired truth conditions? In principle, there are three possible answers to this question: n-words can be existential, universal or negative. All three options have been defended in the literature, and it has also been proposed that n-words can be ambiguous between two or more of these meanings.

The remainder of this section describes the view that n-words are existential in nature. Section 4.4 discusses proposals involving lexical ambiguities. A crucial argument in the debate is provided by fragment answers, as already anticipated in Section 4.2. Therefore, Section 4.5 is devoted to the status of fragment answers in a theory of polarity and negative concord. The conclusion to Section 4 will be that n-words are inherently negative. In line with that conclusion, Section 5 develops a compositional semantics of double negation and negative concord based on the lexical semantics of negative quantifiers (such as the English *nobody*) and n-words (such as the Italian *nessuno*, Greek *KANENAN* and French *personne*) as negative indefinites. But before I can develop that analysis, I need to embed my ideas in the literature.

Laka (1990) takes n-words to denote existential quantifiers (∃) taking narrow scope with respect to negation. This would work well for configurations like (58a), and it would explain the (infrequent, possibly archaic, but existing) existential uses of *nessuno* and *niente* licensed by downward entailing (but not anti-additive) operators in (59) (from Zanuttini 1991).

(59) a. Ha telefonato *nessuno*? [Italian]
Has called nobody
'Did anybody call?'

b. Dubito che venga *nessuno*.
Doubt.1SG that comes nobody
'I doubt that anyone will come.'

The drawback of the proposal is that special syntactic assumptions are required to extend the treatment of *nessuno* and *niente* in terms of existential quantification to sentences like (58b) or fragment answers like (46) through (51). Such assumptions typically involve postulating an implicit negation operator. Such an implicit operator would be syntactically covert, but semantically potent, and contribute the truth-functional connective $\neg$. Laka (1990) locates such an implicit negation operator in a special functional projection, labelled ΣP. Rowlett (1998) exploits Haegeman's (1995) NEG-criterion. Recent versions of the same idea have exploited the feature checking theory of minimalist syntax. Zeijlstra (2004) posits a covert negation operator which provides the interpretable negation feature needed to check the uninterpretable negation feature of the n-word. I discuss Zeijlstra's analysis in Section 5, where I argue that it is difficult to uphold the principle of compositionality of meaning in the presence of covert and empty negation operators.

1.4.4 Lexical Ambiguities

Giannakidou (2000, 2006) analyzes Greek n-words as NPIs that denote universal quantifiers taking a wide scope with respect to negation. Under this analysis, the truth conditions of (60a) involve $\forall x \neg V(x)$ (60b), which is of course logically equivalent to $\neg\exists x V(x)$.

(60) a. Dhen ipe o Pavlos TIPOTA. [Greek]
SN said.3SG the Paul nothing
'Paul said nothing'
b. $\forall x\ [\text{thing}(x) \rightarrow \neg\text{Said}(\text{Paul},x)]$

The analysis is designed to work for Greek and other strict negative concord languages in which the marker of sentential negation always co-occurs with an n-word. Thanks to their emphatic nature, which allows them to undergo topicalization, the Greek n-words escape the usual direct scope requirements, and are also licensed in preverbal position (cf. above 45a).

The non-negative interpretation of n-words predicts that double negation readings are impossible in strict negative concord languages. Section 5 will show that this prediction is not borne out by strict NC languages such as Bulgarian, Romanian and (written) French, which constitutes a problem for Giannakidou's (2000, 2006) analysis.

An extension to the nonstrict variety of negative concord displayed by languages like Italian has to appeal to an implicit negation operator or to lexical ambiguities in order to provide a unified analysis of examples (58a–c). Giannakidou (2000, 2006) defends the view that n-words in natural language come in different types, and she assigns *KANENAN/TIPOTA* and *nessuno/niente* a different lexical semantics.

The emphatic Greek n-words of the *KANENAN/TIPOTA* series in (60a) are uniformly treated as (strong) NPIs denoting a universal quantifier. N-words in nonstrict negative concord languages such as the Italian *nessuno/niente* are ambiguous between an existential and a negative meaning. Romance n-words that occur in non-negative contexts have an existential use (59). When embedded under negation (*nessuno* in 58a) or another n-word (*niente* in 58c), they also get an existential interpretation. In sentences in which they are the sole contributors of negation (*nessuno* in 58b) or the structurally highest n-word (*nessuno* in 58c), they get a negative meaning.

Giannakidou (2000, 2006) is not the first to defend an ambiguity thesis. van der Wouden and Zwarts (1993), Corblin (1996) and Herburger (2001) offer versions of an account under which n-words are underspecified or ambiguous, and denote $\exists$ if embedded under negation or a negative quantifier and $\neg\exists$ if unembedded (cf. Section 5.1). The ambiguity thesis is attractive because of its lack of hidden negation operators. However, the ambiguity thesis is claimed to suffer from lack of independent evidence and testability (cf. Giannakidou 1997: 166–168 and de Swart and Sag 2002 for critical discussion).

The fine-grained lexical approach is meant to provide a flexible semantics of negation and polarity items across languages. Unfortunately, Giannakidou's (2000, 2006) analysis does not go beyond the lexical level, and does not spell out the compositional semantics of negation and n-words in languages other than Greek.

If expressions like *nessuno* and *niente* are ambiguous between an existential meaning ($\exists$) and a negative meaning ($\neg\exists$), we would expect sentences like (58c) to be four-ways ambiguous, and have the readings (i) $\exists\exists$, (ii) $\neg\exists\exists$, (iii) $\exists\neg\exists$ and (iv) $\neg\exists\neg\exists$. Of course, the sentence only has the reading in (ii). In order to eliminate the other readings as potential interpretations of sentences like (58c), a compositional interpretation mechanism for the sentence as a whole is called for.

Such mechanisms are provided in earlier ambiguity analyses (van der Wouden and Zwarts 1993, Corblin 1996, Herberger 2001, cf. Section 5.1), but are only touched upon in rather general terms in Giannakidou (2000, 2006). According to Giannakidou (2006: 357) "the best we can come up with is to stipulate an additional syntactic condition that negation must be expressed at the topmost level of the sentence, and that this can be done either by the SN itself, or by an n-word (which is essentially the proposal in Zanuttini 1991)". This syntactic stipulation is designed to eliminate the unwanted readings (i) and (iii).

Giannakidou (2006: 355) also suggests that the resumptive quantification mechanism defined by de Swart and Sag (2002) might be relevant to the explanation of negative spread as in (53c). If true, this eliminates the unwanted reading (iv). How resumptive quantification, lexical ambiguities and syntactic stipulations are to work together in a compositional semantics remains an open question.

In view of the fact that additional syntactic stipulations and supplementary semantic mechanisms (including resumptive quantification) must be posited for negative concord varieties other than the 'strict' type found in languages like Greek, I conclude that the rich lexical semantics posited by Giannakidou (2006) does not solve the compositionality problems raised by negative concord. In addition, fragment answers in which n-words appear to be 'self-licensing' even in languages such as Greek raise problems for Giannakidou's views. Given the importance of fragment answers in the debate on the quantificational status of n-words in the literature, I devote a separate subsection to this issue.

1.4.5 The Status of Fragment Answers

For many people, including Zanuttini (1991), Ladusaw (1992), Haegeman and Zanuttini (1996), Herburger (2001), de Swart and Sag (2002), Zeijlstra (2004), and Penka (2006, 2007), the 'self-licensing' nature of n-words is most visible in their felicity as negative fragment answers. Fragment answers thus constitute an important context in which n-words behave differently from NPIs. However, according to Giannakidou (1998, 2000, 2006) the fact that n-words are interpreted negatively in the absence of overt negation does not prove that they are negative. In view of these claims, the theoretical implications of the empirical observations made in Section 4.2 have to be reviewed in more detail.

Giannakidou (2000, 2006) defines fragment answers as elliptical structures, and takes elided material to be responsible for the negative meaning. Thus, in response to the question 'Who arrived?' or 'What did you see?', Giannakidou (2000, 2006) spells out the full answer as in (61), where strikethrough indicates the elided material of the fragment answer.

(61) a. KANENAS ~~dhen irthe~~. [Greek]
nobody SN arrived.3SG
b. TIPOTA ~~dhen idha~~.
nothing SN saw.1SG

The "negative meaning in elliptical fragments then arises not as an inherent contribution of the n-words, but rather as the result of their being associated with negation at the level at which ellipsis is resolved" (Giannakidou 2006: 363).

If ellipsis is resolved in the syntax, this route is closed for me, because I adopt a surface-oriented syntax in this book (see Section 4.6). But Giannakidou (2006) follows the semantic approach to ellipsis developed by Merchant (2001): the antecedent proposition must semantically license the elliptical one. The elliptical proposition will be licensed only if it can be inferred by the proposition that serves as its antecedent. A negative answer is part of the denotation of a question, and thus a possible inference from it.

According to Watanabe (2004), a semantic approach to ellipsis does not yield the right results, though.[12] The system of negation in Japanese is closely related to that of Greek. In relation to the Japanese examples in (62), Watanabe points out that Giannakidou's analysis would predict that the representation of the fragment

---

[12] Both Watanabe (2004) and Giannakidou (2006) were circulating as draft versions long before their publication data, which explains the cross-references.

answer in (62b) extends to the one in (62c). Of course, that is not the case, and the answer should be read as in (62d).

(62) a. Nani-o mita no? [Japanese]
what-ACC saw Q
'What did you see?'
b. Nani-mo ~~mi-nak-atta~~.
Nothing see-SN-PAST
'Nothing.'
c. Hebi-o ~~mi-nak-atta~~.
Snake-ACC saw-SN-PAST
'I didn't see a snake.'
d. Hebi-o ~~mita~~.
snake-ACC saw
'I saw a snake.'

Watanabe rejects Giannakidou's inferential approach as a viable solution to the problem of ellipsis resolution. In particular, he argues that under the semantic analysis of ellipsis developed by Merchant (2001), the antecedent is supposed to be a linguistic expression. Accordingly, it is illegitimate to pick up a member from the set of propositions denoted by a wh-question as an antecedent for ellipsis resolution.[13]

Bošković (2008) provides data from Serbo-Croatian that support Watanabe's argumentation. He points out that n-words in this language can be used as fragment answers to affirmative (63), but not negative questions (64).

(63) Q: Šta si kupio? A: Ništa. ~~nisam kupio~~.
what are bought? nothing SN.am bought
Q: 'What did you buy?' A: 'Nothing.'

(64) Context: There was a party yesterday. A knows that John, Mary, and Jane were at the party, but does not know whether Bill, Joan, and Peter were there:

---

[13] Furukawa (2007) points out that 'unembedded' negative indefinites, which are isolated in conversational initial position raise further problems for the ellipsis approach. Consider the context and possible utterances in (i):

(i) Situation: Mary is a TA. Today, since she had to teach, she went into the classroom. When she opened the door, she found that no one was there. Then, with surprise she said:

a. Nobody!	a'. There is nobody there!	[English]
b. Personne!	b'. Il n'y a personne!	[French]
c. KANENAS!	c'. Dhen iparxei KANENAS!	[Greek]
d. #Ni-kogo!	d'. Ni-kogo net!	[Russian]
e. #Dare-mo!	e'. (dare-mo) i-nai!	[Japanese]

In the case of isolation, there is no discourse antecedent for the elided material. On the basis of the felicitous fragment utterances in (ia-c), Furukawa concludes that the ellipsis approach is untenable for English, French and Greek. According to Furukawa, the infelicity of (id) and (ie) does not provide support for the analysis of Russian and Japanese n-words as inherently negative. Given that negative concord languages do not pattern alike in conversation initial position, I will leave this issue open for further investigation, and focus the discussion on fragment answers to an overt question.

Q: Ko nije došao? A: ?*Niko ~~nije došao~~.
who SN.is come ?*Niko ~~nije došao~~.
Q: 'Who didn't come?' A': Niko ~~nije došao~~
nobody SN.is come

The full answer is felicitous in this context, but the fragment answer is not. If anything, negative questions should make it easier to license the n-words approach, for the elided negation can be recovered from the negative question. According to Bošković, these observations pose a serious threat to Giannakidou's approach, and support Watanabe's (2004) claim that the ellipsis analysis of fragment answers is untenable.

Not all languages rule out n-words from fragment answers in response to negative questions. But when they can be felicitously used, they give rise to a double negation reading, i.e. the answers in (65) mean 'nobody did not answer'.

(65) a. Q: Chi non ha risposto? A: Nessuno. [Italian]
Q: Who SN has answered A: nobody
Q: 'Who has not answered?' A: 'Nobody.'
b. Q: Kto nie odpowiedział? A: Nikt. [Polish]
Q: Who SN answered A: nobody
Q: 'Who did not answer?' A: 'Nobody.'

The examples in (65) support the view that negative concord is a clause-bound phenomenon (cf. Chapter 6, Section 1 for discussion). The negation in the question and the n-word in the fragment answer contribute two independent negations, which add up to a double negation reading. Under the ellipsis analysis of n-words in fragment answers in Giannakidou (2000, 2006), the data in (62)–(65) are hard to explain.

Even if the problems related to (62)–(65) can somehow be solved, it remains unclear how to maintain the contrast between NPIs and n-words in fragment answers like (46) through (51). The relevant Greek example is repeated here as (66):

(66) Q: Pjon ihes? A: KANENAN [Greek]
Who did you see? Nobody
A: *kanenan
Anybody

If the n-word *KANENAN* in (66) can take a negative proposition as its antecedent, along the lines of (61), why cannot its NPI counterpart *kanenan* in (66) do the same (pace requirements on the NPI being in the c-command domain of the negation marker in the full answer)? Giannakidou (2006) maintains that emphasis plays a role here. The emphatic n-word *KANENAN* licenses a fragment answer with a negative meaning, but its non-emphatic counterpart *kanenan* does not (66).

Support for this analysis comes from the use of emphatic bare minimizers with a negative meaning in fragment answers, as illustrated in (52), repeated here as (67):

(67) Q: Ipe tipota i Ilectra olo to vradi? [Greek]
said.3SG anything the Electra all the evening
'Did Electra say anything all evening?'
A: LEKSI
word
'Not a word.'

Not all Greek speakers tolerate emphatic bare minimizers in fragment answers.[14] There are two problems with Giannakidou's proposal. On the one hand, emphasis is not sufficient to legitimate a negative reading, for emphatic *LEKSI* gets an existential interpretation in the *if*-clause of the conditional in (42b). It is not clear how emphasis comes into play in the licensing of a negative reading under ellipsis, when it does not in downward entailing contexts more generally.

Second, even if emphasis can be argued to play a role in Greek on the basis of these examples, the cross-linguistic validity of this argumentation remains to be established.[15] Hoyt (2006) confirms that *wɛla*-DPs, Palestinian Arabic n-words which occur in fragment answers (68) are pronounced with strong focal stress, and are more emphatic than NPIs like *ḥada* ('anyone') or *iši* ('anything') in the language.

(68) Q: šu ḳal-l-ak? A: wɛla iši. [Palestinian Arabic]
What said.3MS-to-you? not.even thing
Q: What did he say to you? A: nothing at all.

However, in languages such as French, Spanish and Catalan, NPIs remain unacceptable as fragment answers, even if we add emphasis, because n-words and NPIs belong to different lexical classes in these languages. The examples in (69) and (70) illustrate this (data from Vallduví 1994).

(69) Q: Qu'est-ce que qu'il a dit? A: Rien. [French]
What did he say? Nothing
A: *Quoi que ce soit.
What that it is-SUBJ
A: *(Pas) un mot
*(not) a word

---

[14] Evangelia Vlachou (p.c.) provides the following example which illustrates that minimizers are not generally licit as fragment answers, and usually require the support of the marker wof sentential negation.

(i) Q: Idhes kanenan na pernai? A: *PSIXI.
Saw.2SG anybody SUBJ pass-by.3SG Soul
Did you see anyone passing by? A: PSIXI dhen perase.
Soul SN passed-by.3SG
'Not a soul passed by.'

[15] According to Giannakidou (2008), special intonation also plays a role in Japanese n-words, but she does not discuss the data, so I am leaving this open.

(70) Q: ¿Queda vino? A: Nada [Spanish]
Is there any wine left? Nothing
A:#(Una) gota!
(a) drop!
Ni (una) gota!
Not a drop!

It is impossible to save the infelicity of the NPIs in (69, 70) by means of stress or other means of emphasis. At the same time, there is no evidence that the n-words in (69, 70) are emphatic as opposed to the NPIs. According to Vlachou (2007: 146–147), *quoi que ce soit* requires both contextually relevant and irrelevant values for the variable to be taken into consideration, whereas *rien* quantifies over contextually relevant alternatives. If anything, *rien* is thus less emphatic than *quoi que ce soit*, because it does not involve domain widening (cf. Section 3).

Zeijlstra (2007) also maintains that negative concord is normally not emphatic in Romance, and Italian speakers prefer the use of a negative polarity item to convey emphatic negation. Moreover, Zeijlstra (2007) proposes an analysis of emphatic multiple negative expressions in a double negation language like (standard) Dutch that treats these cases as different from negative concord (cf. Chapter 5, Section 10).[16]

All in all, there is insufficient evidence that emphasis plays a role in negative concord languages in general. It would defeat the purpose to arbitrarily label n-words as emphatic, and NPIs as non-emphatic; so in the absence of a principled explanation of the role of emphasis in languages other than Greek, I do not want to pursue this avenue in a cross-linguistic theory of negative concord.

Zanuttini (1991), Déprez (1999, 2000), de Swart and Sag (2002), Watanabe (2004), Falaus (2008) and Bošković (2008) conclude that a non-negative interpretation of n-words cannot be upheld in fragment answers. These contexts show that n-words are inherently negative, whereas negative polarity items are not. I endorse their conclusions in this book.

---

[16]An anonymous reviewer agrees with Giannakidou that words like *anything, kanenan* and other existential NPIs are not emphatic, hence cannot appear as fragment answers. As pointed out by the reviewer, if *any* becomes emphatic (in its free choice use), it is fine:

(i) Q: Who would you like to talk to?
A: Anybody!

I do not dispute the felicity of (i), but the argument is invalid for two reasons. First, Krifka (1995) establishes a strict distinction between stressed and unstressed *any*, so we can take these to be two different lexical items. Free choice items are not subject to licensing conditions according to Vlachou (2007), so under this analysis, the felicitous occurrence of *anybody* in (i) is tangential to the NPI/n-word distinction. Second, the argumentation does not generalize, for not all NPIs in all languages double as free choice items (in fact, very few do). Other minimizers in English (like *a word, a drop, a soul*) cannot be saved by emphasis.

1.4.6 Toward a Compositional Semantics of Negative Concord

One possible way out of the conclusions drawn in Section 4.5 concerning the status of n-words as inherently negative is for the proponents of a non-negative analysis of n-words to postulate that some languages can express negation covertly, while others have to always realize it overtly. This view seems to underlie several of the approaches discussed here (including, most recently, Penka 2007 and Tubau 2008). Along similar lines, Zeijlstra (2004) and Bošković (2008) suggest that not all markers of sentential negation convey semantic negation.[17]

The ambiguities discussed in Section 5 make it difficult to maintain this view, for the distinction between double negation and negative concord languages is fluid, and intermediate cases are possible. Even if it would be possible to parametrize the languages according to their capacity to realize negation covertly, and deal with the intermediate cases and with the problems raised by negative polarity items in some way, this solution raises two conceptual problems.

My first problem is that, as a semanticist, I find it impossible to defend the view that a compositional analysis of negation and negative concord is to be based on a covert or empty negation operator. If a truth-functional operator like $\neg$ can remain implicit in the sentence, or the negation particle is semantically potent in some sentences, but not others, the distinction between affirmation and negation is blurred.

As long as independent means of establishing when the invisible operator is there, how it is licensed, and in which configurations negation markers do not contribute any meaning are lacking, it is impossible to build a compositional semantic theory of negation. So far, theories exploring this option do not agree on the contexts in which an implicit truth-functional operator $\neg$ occurs, the conditions under which it can and cannot be licensed, and the configurations in which the negation marker is semantically empty.

The second problem is that syntactically invisible but semantically potent negations, and syntactically visible but semantically empty negations are not in line with the view that negation is semantically marked, and therefore universally more complex in form (Horn 1989). The view of negation as the marked member of the pair <affirmation, negation> has been outlined in Section 1, and will be grounded in an evolutionary bidirectional OT model in Chapter 3. Horn's division of pragmatic labor requires that unmarked forms pair up with unmarked meanings, and marked forms with marked meanings. Covert and empty negation operators are not in line with the Horn pattern, which constitutes the communicative underpinning of the bidirectional OT grammar presented in Chapter 2.

In view of these conceptual problems, this book applies the principles of compositionality of meaning to a surface level syntax without hidden levels of

[17] This might be appropriate for instances of so-called 'expletive' negation, which are not discussed in this book. Compare Espinal (1992) for a proposal.

representation and covert operators (as far as negation and negative indefinites are concerned).[18] This means that I cannot adopt a lexical semantics of n-words in terms of existential or universal quantification, or a mixture of those interpretations, as proposed by Laka (1990), Corblin (1996), Herburger (2001) or Giannakidou (2000, 2006), because under such an analysis, an appeal to covert negation is the only way out in cases where the n-word appears to be 'self-licensing' (cf. Sections 4.3–4.5). In line with Zanuttini (1991), Ladusaw (1992), Déprez (1997), Espinal (2000) de Swart and Sag (2002), Watanabe (2004) and Bošković (2008), I analyze n-words as inherently negative. Accordingly, their lexical semantics is closer to that of negative quantifiers like the English *nobody* than to that of NPIs like the English *anybody*, so this position implies a strict distinction between NPIs and n-words.

The association of n-words to negative quantifiers also underlies the analysis of negation in Penka (2006, 2007). Penka categorizes the English *nobody,* German *niemand,* Italian *nessuno,* and Polish *nikt* as members of a broad class of negative indefinites. Their felicity to occur as negative fragment answers opposes negative indefinites to NPIs. Note that the lexical semantics Penka adopts is quite different from the one proposed in this book, because it crucially relies on a covert, abstract negation marker and a minimalist checking approach. Nevertheless, as emphasized by Falaus (2008), whatever proposal turns out to provide the best analysis for negative quantifiers, will also have to apply to negative concord languages. In line with Penka and Falaus, I adopt the same terminology, and label both negative quantifiers (in double negation languages like standard English and standard German) and n-words (in negative concord languages like Italian, Greek and Polish) as negative indefinites.

In the analysis developed in this book, n-words and negative quantifiers are uniformly interpreted as negative indefinites ($\neg\exists$). The principle of compositionality of meaning is at the heart of the semantics. So under a unified lexical semantics of negative indefinites, the distinctions between double negation and negative concord languages reside solely in the grammar.

1.5 A Polyadic Quantifier Analysis of Double Negation and Negative Concord

Most of the approaches presented so far use only the tools of first-order predicate logic. However, this line of analysis runs into a dead end if *nobody, niemand, nessuno, nikt,* and *kanenan* all have the same lexical semantics, and function application constitutes the sole mode of composition in a first-order system. The combination of these assumptions makes it impossible to account for the contrast between double negation and negative concord languages, as outlined in Section 4.

[18] My position is limited to negation operator and negative indefinites, which have basic truth-conditional import. This book is not committed to any claims about the status of empty categories in syntax in general, but does not rely on any.

Alternative analyses go beyond first-order logic (or standard generalized quantifier theory) in one way or another, and expand the inventory of semantic tools. The key is to propose minimal or independently motivated extensions of first-order logic, which pay off by offering a higher explanatory value. Two analyses exploring such ideas were developed around the same time (Section 5.1). The analysis I adopt in Section 5.2 inherits features of both of them, and leads toward a typology of negation (Section 5.3).

1.5.1 Compositionality and Ambiguities

A highly influential proposal concerning the semantics of negative concord was made by Ladusaw (1992), who proposed treating n-words as self-licensing negative polarity items. Thus, in the absence of a trigger, n-words such as *nessuno* and *niente* license themselves, but regular NPIs such as *anybody* do not. Technically, the n-word contributes an existential quantifier $\exists$ to the truth conditions of the sentence. The negative force of the n-word *nessuno* is located in a negative feature that a regular NPI like *anything* lacks. All negative features contributed by sentential negation and n-words percolate up the tree, and get discharged at the top, leading to a single, wide scope negation $\neg$ that has all the existential quantifiers contributed by the n-word(s) in its scope. The extra tool needed in this analysis is a feature percolation and interpretation mechanism, which Ladusaw borrows from the grammatical framework of GPSG (Gazdar et al. 1985).

The spirit of Ladusaw's ideas has been pervasive in much subsequent work, because it highlights the nature of negative concord as an agreement phenomenon: even though negation is expressed in different places in the syntax, it is interpreted only once. What the analysis in this book inherits from Ladusaw's analysis is the nature of n-words as inherently negative, and the idea that negative concord is an instance of agreement. What it opposes to Zeijlstra's (2004) implementation of Ladusaw is that negative concord is viewed as an instance of semantic, not syntactic, agreement.

Zanuttini (1991) and Haegeman and Zanuttini (1996) also emphasize the nature of negative concord as an agreement phenomenon, but in their analysis, n-words denote $\forall\neg$. They define an operation of factorization which reinterprets a sequence of quantifiers $\forall x_1\neg\forall x_2\neg\ldots\forall x_n\neg$ as a new sequence $\forall x_1, x_2\ldots x_n\ \neg$. According to May (1989), factorization fails to respect compositionality, because part of the semantic contribution of the composing elements is simply erased.

As an alternative, May defines an absorption operation which interprets a sequence of negative indefinites $NO_{x1},\ldots NO_{xn}$ as a polyadic quantifier complex $NO_{x1\ldots xn}$ (cf. also van Benthem 1989, Keenan and Westerståhl 1997). May's analysis has also been criticized for its lack of compositionality (e.g. Corblin 1996). Note that absorption requires a mode of composition different from function application, so it does not respect first-order (Fregean) compositionality.

However, absorption as quantifier resumption is embedded in the theory of polyadic quantification (May 1989, van Benthem 1989, Keenan and Westerståhl 1997), so it is one of a series of operations in natural language that goes beyond standard generalized quantifier theory. If the set of operations defined in polyadic generalized quantifier theory constitute permissible combinatoric rules, May's analysis is compositional in a higher order theory of meaning. This view is defended by de Swart and Sag (2002), who propose negative resumption as the interpretation of negative concord (Section 5.2).

Corblin (1996) observes that almost all analyses of negative concord focus exclusively on deriving a single negation reading from a sequence of n-words, and the analyses developed by Ladusaw, Zanuttini and May are no exception. Corblin points out that such analyses do not do justice to the observation that, in certain languages at least, sentences involving two negative indefinites are ambiguous, and allow both a single and a double negation reading, depending on the context. Corblin's French examples are in (71). Corblin and Derzhanski (1997) make similar claims about the Bulgarian example in (72).

(71) a. Personne n' aime personne [French]
nobody SN loves nobody
= No one loves anyone. [NC]
= Everyone loves someone. [DN]
b. Personne n' est l' enfant de personne.
nobody SN is the child of nobody
= No one is the child of anyone. [NC]
= Everyone is the child of someone. [DN]

(72) Nikoj ne običa nikogo [Bulgarian]
nobody.NOM SN loves nobody.ACC
= No one loves anyone. [NC]
= Everyone loves someone. [DN]

The existence of double negation readings in (71) and (72) leads Corblin (1996) to defend an ambiguity thesis at the compositional level. Corblin formulates a construction rule for negative quantifiers in a DRT framework, which introduces a negation and an indefinite in the scope of negation. If a new quantifier shows up when the construction rule has already been applied, one option is to apply just the second half of the rule. This is equivalent to a shift of the n-word to an existential quantifier, and results in the desired negative concord interpretation.

The formulation in terms of a construction rule which optionally applies in a context already containing a negative quantifier strongly suggests that the ambiguity between the single and the double negation reading of examples like (71) and (72) is in the construction, rather than the lexicon. What my analysis inherits from Corblin is the emphasis on a grammatical approach to negative concord. Its shares with Corblin (1996), Herburger (2001) and Falaus (2007a, b) the desire to account for double negation readings in negative concord languages.

1.5.2 The Semantics of Resumptive Negative Quantification

de Swart and Sag (2002) propose an analysis of double negation and negative concord in the framework of polyadic quantifier theory which builds on the ideas advanced by Zanuttini, van Benthem and May. The analysis focuses on the derivation of the single as well as the double negation reading of sentences like (71) and (72).

Polyadic quantifier theory is an elaboration of standard generalized quantifier theory, which deals with interpretations of sequences of quantifiers that cannot be derived by function application. The combination of quantifiers by function application leads to an iteration of quantifiers, corresponding to the scopal order of the nominals.

Iteration of quantifiers leaves a variety of cases unaccounted for. It does not provide the bound reading of the reflexive in (73a), the reading in which the books vary with the students in (73b), the cumulative reading of (73c), or the pair-list reading of (73d).

(73) a. Every student likes himself.
b. Every students bought a different book.
c. Five hundred companies own three thousand computers.
d. Who loves who?

What the cases illustrated in (73) have in common is that a bottom-up interpretation of the sentence in standard generalized quantifier theory fails, because the lower quantifier depends on the higher quantifier for its meaning. Note that it may not be impossible to represent the truth conditions of the sentence in first-order logic, as (73a) illustrates. What is at stake is the derivation of the intended interpretation in a compositional way, namely by formulating the different modes of composition for a sequence of quantifiers.

A number of rules for the interpretation of sequences of quantifiers are formulated by Keenan (1987), May (1989), van Benthem (1989), and Keenan and Westerståhl (1997). In so far as polyadic quantifier theory is motivated by the need to account for a range of constructions that cannot be handled by iteration, Déprez (1997), Espinal (2000) and de Swart and Sag (2002) consider it legitimate to use this framework to account for negative concord, viewed as a configuration in which the interpretation of the lower negative quantifier depends on that of a higher one. Déprez and Espinal take n-words to denote zero cardinality, and rely on cumulativity to derive the single negation reading of negative concord constructions. de Swart and Sag (2002) follow May (1989) and van Benthem (1989) in treating negative concord as an instance of absorption or resumption of negative quantifiers.

Keenan and Westerståhl (1997) define the resumption of a standard quantifier as the polyadic quantifier which results from the application of the original quantifier to *k*-tuples (pairs, triples, etc.), instead of individuals. The binary resumption of a quantifier Q denoted by an NP is the quantifier Q' given by the following rule.

(74) Binary resumption (Keenan and Westerståhl 1997):
$Q'^{A,B}_{E}(R) = Q^{A\times B}_{E^2}(R)$
Where A and B are subsets of the universe of discourse E, and A×B and R are subsets of E^2, i.e. sets of pairs of entities in the universe E.

Suppose the Italian n-words *nessuno* and *niente* are treated as expressions lexically denoting a negative quantifier ¬∃x. This leads to the generalized quantifier representation in NO^{hum}_{E} for *nessuno* or *niente*, with NO being the quantifier interpreted on the universe of discourse E, restricted to the subset of humans (for *nessuno*) or things (for *niente*). The semantics of NO is standard: in set-theoretic terms, it denotes the empty intersection between two sets.

Application of the rule of binary resumption to the sequence of n-words in (40a), repeated here as (75a), leads to the structure in (75b), which has the truth conditions spelled out in (75c) in first-order logic.

(75) a. *Nessuno* ha detto *niente*. [Italian]
Nobody has said nothing.
'Nobody has said anything.'
b. $\text{NO}_{E2}^{\text{hum}\times\text{thing}}$ (SAY)
c. $\neg\exists x \exists y$ Say(x,y)

The resumptive quantifier in (75b) ranges over sets of pairs of humans and things. The empty intersection with the set of pairs in the denotation of *say* requires that there be no pair of a person and a thing such that that pair is a member of the denotation of *say*. Quantification over pairs is equivalent to the first-order representation in (75c). Even though the truth conditions of the sentence can be written in first-order logic, the only way to obtain a compositional interpretation of the sentence based on the lexical semantics $\neg\exists x$ of the n-word is to adopt an interpretation in terms of polyadic quantification.

The resumptive interpretation accounts for negative concord by viewing the two occurrences of the negative indefinite as an instance of semantic agreement, in the spirit of Ladusaw's (1992) analysis. Technically, resumption pairs up the two negative indefinites as two variables bound by a single negative quantifier. In this book, I use negative resumption primarily to provide an interpretation of negative concord, but Chapter 4 (Section 5) discusses Szabolcsi's (2004) extension of the resumption mechanism to negative polarity. Szabolcsi's unification of the two phenomena is particularly relevant for the diachronic development of NPIs into n-words, as shown there. Besides that section, the book concentrates on the mechanism of resumptive negative quantification for sequences of n-words.

Following Keenan and Westerståhl, I generalize the definition of resumptive quantification to a sequence of *k* monadic quantifiers Q' binding just one variable each, and interpreted on the universe of discourse E, with a one-place predicate A as their restrictor, and taking a k-ary relation R as its scope.[19]

(76) Resumption of a k-ary quantifier.
$Q'_{E}^{A1, A2, \ldots Ak}(R) = Q_{Ek}^{A1\times A2\times \ldots Ak}(R)$.

The resumptive quantifier is a polyadic quantifier binding *k* variables, interpreted in the universe of discourse E^k, taking the subset $A_1 \times A_2 \times \ldots A_k$ of E^k as its restrictor, and the k-ary predicate R as its scope. This generalized definition is applied in (77).[20]

[19] Keenan and Westerståhl's definition is slightly more complex than mine, because they want to generalize to the possibility of resumptive quantification with relational nouns. This book is not concerned with relations nouns, so I maintain the easier definition for readability.

[20] For now, I ignore the clitic *ne*, which will be argued to not contribute a semantic negation in Chapter 5 (Section 6).

(77) a. Personne n' a rien dit à personne. [written French]
Nobody SN has nothing said to nobody
'Nobody said anything to anyone.'
b. Nessuno *(non) ha parlato di niente con nessuno. [Italian]
Nobody *(SN) has talked about nothing to nobody.
'Nobody talked to anyone about anything.'
c. $\text{NO}_{E3}^{\text{HUM} \times \text{INAN} \times \text{HUM}}(\text{SAY})$
d. $\neg\exists x \exists y \exists z\ R(x,y,z)$

The sequence of n-words (*personne, rien, personne* in 77a, *nessuno, niente, nessuno* in 77b) provides a series of quantifiers NO, ranging over humans, things, and humans respectively. R is provided by the three-place predicate *say-to* in (77a), and *talk-to* in (77b). The resumptive quantifier reads as $\text{NO}_{E3}^{\text{HUM} \times \text{INAN} \times \text{HUM}}$ (SAY) in (77c), and spells out the semantics of (77a). The truth conditions of the sentence require that there be no triple of a human, a thing and a human such that that triple stands in the 'say-to' relation. The truth conditions of the resumptive quantifier are equivalent to the first-order formula $\neg\exists x \exists y \exists z\ R(x,y,z)$ in (77d), which requires there not to be an individual *x*, a thing *y* and an individual *z* such that *x* says *y* to *z*. (77b) is parallel.

Keenan and Westerståhl take resumption to apply only to a sequence of quantifiers that are somehow 'the same'. In the case of negative concord, resumption applies to a sequence of anti-additive quantifiers provided by negative indefinites (*niente*, *personne*, etc.). Quantifiers like *few, at most two* are monotone decreasing, but not anti-additive. They license negative polarity items, but do not participate in resumptive quantification, and do not lead to negative concord interpretations.

The marker of sentential negation and connectives like *without* are also anti-additive, so they can participate in the construction of the polyadic quantifier as well. Section 6 discusses the status of the marker of sentential negation in various negative concord systems. The participation of sentential negation in the resumptive negative quantifier is spelled out in Section 6.3. Chapter 6 (Section 2) returns to subordinate clauses introduced by *without.*

In this book, the focus is on pronominal negative indefinites such as *nobody, personne,* and *nessuno*, and full DPs such as *no student, aucun étudiant,* and *nessuno degli studenti* are not addressed. Differences between pronominals and full DPs have been referred to occasionally in the literature (cf. Déprez 2000 and Corblin and Tovena 2003 for Romance, Haegeman and Lohndal 2008 for West Flemish). However, a broad cross-linguistic overview of the data for full DPs is missing, which motivates the restriction to pronominal and adverbial negative indefinites in this book.

1.5.3 Ambiguities: Iteration and Resumption

The syntax–semantics interface defines how the DN and NC readings are obtained from the syntax. HPSG uses a notion of Cooper storage in which all quantifiers are collected into a store, and interpreted upon retrieval from the store (cf. Manning, Sag and Iida 1999). This mechanism is generally used to account for scope ambiguities, but de Swart and Sag (2002) extend it to polyadic quantification. All negative (anti-additive) quantifiers are collected into a so-called N-store. Interpretation upon retrieval from the store is by means of iteration of monadic quantifiers (leading to DN) or by resumption, building a polyadic quantifier (leading to NC). I will not elaborate on the retrieval mechanism here, but refer to de Swart and Sag (2002) for details.

Crucially, the HPSG grammar does not distinguish between DN and NC. This accounts for the situation in languages like French, in which both readings are available for a sequence of negative indefinites. Consider the ambiguity of the following sentence in the HPSG analysis of de Swart and Sag (2002).

(78) Personne n'aime personne. [French]
Arg-St<[Store $\{NO_{\{x\}}^{\{Person(x)\}}\}$], [Store $\{NO_{\{y\}}^{\{Person(y)\}}\}$]>
Content Quants <$NO_{\{x\}}^{\{Person(x)\}}$, $NO_{\{y\}}^{\{Person(y)\}}$> Nucleus *Love(x,y)*
Semantic interpretation (iteration): NO(HUM, {x|NO(HUM, {y|x loves y})})
In first-order logic: $\neg\exists x\neg\exists y$ Love(x,y) [DN]

(79) Personne n'aime personne. [French]
Arg-St<[Store $\{NO_{\{x\}}^{\{Person(x)\}}\}$], [Store $\{NO_{\{y\}}^{\{Person(y)\}}\}$]>
Content Quants <$NO_{\{x,y\}}^{\{Person(x),\ Person(y)\}}$> Nucleus *Love(x,y)*
Semantic interpretation (resumption): $NO_{E2}^{HUM\times HUM}$(LOVE)
In first-order logic: $\neg\exists x\exists y$ Love(x,y) [NC]

The representations in (78) and (79) are identical as far as the argument structure, the storing mechanism, and the relational interpretation of the word *love* are concerned. The difference resides in the interpretation of the polyadic quantifier upon retrieval from the N-store: iteration in (78), and resumption in (79). The iteration of quantifiers in (78) requires that there be an empty intersection between the set of persons, and the set of individuals that love no one. This is equivalent to the double negation reading represented in first-order representation. The resumptive interpretation in (79) creates a negative quantifier ranging over pairs of individuals, which excludes all pairs of humans from the denotation of the love relation. This amounts to the single negation reading in a first-order formula.

The main insights of this analysis are the following. The HPSG grammar assumes no lexical difference between negative quantifiers and n-words: both contribute a negative existential quantifier. In line with the unified lexical semantics, the rest of this book uses the terms 'negative indefinite' and 'Neg-expression' to designate both negative quantifiers (like *nothing*) and n-words (like *personne, nessuno*, KANENAN). The analysis works for n-words in argument and adjunct position alike (so *nobody* and *nothing,* as well as *never* and *nowhere*).

Finally, it does not involve covert or empty negations (i.e. syntactically invisible but semantically potent negations, or syntactically visible but semantically inoperative negations). Although both readings of the sentence can be spelled out by means of a first-order logical formula, resumption is not dispensable, for the resumptive polyadic quantifier provides a higher-order compositional interpretation in a surface-oriented syntax.

1.5.4 Toward a Typology of Negation

The ambiguity of French and Bulgarian examples like (71) and (72) is real, and constitutes a problem for most analyses of negative concord, which exclusively focus on deriving the single negation reading. The polyadic quantifier analysis has an advantage here, as it spells out the two readings in terms of iteration and resumption (cf. 78 and 79). At the same time, it is quite clear that the double negation reading of these sentences is highly marked, and that most instances of a sequence of two n-words in French or Bulgarian lead to a single negation reading.

An important question raised by the analysis proposed by de Swart and Sag (2002), and first pointed out by Zeijlstra (2004: 207) is why certain languages are predominantly negative concord languages (French, other Romance languages, Slavic, Greek, Afrikaans, etc.), whereas other languages normally interpret a sequence of negative indefinites in terms of double negation (standard English, Dutch, German, Swedish, etc.) The HPSG analysis developed by de Swart and Sag (2002) provides the space of possible meanings created by the grammar, but does not predict cross-linguistic variation where it arises.

Unlike Zeijlstra, I do not take this as a decisive argument against the HPSG analysis. I have two reasons for it. First, double negation readings are attested for concord languages like French and Bulgarian (see above), but analyses other than the polyadic quantifier analysis do not offer a proper account of these ambiguities. Second, resumptive readings are marginal in double negation languages, but they are not excluded, and we need a theory that can handle them.

As far as double negation languages are concerned, van Benthem (1989) claims that the English sentence (80) has the same two readings as its French counterpart (78/79).

(80) Nobody loves nobody.

Not everyone I consulted finds the ambiguity of (80) easy to access. The attested example (81), taken from an internet source, might be a better example. It illustrates the resumptive reading of the sequence *nobody-nothing* in (standard) English, as opposed to the double negation reading of (40d).

(81) When *nobody* knows *nothing*, everybody is an expert. Nobody can seriously claim to be an expert on the collapse of the World Trade Center, simply because nobody had a chance to study the rubble. Everybody who has looked at the photographs and television news video knows as much about the collapse as the most knowledgeable scientists. Therefore, everybody who has viewed the photographs and videos can claim to be an expert.

The first line of (81) contains a claim that is elaborated by the following sentences. The elaboration establishes the resumptive reading of the sequence *nobody-nothing* as the contextually relevant interpretation. The intended reading of the sentence is that there is no pair of an individual *x* and a thing *y*, such that *x* saw *y*. The first line of (81) then has the same interpretation as 'nobody knows anything.'

Similarly, the example (82b) was used as a slogan by Amnesty International in the seventies, and supports a marginal use of the resumptive reading in (standard) Dutch, another typical double negation language (82a):

(82) a. *Niemand* hoeft voor *niets* te werken.
Nobody needs for nothing to work
'Nobody needs to work for free.'
b. Als *niemand* luistert naar *niemand* vallen er doden in plaats van woorden.
If nobody listens to nobody fall there deaths in stead of words
'If nobody listens to anybody, the conversation doesn't die, people do.'

The examples in (81) and (82b) involve resumption of a sequence of negative quantifiers.

However, the examples are infrequent, and resumption seems to be a marginal phenomenon in typical double negation languages like English and Dutch. Even if these examples are analyzed in terms of emphatic negation along the lines of van der Wouden (1994) and Zeijlstra (2007), the semantics will have to rely on resumptive quantification (at least in a surface-oriented syntax).

I conclude that the polyadic quantifier analysis has so many advantages that it is worth upholding, even if it is unable to account for the cross-linguistic variations found in the availability of negative concord and double negation readings. However, it needs to be enriched with a typological dimension. The approach adopted in this book allows me to distinguish two classes of languages in terms of the optimality theoretic grammar they adopt. The OT analysis developed in Chapter 4 is built on top of the polyadic quantifier analysis, so it should be viewed as an elaboration of the earlier proposal made by de Swart and Sag (2002) along a typological dimension.

1.6 Negation and Negative Indefinites

Section 2 of this chapter focused on the marker of sentential negation. Sections 4 and 5 focused on n-words participating in negative concord. This section brings the two issues together and presents the main systems of negative concord found in languages.

1.6.1 Varieties of Negative Concord

Den Besten (1986) and Haspelmath (1997) distinguish three types of negative concord systems. Here I use the labels *strict negative concord, nonstrict negative concord* and *negative spread* introduced by Giannakidou (1997, 1998) to describe them.

In strict negative concord varieties, the marker of sentential negation is obligatorily present in all sentences containing an n-word. Polish, Greek, Hungarian,

Romanian, Japanese and Slavic exemplify this system (with Polish examples from Haspelmath 1997: 201, Romanian ones from Corblin and Tovena 2003, Greek ones from Giannakidou (2006), and Japanese ones from Watanabe 2004).

(83) a. Nikt *(nie) przyszedł. [Polish]
nobody *(SN) came.
'Nobody came.'
b. *(Nie) widziałam nikogo.
*(SN) saw nobody.
'I saw nobody.'

(84) a. Nimeni *(nu) a venit. [Romanian]
nobody *(SN) has come.
'Nobody came'
b. *(Nu) a venit nimeni.
*(SN) has come nobody.
'Nobody came'

(85) a. KANENAS *(dhen) ipe TIPOTA. [Greek]
nobody *(SN) said.3SG nothing
'Nobody said anything.'
b. O Petros *(dhen) idhe TIPOTA.
the Peter *(SN) saw.3SG nothing
'Peter didn't see anything.'

(86) a. Dare-mo John-o hihanshi- *(nak)-atta. [Japanese]
Who-mo John-ACC criticize- *(SN)-PAST
'Nobody criticized John.'
b. John-wa nani-mo tabe-*(nak)-atta.
John-TOP what-mo eat- *(SN)-PAST
'John didn't eat anything.'

In contrast to the instances of strict negative concord in (83)–(86), Spanish, Italian and European Portuguese exemplify nonstrict negative concord. The examples in (87a) and (88a) illustrate that a postverbal n-word requires the presence of a preverbal marker of sentential negation. However, when the n-word is in preverbal position, the negation marker is not used in the expression of a single negation reading (87b), (88b) (examples from Zanuttini 1991, Herburger 2001).

(87) a. Mario *(non) ha parlato di niente con nessuno. [Italian]
Mario *(SN) has talked about nothing to nobody.
'Mario didn't talk to anyone about anything.'
b. Nessuno (*?non) ha parlato con nessuno.
Nobody (*?SN) has talked with nobody.
'Nobody talked to anyone.'

(88) a. *(No) he visto a nadie. [Spanish]
*(SN) has seen nobody
'He hasn't seen anybody.'

b. Nadie (*?no) ha dicho nada.
Nobody (*?SN) has said nothing
'Nobody said anything.'

The phenomenon whereby the negative concord relation is established exclusively between n-words is called negative spread. The examples (87b) and (88b) exemplify negative spread in a nonstrict negative concord language, because the expression of a single negation relies on a sequence of negative indefinites, without the support of a marker of sentential negation.

In nonstrict negative concord languages, negative spread is found in certain constructions, but not others. Systematic negative spread is exemplified by spoken French (89a). The combination of an n-word with the marker of sentential negation *pas* always leads to double negation readings (89b).

(89) a. Personne a rien dit. [Spoken French]
Nobody has nothing said
'Nobody said anything.'
b. Il est pas venu pour rien.
He is SN come for nothing
≠ He didn't come for anything. [NC]
= 'He didn't come for nothing.' [DN]

In the remainder of this book, I will reserve the term negative spread for languages such as spoken French, in which the marker of sentential negation is always incompatible with n-words in the expression of a single negation reading. I will use the term nonstrict NC for languages such as Italian and Spanish that require the support of a marker of sentential negation in postverbal, but not in preverbal, position.

Any typological theory of negative concord needs to provide an analysis of the three main systems of strict NC, nonstrict NC and negative spread. Furthermore, languages do not always clearly fall into one of these three categories. Optional instances of the negation marker and mixed patterns are attested, and need to be integrated in such a typological theory of negation. Chapter 5 provides details on many languages, and develops a range of grammars to handle the observations.

1.6.2 The Marker of Sentential Negation in the Debate on Negative Concord

As pointed out by de Swart and Sag (2002: 401), the fact that the role of the marker of sentential negation in negative concord is subject to considerable cross-linguistic variation constitutes a significant problem for approaches to negative concord in which sentential negation plays an important role as the licensor of the n-word (Laka 1990, Ladusaw 1992, Przepiórkowski and Kupść 1999, Giannakidou 1998, 2000, Zeijlstra 2004 and others).

Licensing conditions on negative polarity items like (*any,* Dutch *hoeven*, etc.) and minimizers (*a red cent, a drop*) are by and large the same across languages. Variation occurs among weak, medium and strong NPIs, but this variation is cross-linguistically stable. If negative concord involves licensing of the n-word by a marker of sentential negation or a negative head, the strict NC, non strict NC and negative spread languages require different sets of licensing conditions on n-words. According to Ladusaw (1992: footnotes 10 and 11), a proliferation of licensing conditions is not very attractive.

Zeijlstra (2004, Chapter 7) fully endorses the consequences of the licensing approach, and claims that negative markers in different types of negative concord languages have different negation features. Variation thus resides in the lexicon. Zeijlstra exploits the distinction the minimalist framework establishes between interpretable and noninterpretable features to this end.

The negation marker in a nonstrict negative concord language such as Italian or Spanish (87 and 88 above) has an interpretable Neg-feature, but the negation marker in strict negative concord languages such as Polish, Greek, Japanese or Romanian (83–86) has an uninterpretable Neg-feature. N-words and negative markers in strict NC languages participate in a feature-checking relation with an abstract, i.e. phonologically empty but semantically potent operator that takes clausal scope. In nonstrict NC languages, the uninterpretable feature of the postverbal n-word is checked against the interpretable feature of the marker of negation, whereas the uninterpretable feature of the preverbal n-word is checked against the interpretable Neg-feature of an abstract negation operator.

Within the overall set-up of the minimalist framework, the covert negative operator is motivated by the unified treatment of negative concord in terms of syntactic rather than semantic agreement (Zeijlstra 2004: 246). In a more surface-oriented syntax, an approach which does not need negations that are semantically potent, but syntactically 'hidden', would be preferred, as outlined in Section 4.6. Zanuttini (1991: 126 sqq) and Ladusaw (1992) have already made this point in relation to Laka's (1990) postulation of a ΣP that hosts semantic negation, but is not always filled with lexical material (cf. Section 4). Although Zeijlstra uses a more recent version of the Chomskian paradigm, he is vulnerable to the same criticism. Under the assumption that covert negation operators are not allowed, it is impossible to reduce negative concord to syntactic agreement, and we are back to where Zanuttini and Ladusaw were in the early nineties.

Watanabe (2004) and Bošković (2008) provide an alternative which also relies on the notion of feature checking in the minimalist framework. Unlike Zeijlstra, Watanabe and Bošković take n-words to be inherently negative (cf. Section 4.5). This raises the question of how the combination of a marker of sentential negation and an n-word can express a single rather than a double negation in contexts like (83)–(86). In double negation as well as negative concord languages, the negative head also contributes a negation.

Watanabe proposes an indirect account in terms of checking focus features, which leads to the copying of neg-features in negative concord languages. Feature copying guarantees the presence of two neg-features on the negative head. The two neg-features cancel each other out, so that the negative head in (83)–(86) denotes the identity function, rather than negation. Bošković proposes two projections for negation, one with an interpretation, and the other with an uninterpretable feature for negation.

Although Watanabe and Bošković account for the doubling of an n-word by a marker of sentential negation in strict and non strict NC languages along these lines, they do not account for the possibility of sequences of multiple n-words in such languages, as illustrated in (90). Watanabe (2004) only discusses the Italian examples in (90a), but data from Hungarian and Greek can be added (examples from Surányi 2006a, b, Giannakidou 2000).

(90) a. Mario non ha parlato di niente con nessuno. [Italian]
Mario SN has talked about nothing to nobody
'Mario didn't talk to anyone about anything.'
b. Sehol nem lát-t-am senki-t. [Hungarian]
Nowhere SN see-PAST-1SG nobody-ACC
'I did not see anybody anywhere.'
c. KANENAS dhen ipe POTE TIPOTA se KANENAN. [Greek]
nobody SN said.3SG never nothing to nobody
'Nobody said anything to anyone.'

The mechanism of feature copying applies to the negative head *non/nem/dhen*, but leaves the negative value of *niente/nessuno* as well as *sehol/senki* and *KANENAS/POTE/TIPOTA* intact. In order to obtain the single rather than the double negation reading of examples like (90), Watanabe suggests that the polyadic quantifier treatment proposed by de Swart and Sag provides a good analysis of negative spread. Watanabe (2004) claims that the extension of this account to negative doubling, as proposed by de Swart and Sag, is not justified, given his treatment of negative doubling.

However, it is just as easy to turn this argument around, and defend the view that a unified analysis, if possible, is preferred. There is no need for an account of negative doubling separate from negative spread under the polyadic quantifier analysis, as the mechanism of resumption of negative quantifiers can account for both phenomena.

1.6.3 Sentential Negation in the Polyadic Quantifier Approach

de Swart and Sag (2002) extend the construction of the resumptive quantifier to include mixed cases in which a sequence of Neg-expressions combines with a marker of sentential negation. Of course, sentential negation is a propositional operator, not a variable binding operator. In terms of the polyadic quantifier theory, this means that it is an expression of a different type. Full NPs (or DPs) denote functions from the power set of the universe of discourse provided by a one-place predicate to truth values; they are defined as type <1> quantifiers in the Lindström type system used by Keenan and Westerståhl (1997). Determiners map a one-place predicate onto a DP, so they denote functions from the power set of the universe of discourse to type <1> quantifiers; they are defined as type <1,1> quantifiers.

A sentential operator like negation is a function from propositional entities into truth-values. Propositions correspond to zero-place predicates, because they denote truth values. This opens the way for the treatment of a nonvariable binding operator such as negation as a quantifier with adicity zero, or a quantifier of type <0>. This proposal is fleshed out in (91).

(91) Non-variable binding, propositional operators such as negation are treated as quantifiers of type <0>.

Once the treatment of sentential negation as a quantifier of type <0> is in place, the definition of resumption can be extended to allow resumption of quantifiers of different types.

Recall that a resumptive negative quantifier interprets a sequence of anti-additive quantifiers $Q^1 \ldots Q^k$ of type <1,1> as one complex negative quantifier Res_Q of type $<1^k, k>$ (cf. Section 5). This means that the resumptive quantifier maps a series of *k* one-place predicates and one *k*-ary predicate onto a proposition. As such, it binds the sum of all the variables of the composing quantifiers. Given that sentential negation does not bind any variables, it does not add any variables to the sum of bound variables, and it does not change the type of the resumptive quantifier.

The extension of the rule for resumptive quantification to a sequence of negative quantifiers that involves a mixture of type <1,1> and type <0> quantifiers is defined in (92).

(92) Resumption of a sequence of *k* type <1,1> quantifiers *Q* and *l* type <0> quantifiers *Q'* leads to the construction of a resumptive quantifier *Q"* of type $<1^k, k>$, such that:

$$Q''^{A1 \ldots Ak}_{E}(R) = Q_{Ek}^{A1 \times A2 \times \ldots Ak}(R)$$

Where $A_1 \ldots A_k$ are subsets of the universe of discourse E, and $A_1 \times A_2 \times \ldots A_k$ and R are subsets of E^k.

As before, resumption is defined only for quantifiers that are somehow 'the same'. The resumptive negative quantifier is defined only for anti-additive quantifiers such as *nobody*, *nessuno*, *personne*, etc. As an antimorphic operator, the semantics of *not, non, nem, dhen* subsumes anti-additivity, so the marker of sentential negation is sufficiently similar to that of the negative indefinite to participate in the resumptive negative quantifier. As a type <0> quantifier, however, it does not affect the type of the resumptive quantifier, and does not change the number of variables bound by the polyadic negative quantifier.

The Italian example in (90a), repeated in (93), and analyzed by means of the extended definition of resumption of negative quantifiers illustrates that this process leads to the desired truth conditions for the sentence.

(93) a. Mario *non* ha parlato di *niente* con *nessuno*. [Italian]
Mario SN has talked about *nothing* to *nobody*
'Mario didn't talk to anyone about anything.'

b. $\text{NO}_{E2}^{\text{INAN} \times \text{HUM}}(\text{TALK\_ABOUT}_M)$
c. $\neg\exists x\exists y$ talk_about(m,x,y)

The two n-words *niente* and *nessuno* in (93a) provide two type <1> quantifiers, the negation marker *non* provides a type <0> quantifier, and the verb applied to the subject (written as TALK_ABOUT_M) denotes a two-place relation. The resumptive quantifier is spelled out in (93b) as $\text{NO}_{E2}^{\text{INAN} \times \text{HUM}}$ (TALK_ABOUT_M). The polyadic negative quantifier binds two individual variables, and requires no pair of a thing and a human to stand in the relation of be-talked-about-by-Mario. This corresponds with the truth conditions spelled out by means of the first-order formula $\neg\exists x\exists y$ talk_about (m,x,y) in (93c), which requires that there not be a thing *x* and an individual *y* such that Mario talked about *x* to *y*. The negation marker *non* has been absorbed in the mixed resumptive quantifier, and leaves no separate reflection in the truth conditions.

The extended definition of resumptive negation quantification emphasizes that the polyadic quantifier analysis relies on n-words, not on sentential negation to express a negative proposition involving multiple indefinites. Accordingly, de Swart and Sag (2002: 401) conclude that the marker of sentential negation is semantically redundant in a negative concord context.

The conclusion drawn by de Swart and Sag should not be misunderstood. Of course, concord languages have a marker of sentential negation, just like any other language. Furthermore, this marker contributes the meaning of the truth-conditional connective ¬, and is employed in the language to convey propositional negation in environments like (94).

(94) a. Gianni *non* mangia. [Italian]
Gianni SN eat
'Gianni doesn't eat.'
b. János *nem* dohányz-ik. [Hungarian]
János SN smoke.3SG
'János doesn't smoke.'
c. *dhen* tha tu to ksana dhósi [Greek]
SN will to.him it again give.PERF.3SG
'He will not give it to him again.'

In examples like (94), the negation marker is responsible for the syntactic marking of negation, as well as for the semantic interpretation of the sentence as expressing a negative proposition. So there is no doubt about the marker of sentential negation actually having the semantics of a negation operator in double negation and negative concord languages alike. The syntactic and semantic status of the markers *non, nem* and *dhen* in (90) is strictly the same as that of *non, nem* and *dhen* in (94).

In fact, *non, nem* and *dhen* in (90) must have the semantics of a truth-functional negation operator; otherwise they cannot participate in the resumption of a sequence of anti-additive quantifiers (93). It is just that in the course of the resumption process, the negation contributed by the sentential negation marker is absorbed in the polyadic negative quantifier, so no separate contribution of *non* is spelled out in the truth conditions in (93c). In the absence of an n-word, no resumptive negative quantifier is built, so the semantic contribution of *non* and *nem* is directly reflected in the truth conditions of (94). Thus the claim that negation is semantically redundant is restricted to sentences involving one or more n-words.

The polyadic quantifier analysis provides a straightforward explanation for the fact that sentential negation is semantically redundant in contexts of resumption, although not in contexts involving just propositional negation. de Swart and Sag (2002) draw on this insight, and predict that languages are free to exclude the marker of negation from concord constructions (as observed for *pas* in French, (89a)) or include the negation marker in the concord system, and exploit it for syntactic purposes, as in nonstrict and strict concord languages. The three varieties of negative concord classified as strict NC, non strict NC and negative spread in Section 6.1 reflect the main typological patterns attested in natural language.

de Swart and Sag (2002) do not offer a typological theory that accounts for the different uses languages make of the marker of sentential negation in contexts involving n-words. The HPSG analysis provides the syntax–semantic interface of natural language grammars in general, and does not predict which language works out which option. Chapter 5 couples the polyadic quantifier analysis proposed by de Swart and Sag (2002) with an OT grammar that accounts for the three main varieties of negative concord, as an extension of the analysis developed in Chapter 4.

1.7 Outline of the Book

The analysis of the expression and interpretation of negation in this book is formulated in the framework of OT. An early case study of negation in OT by Newson (1998) suggests that cross-linguistic variation in the expression of negation can be accounted for in terms of different rankings of constraints. Newson's paper deals mostly with English and Hungarian, and his analysis relies on specific syntactic assumptions from the Minimalist Program. Morimoto (2001) presents an OT-LFG analysis of the placement of negation in the sentence.

There are clear similarities between these early OT accounts, and the analysis developed in this book. In all proposals, the constraint rankings seek a balance between the proliferation of negative expressions in some languages, versus a ban on multiplication of negation in others. My work shares with Morimoto's a concern with the placement of negation in relation to the verb. With Newson, I intend to connect the formal realization of negation to its interpretation. However, the analysis developed in this book is more general than its precursors in four respects.

First, it expands the empirical domain of the study to a larger number of languages, so that a broader typological perspective on negation in natural language can be developed. Second, this book explores not only the marker of sentential negation corresponding to *not* in English, but also negative quantifiers such as the English *nobody* in relation to the n-words characteristic for negative concord languages (such as the Italian *nessuno* and Greek *KANENAN*). Third, I investigate not only the syntax of negation and negative indefinites, but also their semantics, and the way form and meaning hang together in the syntax–semantics interface of negation. Fourth and finally, the analysis is mostly neutral with respect to the syntactic theory in which the syntactic constraints are formulated, and relies on fairly general assumptions about phrase structure and word order. This means that the analysis is compatible with different grammatical frameworks.

The analysis makes two specific assumptions that are not necessarily shared by all syntactic theories. First, it is exclusively surface oriented, and does not account for semantic effects (scope, licensing, etc.) in terms of syntactic movement. Second, it does not rely on empty categories. These two assumptions are shared by grammatical theories such as HPSG and LFG, but typically not by the Minimalist Program (or Principles and Parameters). The restrictions I impose on the general format of the grammar have important consequences for my analysis of negation. In particular, I shy away from covert negation operators (syntactically invisible, but semantically potent negations), and empty negations (syntactically visible, but semantically inactive negations).

Both covert and empty negations are widely used in current analyses of negation discussed in this chapter. My analysis will be different from some of the influential proposals in the literature because of the severe restrictions imposed on the syntax–semantics interface. However, I believe that the limitations provide a more insightful perspective on cross-linguistic variation. If it is possible to develop a typology of the expression and interpretation of negation without covert and empty negations, it would provide a more economical and explanatory theory of cross-linguistic variation.

In order to set the stage for the analysis to be developed in later chapters, Chapter 2 offers an introduction to OT. This chapter motivates the use of OT to provide the grammar of individual languages, and shows how it functions as a theory of linguistic variation, both in a synchronic (typology) and diachronic perspective (language change). The basic assumptions underlying OT syntax and OT semantics are spelled out, and bidirectional OT is offered as a theory of the syntax-semantics interface.

Chapter 3 focuses on the expression and interpretation of propositional negation. The chapter takes its starting point in the markedness of negation, and derives the basic constraints of the OT system used in this book from the asymmetry between assertion and negation. The conflict between the faithfulness constraint FNEG (be faithful to negation in the input) and the markedness constraint *NEG (avoid negation) is resolved by ranking FNEG above *NEG in all languages. This derives Dahl's (1979) generalization that negation is a universal category of natural language.

The chapter further discusses the different positions and realizations of negation in a typological perspective. The different systems are related to diachronic change. The Jespersen cycle is modeled as a series of constraint re-rankings.

Chapter 4 works out a bidirectional OT typology on top of the polyadic quantifier analysis developed by de Swart and Sag (2002). Double negation and negative concord languages strike a different balance between two opposing tendencies. On the one hand, there is strong motivation in favor of the marking of "negative variables" (Corblin and Tovena 2003), which drives the use of n-words in negative concord languages. On the other hand, languages prefer (first-order) iteration over a (second-order) resumptive interpretation, and this preference wins in double negation languages. The syntactic and semantic needs are balanced by the economy constraint *NEG.

Chapter 5 puts Chapters 3 and 4 together in a study of the interaction of negation and negative indefinites in double negation and negative concord languages. The OT analysis accounts for the different negative concord systems one finds in typology and diachrony. The ranking of constraints targeting scope marking of negation governs the use of the negation marker in strict and nonstrict negative concord. Negative spread results under rankings in which the scope constraints rank below the economy constraint *NEG.

Chapter 6 (Section 4) takes another look at the interaction of n-words and the negation marker. No negative concord is established with constituent negation or across clause boundaries. This confirms that double negation is not a conceptual problem in negative concord languages, but a grammatical phenomenon concerning sequences of negative indefinites within a single argument structure. The double negation readings arising with the combination of an n-word and *pas* in spoken French, exemplified in (89b) will be shown to be part of a systematic pattern. The negation marker is not needed in (89b) in order to express a single negation reading, as indicated by (89a). For economy reasons, it should therefore be left out. If it is inserted anyway, its presence needs to be justified for interpretive reasons. The combination of a syntactically marked expression with the semantically marked double negation interpretation is accounted for in a weak bidirectional OT framework.

Double negation readings do not arise in the interaction of the negation marker and n-words in strict negative concord languages with a single negation marker, as observed by Giannakidou (2006). In the OT analysis, this is the result of the negation marker being licensed as a scope marker, which leaves no room for weak bidirectionality. However, double negation readings do arise in strict negative concord languages with discontinuous negation, in nonstrict negative concord languages and in languages exemplifying negative spread, as Chapter 6 will show. Although the examples are rare, and sometimes conceived as marginal, their existence provides independent support for the OT analysis advanced in this book.

Chapter 7 summarizes the main conclusions to be drawn from the proposals made in the book, and sketches perspectives for further typological and theoretical research.

Chapter 2
Expressive and Interpretive Optimization

Introduction and overview This chapter is an outline of Optimality Theory (OT) as a model of grammar. OT is a linguistic theory that is explicitly embedded in a broader cognitive architecture. Expressive optimization is used as a theory of syntax, and interpretive optimization as a theory of semantics. In bidirectional OT, the two come together in a theory of the syntax–semantics interface. As far as the empirical coverage is concerned, this book focuses on the use of OT in typology, with stochastic extensions for language variation and language change.

2.1 Fundamentals of OT as a Model of Grammar

Prince and Smolensky (1997) explore the implications of neural computation as optimization for the theory of grammar. Optimization over symbolic linguistic structures provides the core of a new grammatical architecture, called Optimality Theory. 'The proposition that grammaticality equals optimality sheds light on a wide range of phenomena, from the gulf between production and comprehension in child language, to language learnability, to the fundamental questions of linguistic theory: What is it that the grammars of all languages share, and how may they differ?' (Prince and Smolensky 1997: 1604).

Prince and Smolensky's conceptualization of linguistic theory through optimization principles is embedded in a broader theory of the mind, most recently explored in Smolensky and Legendre (2006). In this work, the authors develop a cognitive architecture based on neural computation, but supporting formally explicit higher-level symbolic descriptions.

According to Smolensky and Legendre (2006: 209), the basic idea is that mental representations are instantiated in the activation values of connectionist units. When analyzed at a higher level as distributed patterns of activity, these same representations are seen as realizations of symbolic structures. In this way, cognitive theories based on neural computation and linguistic theories employing symbolic computation can be integrated, and strengthen each other.

H. de Swart, *Expression and Interpretation of Negation: An OT Typology*,
Studies in Natural Language and Linguistic Theory,
DOI 10.1007/978-90-481-3162-4_2,

In this chapter, I briefly discuss the grounding of OT in a connectionist cognitive architecture, and present the formal organization of the theory (Section 1). Section 2 discusses applications of the OT model in syntax. Section 3 does the same for semantics. Section 4 brings syntax and semantics together in bidirectional OT. Sections 5 and 6 discuss issues in language typology and language change that are relevant to the concerns of this book. The discussion on the fundamentals in this section is based mostly on Smolensky and Legendre (2006: Chapter 1).

Following widely accepted views in cognitive neuroscience, Smolensky and Legendre (2006) adopt a connectionist cognitive architecture. They consider the brain to be a massively parallel computer consisting of billions of processors (neurons). These processors manipulate numbers (neural activation levels). The quantitative internal interactions within the computer (the efficacy of synaptic connections between neurons) change in response to the statistical properties of the computer's experience. The study of the way complex cognitive functions are computed by the brain exploits mathematical models of neural computation known as connectionist networks.

Connectionist networks are collections of simple, parallel computing elements, each of which carries a numerical activation value that it computes from the values of neighboring elements in the network. Each connection carries a numerical strength or weight. The network elements (units) influence each other's values through connections. In a typical connectionist network, input to the system is provided by imposing activation values on the input units of the network. The activation on the input units propagates along the connections until some set of activation values emerges on the output units. These activation values encode the output the system has computed from the input. Mediating between the input and output units, there may be hidden units that do not participate directly in the representation of either the input or the output.

The computation performed by the network in transforming the input pattern of activity to the output pattern depends on the set of connection strengths. These weights are regarded as encoding the system's knowledge. Many connectionist networks perform optimization: they compute those activation values for hidden and output units that, together with the given activation values of the input units, maximize a measure of well-formedness, called harmony. The harmony of a network is interpreted as the degree to which the state satisfies a set of 'soft' constraints implemented in the network's connections. Thus, when the network achieves a state of maximal harmony, it has optimally satisfied these constraints.

In the field of cognitive science, connectionist networks are used to model a wide variety of cognitive tasks. OT is an application of the connectionist view to language. A possible linguistic structure is evaluated by a set of well-formedness constraints, each of which defines one desirable aspect of an ideal linguistic representation. These constraints are highly general, and frequently conflicting. Typically, no structure meets all the constraints, and a mechanism is needed for deciding which constraints are the most important. The well-formed or grammatical structures are the ones that optimally satisfy the constraints, taking into account differing strength or priority of constraints.

In ordinal OT, the constraints are ranked in a strict domination hierarchy. This means that each constraint has complete priority over all the constraints that are lower in the hierarchy. An optimal structure may violate a given constraint C, but only if that permits the structure to better satisfy some constraint C′ ranked above C. The constraints of OT are in this sense minimally violable, and grammaticality is defined in terms of maximal harmony. An ordinal OT grammar is a set of constraints defining the preferred characteristics of linguistic representations, priority-ranked in a strict domination hierarchy. A fundamental hypothesis of OT is that human grammars differ only in ranking, that is, in the way conflicts among constraints are solved. Given that the ranking varies across languages, it must be learned. The constraints themselves are the same across languages – they are strictly universal.

The origin of these universal constraints is very much an open question, one on which OT itself is silent. Many OT constraints are grounded in general cognitive or functional principles. In this book, I will not be committed to the view that the constraints adopted are innate, but I do not exclude this as a possibility either. I will come back to the grounding of the constraints governing the expression and interpretation of affirmation and negation in an evolutionary perspective in Chapter 3.

The constraints are universal, but the constraint ranking is language specific. Grammatical knowledge of a particular language is knowledge of the constraint hierarchy. Use of that knowledge then consists in determining, under various conditions, which linguistic structures optimally satisfy the constraint hierarchy of a particular language. Grammatical knowledge determines the expression of a given meaning by the speaker (production) as well as the interpretation of a given expression by the hearer (comprehension). The two directions of expressive optimization (from meaning to form) and interpretive optimization (from form to meaning) are central to this book.

Despite its embedding in a broader theory of cognitive science, the developments and applications of OT have first and foremost been centered in theoretical linguistics. This is how I will use OT in this book. The earliest applications of OT were concerned with phonology (Prince and Smolensky 1993/2004, McCarthy 2002). Applications to syntax followed quite quickly (Aissen 1999, 2003, Grimshaw 1997, Sells 2001, the volumes edited by Barbosa et al. 1998 and Legendre et al. 2001). The OT study of semantics and pragmatics took shape in the works of Hendriks and de Hoop (2001) and de Hoop and de Swart (2000). Bidirectionality first emerged in Blutner (1998, 2000), and in the volume edited by Blutner and Zeevat (2003), followed by Blutner et al. (2006) and Hendriks et al. (2009).

Given that this book is concerned with the marking and interpretation of negation, bidirectionality is a central notion. Section 2 focuses on expressive optimization as a theory of syntax, Section 3 develops the notion of interpretive optimization as a theory of semantics, and Section 4 presents bidirectional OT as a theory of the syntax–semantics interface.

2.2 Fundamentals of OT Syntax

Smolensky and Legendre (2006: Chapter 12) characterize grammatical knowledge as a system of universal violable constraints on well-formed linguistic combinations, ranked in a language-particular hierarchy. OT is a framework for stating theories of linguistic phenomena; it is not itself such a theory. In other words, OT is a theory of the structure of universal grammar, not of its content.

In OT syntax, the input consists of a meaning or interpretation, and the output units are forms or expressions. A meaning is a dynamic semantic structure, typically a logical representation of predicate–argument and operator–variable structure, possibly including the discourse status of elements (such as topic or new information). A form is a sequence of words structured into syntactic constituents (phrases), possibly containing other syntactic information. The function GEN specifies the set of candidate expressions for a particular interpretation.

GEN is constrained by a correspondence function between interpretations and forms. A correspondence relation connects the entities of the meaning (predicates, arguments, operators, variables) with the elements that express them in the syntactic form in such a way that parts of the linguistic form are related to parts of the meaning. Wholes derive their meanings from their parts and the way these parts are combined, although not necessarily in the strict way the principle of compositionality of meaning is often conceived (cf. Blutner et al. 2003).

In practice, the output candidates generated by GEN are required to consist of licit elements from the universal vocabularies of linguistic representation, respecting X-bar structure, heads/complementizer configurations, etc. (Kager 1999: 20). The grammatical expression of a particular input meaning is the candidate generated by GEN that is evaluated as the 'best', 'least marked', most harmonic, or most optimal according to the constraint ranking in the language. GEN generates a potentially infinite set of candidate forms.

Legendre (2001) discusses a simple example involving expressive optimization. In certain languages (e.g., English), weather verbs take an expletive subject (1a). In other languages (e.g., Italian), a subjectless sentence is used (1b).

(1) a. It is raining. [English]
b. Piove. [Italian]
Rain.3SG

Obviously, English and Italian use different lexical items to represent the meaning 'rain', but that is not the issue here; both use a weather verb. Semantically, weather predicates are zero-place predicates. One-place predicates such as *sing* or *laugh* predicate singing or laughing of someone. But rain is not predicated of anything. In the absence of an argument position in the lexical semantics of the verb, the two languages make a different choice with respect to the tendency of well-formed sentences of natural language to have clauses with an explicit subject (favoring the English structure in 1a), and the desire to give content to all the expressions used in the sentence. Expletive subjects do not have content, so the pressure to use only meaningful expressions favors the Italian structure in (1b).

In OT, violable constraints model such conflicting tendencies. Legendre proposes the two constraints Subject and Full-Interpretation, as defined in (2).

(2) a. Subject: all clauses must have a subject.
  b. Full-Interpretation: all constituents in the sentence must be interpreted.

These constraints come into play when the speaker builds a well-formed sentence based on the input of the weather verb. The content 'rain-here-and-now', represented as ☔ constitutes the message that the speaker intends to convey. Given that weather predicates are zero-place predicates, the two constraints Subject and Full-Interpretation are giving conflicting instructions for the best form to use. The constraint Subject favors a sentence with an expletive subject (as in 1a), whereas the constraint Full-Interpretation prefers a subjectless sentence (as in 1b).

Crucially, the choice between the two constructions is not free. The English sentence would be ungrammatical without the expletive subject, and Italian does not have an expletive form that could take the place of *it* in (1b). What is the grammar that accounts for the two languages?

Suppose Gen produces two possible sentences: one with an expletive subject and the other without a subject as the possible outputs for an input containing a weather verb. The grammatical contrast between English and Italian can then be modeled as a difference between the importance or strength of the two constraints. In English, Subject is the dominant constraint, and a violation of Full-Interpretation is accepted when the input contains a zero-place predicate. In Italian, Full-Interpretation is the dominant constraint, and a violation of Subject is accepted in such cases.

The ranking and the process of optimization is modeled in Tableaux 1 and 2, which provide the English and Italian patterns with weather verbs, respectively.

The top left-hand cell in the two tableaux represents the input meaning ☔. There are infinitely many ways in which the speaker could convey the meaning ☔, many of them nonlinguistic (e.g., by pointing at the sky, sighing while getting out an umbrella, performing a rain dance, etc.). Linguistic expressions are the only possible outputs considered in this book. Even so, an infinite number of options remain, some banal, others poetic, ironic, or bizarre. For simplicity, the only forms listed as possible candidates for the expression of the meaning ☔ are simple sentences using a weather verb. Thus, the candidates in Tableaux 1 and 2 list a finite subset of a possibly infinite set of output candidates.

Tableau 1 Weather verbs in English (production)

Meaning ☔	Form	Subject	Full-Int
	Rains	*	
☞	It rains		*

Tableau 2 Weather verbs in Italian (production)

Meaning ☔	Form	Full-Int	Subject
☞	Piove		*
	expl piove	*	

Tableaux 1 and 2 are concerned with the choice between a sentence with an expletive subject and a subjectless sentence. The well-formedness of the sentence depends on the ranking of the two violable constraints, SUBJECT and FULL-INTERPRETATION. The constraints are ranked across the top, going from the highest ranked constraint on the left to the lowest ranked constraint on the right.

An asterisk (*) in a cell indicates a violation of the constraint. An expletive subject violates the constraint FULL-INTERPRETATION: the word *it* in *it is raining* does not have a meaning. Subjectless sentences such as *piove* violate the constraint SUBJECT. The little hand (☞) points at the optimal candidate. According to the strict domination hierarchy in ordinal OT, the optimal candidate is the grammatical sentence in the language at hand. Suboptimal candidates are not simply less good, but plain ungrammatical. This is known as the principle that 'the winner takes all.'

Tableaux 1 and 2 reflect that in languages like English it is more important to have a subject than to avoid meaningless words, whereas in languages like Italian, it is more important to have only meaningful words in the sentence than to have a subject. In running texts, the two grammars are written as SUBJECT >> FULL-INTERPRETATION for English and FULL-INTERPRETATION >> SUBJECT for Italian. In general, $C_1 >> C_2$ indicates that constraint C_1 is ranked above C_2.

The well-formedness constraints used in OT are of two general types: markedness constraints and faithfulness constraints. Faithfulness constraints evaluate the relation between input and output. The constraint FULL-INTERPRETATION in (2b) is a faithfulness constraint: it relates parts of the form to a correspondent in the input meaning. A markedness constraint is output oriented. Markedness constraints in OT syntax exclusively concern constraints on form. The constraint SUBJECT defined in (2a) is a markedness constraint: the requirement that clauses must have a subject is a requirement on forms that is unrelated to the input meaning.

The notion of markedness goes back to the Prague school of linguistics (Trubetzkoy 1931, 1939, Jakobson 1962, 1971). In later developments of linguistics, the notion of markedness was conceived as problematic for formal linguistic theory, because the concept proved difficult to define (cf. Haspelmath 2006 for an overview). In OT, markedness theory plays a central role. Marked structures are identified as those that violate a universal constraint in OT. Given that constraints are violable, and candidates compete, OT can maintain a notion of markedness, even if we are dealing with tendencies, and degrees of markedness.

As already outlined in Chapter 1, the markedness of negation with respect to affirmation is the starting point of the analysis. In Chapter 3, the core faithfulness and markedness constraints concerning negation will be derived in an evolutionary OT approach. The markedness of negation carries over to the next level of complexity: Chapters 4, 5, and 6 investigate the markedness of double negation with respect to single negation in the syntax as well as in the semantics. Throughout the book, new faithfulness constraints are grounded in cognitive and functional principles.

The analysis of the contrast in (1) shows that differences between languages can be explained through a different ranking of the same set of constraints. This is called typology by reranking. The space of all possible human grammars is formally specified by the factorial ranking possibilities of the set of constraints. So a set of two constraints allows two possible rankings, a set of three constraints allows eight possible rankings, and so on. In practice, the number of possible languages is smaller than the number of possible rankings within the factorial typology. Constraint rankings implementing markedness hierarchies may be universal, or permit only restricted reranking because of implicational hierarchies. In other cases, several different constraint rankings define the same language because two or more constraints do not interact, so their ranking with respect to each other does not lead to a different grammar. Nevertheless, typology by reranking remains an important tool for the description of cross-linguistic variation.

In this book, typology by reranking is explored at the syntax–semantics interface, and exploited to define classes of languages with respect to the expression and interpretation of negation.

2.3 Fundamentals of OT Semantics

OT semantics is a mirror image of OT syntax, and spells out a process of interpretive optimization. The input is a given form, and the output involves a set of candidate meanings. The form is a sequence of words structured into syntactic constituents (phrases). A meaning is a dynamic semantic structure, including a logical representation of predicate–argument and operator–variable structure, and temporal–information structure, if applicable. The function INT specifies the set of candidate meanings for a particular expression. INT is constrained by a correspondence function between forms and interpretations. INT generates a potentially infinite set of meanings (Hendriks and de Hoop 2001).

The candidates generated by INT are constrained by standard semantic theories such as type theory, the lambda calculus, and generalized quantifier theory. The interpretation of a particular input expression is now the candidate generated by INT that is evaluated as the 'best', 'least marked', most harmonic, or most optimal according to the constraint ranking in the language at hand.

Analyses in terms of interpretive optimization have been formulated for issues involving anaphora resolution of reflexives and pronouns, discourse anaphora, temporal structure, focus, the conceptualization of color terms, and lexical semantics (cf. Hendriks and de Hoop 2001; de Hoop and de Swart 2000; Zwarts 2003, 2004; Hendriks 2004a, b; Blutner et al. 2006; Jäger and van Rooy 2007; Hendriks et al. 2009). In this section, I discuss the temporal structure of *when*-clauses (based on de Hoop and de Swart 2000) as an example of interpretive optimization.

Temporal adjunct clauses introduced by *when*, *before*, *after*, etc. come with tense and aspect. Heinämäki (1978) points out that the temporal relation established by *when* depends on the aspectual features of the main and the subordinate clause. Intervals or moments denoted by event predicates are included in the intervals referred to by durative sentences (3a). Two durative sentences overlap in time (3b), and two event predicates describe events happening in succession (3c).

(3) a. Everybody was away when Jane destroyed the documents.
b. It was raining in San Francisco when we were there.
c. When Robert wrecked the car, Jane fixed it.

Heinämäki's claim that two events related by *when* happen in succession (3c) has been challenged. According to de Swart (1999), neither preposed nor postponed *when*-clauses express a succession of events in which the event described by the subordinate clause follows the main clause event, even if this is strongly suggested by world knowledge, compare (4a–d).

(4) a. When the president asked who would support her, Robert raised his hand.
b. Robert raised his hand when the president asked who would support her.
c. When Robert raised his hand, the president asked who would support her.
d. The president asked who would support her when Robert raised his hand.
e. The president asked who would support her. Robert raised his hand.

The preposed *when*-clauses in (4a, c) allow for only one reading. In both cases, the main clause event is located shortly after the event described by the subordinate clause. The main clause event is caused by or otherwise made possible by the subordinate clause event. The preferred reading of (4b) is the same as that of (4a). However, an alternative interpretation is available in which Robert raises his hand just at the moment at which the president asks who would support her. Under this reading, there is no causal connection between the two actions; there is just a relation of temporal overlap.

The puzzle is (4d). If two events related by *when* could describe the two events as happening in succession independently of subordination, then the prediction would be that Robert's raising of his hand is located after the president's request for support, just like in the sequence of two independent sentences (4e). Even in the presence of strong rhetorical support, this reading is unavailable in (4d): no causal or enablement relation leading from a request for support to a raising of the hand can be established; there is just a temporal relation between the two events.

De Swart (1999) appeals to topic-focus articulation and the difference in anaphoric behavior between main and subordinate clauses to explain the paradigm in (4). Main clauses are anaphoric just like independent clauses. This means that their temporal anchoring is determined by the relation with the preceding discourse. In line with Lascarides and Asher's (1993) claim that temporal relations are derived from the rhetorical structure of the discourse, an independent clause β seeks to establish a rhetorical relation $R(\alpha,\beta)$ with an earlier sentence α in the discourse. In an OT setting, the preference for an anaphoric discourse structure is captured by means of a temporal version of the constraint DOAP proposed by Hendriks and de Hoop (2001):

(5) Don't overlook anaphoric possibilities (DOAP): opportunities to establish a rhetorical relation must be seized.

Following DOAP, a main or independent clause β tries to establish a rhetorical relation R with a clause α that is already part of the discourse representation structure built up so far.

Time adverbials do not build such an anaphoric relationship. Time adverbials are presuppositional (Heinämäki 1978), which means that their location in time is taken to be determined independently of the local context. Accordingly, time adverbials cannot fulfill the role of β in a rhetorical relation R(α,β). De Hoop and de Swart (2000) define a constraint TA on temporal adjuncts that captures their non-anaphoric behavior (6).

(6) TA: temporal adjuncts do not function as β in a rhetorical relation R(α,β).

Topic-focus articulation is related to clause order. Preposed temporal clauses as in (4a, c) are topicalized, and always provide the rhetorical antecedent of the main clause.[1] Postponed temporal clauses can be either topic or focus. Thus they provide the rhetorical antecedent of the main clause (functioning as α in R(α,β)), or they establish a relation of temporal overlap in the absence of a rhetorical relation. The MIRROR PRINCIPLE in (7) relates clause order, information structure, and rhetorical structure.

(7) MIRROR PRINCIPLE (α < β): α < β: R(α,β): topic < focus.
The linear order of two syntactic constituents corresponds to the order antecedent-anaphor in a rhetorical relation, which mirrors the order topic-focus in the information structure.

Tableau 3 shows that the optimal interpretation for the input sequence of a preposed *when*-clause is the result of the satisfaction of all three constraints.

Tableau 3 Preposed *when*-clause (e.g., 4a, c) (interpretation)

Form *when*-clause(e_1) < main-clause (e_2)	Meaning	TA	DOAP	α < β
☞	R(e_1, e_2) *when*-clause$_{top}$, main clause$_{foc}$			
	R(e_1, e_2) main clause$_{top}$, *when*-clause$_{foc}$			*
	R(e_2, e_1) *when*-clause$_{top}$, main clause$_{foc}$	*		*
	R(e_2, e_1) main clause$_{top}$, *when*-clause$_{foc}$	*		**
	¬ R(e_1, e_2) *when*-clause$_{top}$, main clause$_{foc}$		*	
	¬ R(e_2, e_1) main clause$_{top}$, *when*-clause$_{foc}$		*	*

[1]The term 'topic' is used here in the information structuring sense (cf. de Swart 1999) and not in the syntactic sense (e.g., Rizzi 1997). Compare Haegeman (2001, 2003) for a more general discussion of fronted adverbial adjuncts in syntax.

There is one interpretation in Tableau 3 that satisfies all three constraints. In this optimal interpretation, a rhetorical relation is established between the two clauses (satisfaction of DOAP), with the *when*-clause as the rhetorical antecedent α (satisfaction of TA). Furthermore, the rhetorical structure satisfies both clause order and topic-focus articulation (satisfaction of the MIRROR PRINCIPLE $\alpha < \beta$). In the optimal interpretation of (4a), the president's request triggers Robert's raising of his hand as a natural response, whereas in (4c), Robert's raising of his hand triggers the president's request.

Given that the preposed *when*-clause satisfies all three constraints, this example does not tell us anything about the ranking of the constraints. Postponed *when*-clauses are more informative in this respect. Consider the representation in Tableau 4 for examples like (4b, d).

Postponed *when*-clauses generally allow two interpretations, as the discussion of examples (4b, d) made clear. One is the same as the construction with the preposed *when*-clause, namely $R(e_1, e_2)$, with the *when*-clause providing the topic of the construction, and the antecedent of the rhetorical relation. In Tableau 3, with the preposed *when*-clause as input, this interpretation did not violate any constraints, but in Tableau 4, with the postponed *when*-clause, the interpretation violates the MIRROR PRINCIPLE, because the linear order of the main clause and the *when*-clause do not correspond to the order antecedent-anaphor in the rhetorical relation. The MIRROR PRINCIPLE is satisfied by the third candidate, but this candidate violates TA, which is a higher ranked constraint.

In the final two candidates, no rhetorical relation is established, and *when* denotes a relation of temporal overlap between two events. This candidate violates DOAP, but satisfies the other two. Note that this interpretation requires the *when*-clause to be in focus. The candidate in which no rhetorical relation is established, but the *when*-clause is interpreted as the topic, incurs an additional violation of the MIRROR PRINCIPLE, which makes this a suboptimal candidate.

Tableau 4 Postponed *when*-clause (e.g., 4b, d) (interpretation)

Form main-clause (e_2) < *when*-clause(e_1)	Meaning	TA	DOAP	$\alpha < \beta$
	$R(e_1, e_2)$ main clause$_{top}$, *when*-clause$_{foc}$			**
☞	$R(e_1, e_2)$ *when*-clause$_{top}$, main clause$_{foc}$			*
	$R(e_2, e_1)$ main clause$_{top}$, *when*-clause$_{foc}$	*		
	$R(e_2, e_1)$ *when*-clause$_{top}$, main clause$_{foc}$	*		*
	$\neg R(e_1, e_2)$ *when*-clause$_{top}$, main clause$_{foc}$		*	*
☞	$\neg R(e_2, e_1)$ main clause$_{top}$, *when*-clause$_{foc}$		*	

The fact that the second and the last candidate both arise as optimal interpretations suggests that the constraints DOAP and $\alpha < \beta$ are equally strong. In the tableau, this is indicated by the dotted line between the two columns. In running text, $C_1 >> \{C_2, C_3\}$ represents a grammar in which C_2 and C_3 are ranked equally high, and the two constraints are outranked by C_1.

The contrast between the second and last candidates on the one hand and the remaining candidates on the other emphasizes that the grammatical candidate violates the constraints minimally, with respect to the constraint ranking. The interaction of the three constraints under the ranking TA >> {DOAP, $\alpha < \beta$} explains how word order, information structure, and constraints on anaphoric relations work together in the selection of the optimal interpretation of *when*-clause constructions. OT thus serves as a theory of temporal anaphora resolution.

Two concerns have been raised with respect to OT semantics that I address in this section, and more in detail in the remainder of this book. The first concern is that processes such as anaphora resolution are not really part of the semantics, because the temporal and information structure of the examples in (4) relies on pragmatic principles like DOAP and the MIRROR PRINCIPLE $\alpha < \beta$. The underlying idea of this objection is that hard-core semantic rules could not be subject to optimization processes, but this is possible for the anchoring of utterances to their context and situation of use. The lexical–semantic analysis Zwarts (2003, 2004) develops for the preposition *round* in English provides strong counterevidence against this idea.

Zwarts adopts a formal semantic analysis of *round* in terms of the vector-space semantics developed by Winter and Zwarts (2000). However, he shows that the use of *round* in a particular context involves the interaction of the prototypical interpretation of *round* as denoting a full circle with the lexical semantics of its environment. In this interaction, the interpretation of *round* can be weakened to a half circle (8a), a quarter circle (8b), an oval (8c), or a criss-cross movement (8d), but the interpretation is always as strong as the context allows.

(8) a. He went round the barrier.
   b. The postman went round the corner.
   c. The earth goes round the sun.
   d. The tourists went round the city centre.

The context-dependency of *round* in examples like (8) is not reducible to pragmatics, because the outcome depends on the way lexical features of *round* interact with the semantics of other words in the sentence. If principles of OT semantics are applicable in lexical semantics, especially to expressions that have a clear algebraic structure such as spatial prepositions, optimization over meanings cannot be relegated to the pragmatic module.

Additional evidence that interpretive optimization is not limited to the domain of pragmatics is provided by the treatment of negation in this book. If truth-conditional operators such as negation are subject to optimization processes, this provides strong evidence in favor of an application of OT principles to domains of hard-core semantics. As part of the emphasis on the role of optimization processes in truth-conditional semantics and the syntax–semantics interface, the pragmatics of negation is marginal to my concerns (cf. some remarks in Chapter 1, Section 1 and Chapter 3, Section 4, though).

The second concern voiced about OT semantics is that there is no crosslinguistic variation in meaning on a par with syntactic variation. The example in (1) was easy to account for in terms of reranking of two syntactic constraints. Pace universal markedness hierarchies, the reranking of constraints always leads to a new OT grammar that should correspond to a possible language. In the examples discussed in this section, it seems hard to come up with a language that would rank the semantic constraints in a different order. The interpretive principles involved seem to depend on universal markedness hierarchies, rather than be subject to cross-linguistic variation. If semantic constraints are always universally ranked, one of the important advantages of OT as a typological theory, accounting for cross-linguistic variation in terms of different constraint rankings seems to be lost as far as interpretive optimization is concerned. But this would be a misconception.

Although semantic variation is certainly more constrained than syntactic variation, true semantic variation does exist, and the domain of negation provides an important illustration. Two key contrasts from Chapter 1 are here repeated under (9) and (10) (from Herburger 2001). In elliptical contexts, a negative answer must be provided by a truly negative expression, such as *nothing* in English. A negative polarity item like *anything* is not felicitous as an answer to the question in (9a), because it needs to be in construction with a licensor with particular semantic properties (negation, a negative quantifier, etc.), as argued in Chapter 1 (Section 3).

(9) a. Q: What did you see? A: Nothing. [English]
A: *Anything

b. Q: A quién viste? A: A nadie. [Spanish]
Q: whom saw.2SG A: nobody
Q: 'Who did you see?' A: *A un alma
a soul

In Spanish, we observe the same contrast between *nadie* and the negative polarity item *a un alma* ('a soul') (9b). This suggests that expressions like *nothing* and *nadie* have the same semantics. However, the situation changes with sentences that involve two instances of expressions like *nothing* or *nadie*. In English, the combination of *nobody* with *nothing* in (10b) leads to a double negation reading, whereas the two instances of *nadie* in (10a) express a single negation.

(10) a. Nadie miraba a nadie. [Spanish]
nobody looked at nobody.
'Nobody looked at anybody.'
$\neg\exists x\exists y$ Look-at(x,y)

b. Nobody said nothing. [English]
$\neg\exists x\neg\exists y$ Say(x,y)

Many analyses of these contrasts have been proposed, and the most important ones have been reviewed in Chapter 1. The outcome of that discussion is that a lexical analysis of the contrast in (9) and (10) is doomed to fail. The similarities between

(9) and (10) support the view that there is no lexical distinction between negative quantifiers such as English *nobody* and n-words that participate in negative concord like Spanish *nadie*. If there is no lexical difference between *nobody* and *nadie*, the contrast between (10a) and (10b) must reside in the grammar.

In Chapters 3 and 4, I propose three constraints governing the expression and interpretation of negation in natural language, and claim that English and Spanish involve two different grammars, and crucially rank the syntactic and semantic constraints on negation in two different orders. The constraint ranking shows a balance between expressive and interpretive optimization, but crucially for the argumentation in this section, there is reranking of constraints in the semantic component as well as in the syntactic component. The typology of double negation and negative concord languages I propose in Chapter 4 is thus a true instance of semantic variation as reranking of interpretive constraints.

The argument that optimization principles do not apply in the semantic domain because there is no reranking of interpretive constraints across languages is refuted by the results presented in this book. This result opens up the possibility of fruitful investigation of other phenomena in the area of cross-linguistic semantics as well (cf. Hendriks et al. 2009 for proposals). Such a typological line of research complements the search for semantic universals put forward in Von Fintel and Matthewson (2008).

2.4 Bidirectional Optimality Theory

So far, I presented unidirectional versions of OT. Expressive optimization is speaker oriented. It takes meanings as input and selects the optimal form for the message to be expressed. Interpretive optimization is hearer oriented. It takes forms as input and selects the optimal interpretation for the given expression. Under the view that language serves a communicative purpose, these two directions of optimization should be connected. After all, the speaker wants the message not only to be transferred to the hearer, but also to be understood. The optimal form is the one that is understood by the hearer to convey the message that the speaker has in mind.

Communication requires the speaker and the hearer to take each other's perspectives into account. Optimization over pairs of forms and meanings is the domain of bidirectional OT (Blutner 1998, 2000, 2004; Blutner et al. 2006; Hendriks et al. 2009). In this book, bidirectional OT provides the syntax–semantics interface of negation across languages.

Hendriks et al. (2009: Chapter 1) illustrates the optimization process underlying bidirectional OT with the nonlinguistic example of a dance. Imagine a situation in which men and women dance in pairs. Men and women are free to choose their preferred partner, but they can have only one. Men prefer better female dancers to less good female dancers, and women prefer better male dancers to less good male dancers. If we want to match the dancers to get the best pair, for example, to win a dance competition, the best pair is the pair consisting of the best female dancer and the best male dancer.

Strong bidirectional optimization (adapted from Blutner 2000) uses the same intuition to pair up the best form (f) with the best meaning (m) as the winner of the linguistic competition. The definition of strong bidirectional optimization is given in (11):

(11) Strong bidirectional optimization:
A form–meaning pair <f,m> is bidirectionally optimal iff:
a. there is no other pair <f′,m> such that <f′,m> is more harmonic than <f,m>.
b. there is no other pair <f,m′> such that <f,m′> is more harmonic than <f,m>.

Under this definition, forms and meanings are not considered separately. Instead, optimization is defined over pairs consisting of forms and their corresponding meanings. A form–meaning pair is an optimal pair if and only if there is no pair with a better form or a better meaning. Such optimal pairs block all other pairs in the same competition.

The notion of strong optimality is illustrated in Figure 1 with the two forms f_1 and f_2 and the two meanings m_1 and m_2. The arrows indicate preference relations.

$<f_1, m_1>$ and $<f_2, m_2>$ are strongly optimal form–meaning pairs, because both the horizontal and vertical arrows point to these pairs. For the interpretation of f_1, the meaning m_1 is preferred over m_2, and for the expression of m_1, the form f_1 is preferred over f_2. Similarly, for the interpretation of f_2, the meaning m_2 is preferred over m_1, and for the expression of m_2, the form f_2 is preferred over f_1. In strong bidirectional OT, optimization over forms and meanings converges. Pairs that are suboptimal in one or the other direction of optimization are blocked.

Blocking in natural language occurs in situations where a meaning can be expressed by two different forms, but one of these forms is simpler, shorter, or otherwise preferred, so the other form is blocked for this meaning. For example, there are two possible ways to realize the comparative form of *good*, namely by means of the regular form *gooder*, or the irregular form *better*. Because the irregular form *better* is preferred, the regular form *gooder* is blocked as the comparative form of *good*.

The blocking of *gooder* by *better* is an instance of total blocking: *gooder* is a nonexisting form in English. Total blocking fits in with the notion of 'the winner takes all' underlying OT (cf. Section 2). However, natural language also presents instances of partial blocking. Again, the underlying idea can be illustrated with the nonlinguistic example of the dance.

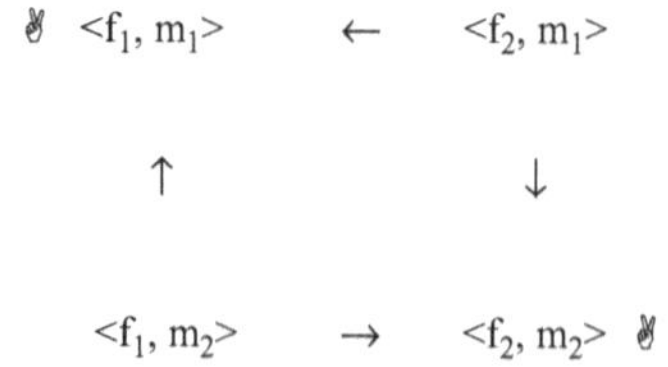

Figure 1 Strong bidirectional OT

In order to win the dance competition, the best female dancer pairs up with the best male dancer. Now imagine that other dancers besides the best pair are allowed to dance as well. The best dance partner for the one-but-best female dancer, as for all female dancers, would be the best male dancer. Given that the best male dancer already forms a pair with the best female dancer, he is no longer available to dance with anyone else. So all imaginable pairs in which the best male dancer dances with someone else than the best female dancer are blocked. Similarly, all imaginable pairs in which the best female dancer dances with someone else than the best male dancer are blocked. But in a second round of optimization, the one-but-best female dancer will end up with the one-but-best male dancer. And if even more dancing pairs are allowed to be formed, the two-but-best female dancer ends up with the two-but-best male dancer, and so on. Crucially, pairs which do not consist of either the best female dancer or the best male dancer will not be blocked, even if they do not constitute the absolutely best possible pair.

In language, this situation can be modeled with the recursive definition of bidirectional optimality, which is called superoptimality (adapted from Blutner 2000):

(12) Weak bidirectional optimization (adapted from Blutner 2000)
A form–meaning pair <f,m> is superoptimal iff:
a. there is no superoptimal pair <f′,m> such that <f′,m> is more harmonic than <f,m>.
b. there is no superoptimal pair <f,m′> such that <f,m′> is more harmonic than <f,m>.

Strong pairs are superoptimal, but not all superoptimal pairs are strong. Figure 2 illustrates the notion of superoptimality with the two forms f_1 and f_2 and the two meanings m_1 and m_2. Again, the arrows indicate preference relations.

According to the preference relations in Figure 2, f_1 is always preferred over f_2, and m_1 is always preferred over m_2. As a result, two arrows are pointing toward the pair <f_1, m_1>, and two arrows are pointing away from the pair <f_2, m_2>. The fact that two arrows are pointing toward the pair <f_1, m_1> indicates that this is a strongly optimal pair. In the dance example, this would be the pair consisting of the best female dancer and the best male dancer.

The pairs <f_1, m_2> and <f_2, m_1> lose against this strongly optimal pair, because <f_1, m_1> has a better form for the same meaning, or a better meaning for the same form. In the dance example, these would be pairs consisting of the best female dancer with the second-best male dancer, or the best male dancer with the second-best female dancer. The pairs <f_1, m_2> and <f_2, m_1> are neither strong nor superoptimal pairs.

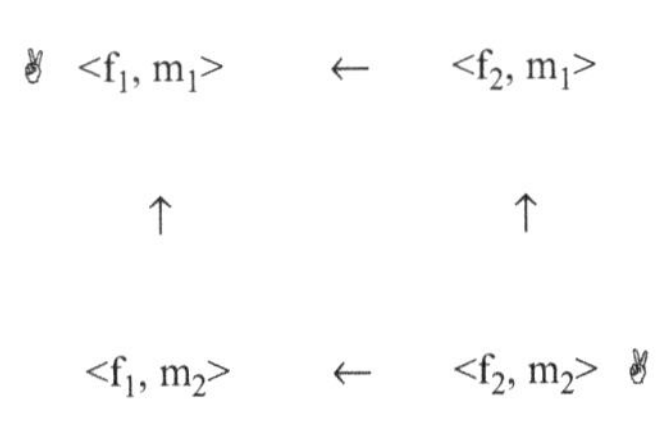

Figure 2 Weak bidirectional optimization

Under strong bidirectional optimization, the pair $<f_2, m_2>$ is blocked, because there are better form–meaning pairs available, as the arrows indicate. Under weak bidirectional optimization, the pair $<f_2, m_2>$ arises as a weakly optimal (or 'superoptimal') form–meaning pair, because there is no superoptimal pair that has either a better form or a better meaning. In the dance example, this would be the pair consisting of the second-best female dancer and the second-best male dancer.

In Figure 2, $<f_2, m_2>$ is not in direct competition with the only other superoptimal pair $<f_1, m_1>$, because the two pairs differ in both their form and their meaning component. Figure 2 shows that weakly optimal pairs involve forms that are suboptimal in unidirectional generation, and meanings that are suboptimal in unidirectional interpretation, but that do not compete with a strongly optimal pair. As a result, superoptimality pairs up marked forms and marked meanings that would not otherwise be available. Superoptimality is used to model instances of partial blocking. Examples of partial blocking are discussed in Blutner (1998, 2000).

A well-known example of partial blocking discussed by Blutner is the relation between *kill* and *cause to die*. *Kill* is the shorter, less complex, thus unmarked form. *Cause to die* is the longer, more complex, hence marked form. Two closely related meanings are available for these forms, varying only in whether the action is accomplished in a direct or an indirect way. Two markedness constraints F1 and M2 can model this situation. F1 penalizes complex structure in the form, and M2 penalizes complex meanings in the interpretation. The ranking is irrelevant in this particular example. Tableau 5 spells out the possible form–meaning pairs and their violation patterns on the basis of this input. The victory sign (✌) indicates the superoptimal pairs.

The combination of the form *kill* and the direct meaning constitutes a strongly optimal pair, because it does not violate any constraint. The combination of *cause to die* with the indirect meaning comes out as a superoptimal pair. This pair violates both markedness constraints, but wins in a second round of optimization, because the competing pairs of [kill, indirect] and [cause to die, direct] lose against the strongly optimal pair [kill, direct]. Given that there are no better superoptimal pairs, the pair [cause to die, indirect] is itself a superoptimal pair.

Weak bidirectional OT is not exclusively operative in the lexicon. De Swart and Zwarts (2009) use it to model how bare singular nominals without an article get default, stereotypical, idiomatic meanings in constructions like incorporation, predication, embedding under certain prepositions, and so on. In those same contexts, full nominals with an article get less idiomatic interpretations. Consider the contrast between (13a) and (13b), as described by Horn (1984), Stvan (1998), van Rooy (2004), and others.

Tableau 5 Weak bidirectional optimization

Input [f,m] f_1: kill, f_2: cause to die m_1: direct, m_2: indirect		F1	M2
[kill, direct]	✌		
[kill, indirect]			*
[cause to die, direct]		*	
[cause to die, indirect]	✌	*	*

(13) a. George is in jail.
b. George is in the jail.

Sentence (13a) is understood as the qualification of George as a prisoner. A speaker who uses (13b) normally conveys that George is in the building described as the jail, where he may be a visitor, a priest, a volunteer, a repairman, or whatever, but not a prisoner. The preference relations between the four possible form meaning pairs are indicated in Figure 3.

De Swart and Zwarts (2009) set up an OT typology of article use to motivate that the bare nominal constitutes the unmarked member of the pair *in jail* and *in the jail*. The constraint *ARTICLE prefers articleless nominals. They use the strongest meaning hypothesis (Dalrymple et al. 1998) to argue that the incarcerated meaning is the unmarked interpretation. The constraint STRENGTH favors idiomatic interpretations over less stereotypical interpretations. Under these assumptions, the pair [in jail, incarcerated] is a strongly optimal pair. The pair [in the jail, just visiting] emerges as a superoptimal pair in a second round of optimization, as shown in Tableau 6.

Weak bidirectional OT is also operative in language change (Blutner et al. 2006) and language learning (Hendriks et al. 2009, Chapters 4 and 5). In a more general perspective, weak bidirectional OT is a way to model Horn's division of pragmatic labor, where unmarked forms are used to express unmarked meanings, and marked forms are used for marked meanings (Horn 1984, 2001). Levinson also models this idea in his M-heuristics (Levinson 2000).

In this book, bidirectional OT models the syntax–semantics interface both in a static (synchronic) analysis of typological variation, and in a dynamic analysis of language change. The expressive and interpretive optimization procedures in Chapter 4 are linked in such a way that the syntactic and semantic components of the analysis converge on the optimal status of form–meaning pairs. The analysis in

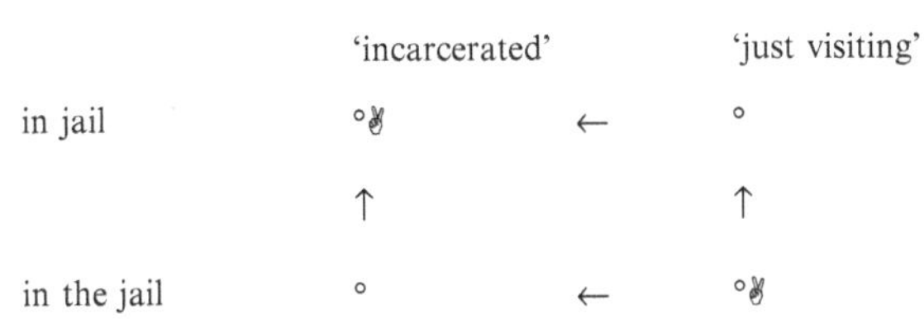

Figure 3 Weak bidirectional optimization over bare nominals

Tableau 6 Weak bidirectional optimization over bare nominals

		*ART	STRENGTH
in jail, λ*x* [IN(*x*,*y*) & JAIL(*y*) & IMPRIS(*y*,*x*)]	✌	✓	✓
in the jail, λ*x* [IN(*x*,*y*) & JAIL(*y*) & IMPRIS(*y*,*x*)]		*	✓
in jail, λ*x* [IN(*x*,*y*) & JAIL(*y*)]		✓	*
in the jail, λ*x* [IN(*x*,*y*) & JAIL(*y*)]	✌	*	*

that chapter is thus an instantiation of strong bidirectional OT. Chapter 6 shows that double negation readings in negative concord languages cannot be accounted for by exploiting the mechanism of strong bidirectional OT. A weak bidirectional OT extension of the analysis is developed for those special cases.

A number of studies have argued against the recursive mechanism of weak bidirectional optimization as an online mechanism of linguistic processing (Zeevat 2000, Beaver and Lee 2004). Beaver and Lee's main objection concerns its property of recursion, which allows, in principle, for an infinite number of rounds of optimization. Because suboptimal candidates can become winners in a second or later round of optimization, 'in weak OT, everyone is a winner', as Beaver and Lee (2004: 126) put it. In the nonlinguistic dance example, multiple rounds of optimization are unproblematic, and recursion is the right strategy if everyone is allowed to dance. But in natural language, recursion results in an overgeneration of form–meaning pairs.

Blutner et al. (2006: 149) suggest that general cognitive limitations on recursion limit recursion in linguistic applications of superoptimality. They propose to limit bidirectional optimization to at most two rounds, in agreement with the bounds that can be observed for higher order epistemic reasoning required for playing strategic games. Beaver and Lee (2004) use a special constraint *BLOCK which ensures the restriction of the optimization over form–meaning pairs to a single recursion step. The analysis developed in Chapter 6 is in line with these views in that it involves just two rounds of bidirectional optimization.

Both strong and weak bidirectional optimization are symmetric, in that they rely on the intuition that speakers take into account the hearer's perspective, and hearers the speaker's perspective. However, some authors have defended the need for asymmetric versions of bidirectionality. Zeevat (2000, 2006) develops an asymmetrical version of OT in which a unidirectional OT model for production forms the basic system. In comprehension, the set of candidate meanings is restricted to the results of the production step. The opposite view has been defended by Wilson (2001), who argues that the candidate set for production should be restricted by using results of comprehension.

Most of this book is based on a symmetric view, in which expressive and interpretive optimization are treated on a par (cf. als Hendriks et al. 2009). The only exception is the appeal to the evolutionary bidirectional learning algorithm developed by Zeevat and Jäger (2002), Jäger (2003) and Mattausch (2005, 2007), which I use in Chapter 2 to derive the marked status of negation in language. This model incorporates an asymmetric version of bidirectionality in which forms compete in the optimal recoverability of the intended meaning, but there is no similar competition among meanings. An asymmetry in the frequency distribution of marked and unmarked meanings is sufficient to obtain a stable system in which unmarked forms pair up with unmarked meanings, and marked forms pair up with marked meanings. The fact that language evolution relies on an asymmetric version of bidirectional optimization is not necessarily incompatible with a model that uses a symmetric version of the theory to model typological variation.

2.5 Language Variation and Language Change in Stochastic OT

OT works with a universal set of constraints, and a language-specific ranking. This allows an account of crosslinguistic variation in terms of constraint reranking. A simple example is the expletive subject in English, versus the subjectless clauses found in pro-drop languages like Italian in weather statements ('it rains/piove') (cf. Section 2).

Reranking is also used to model diachronic change. Vincent (1999) is an early example of an OT analysis of patterns of change in the pronominal system between Latin and Romance. In Chapter 3, I will show how the three main phases of the Jespersen cycle (the diachronic development of negation introduced in Chapter 1) are accounted for by three rankings of three constraints in ordinal OT. Chapter 4 (Section 5) sketches the diachronic development of negative polarity into negative concord.

Both in language variation and in language change, there are situations that can be classified as combining features from two systems or being in between two stages. Stochastic versions of OT can be used to model gradience. Clark (2004) uses stochastic OT to model patterns of syntactic change from Old to Middle English in the domain of headedness (e.g., OV structures) and the syntax of subjects.

The main difference between standard (ordinal) OT and stochastic OT involves the ranking of the constraints. According to ordinal OT, in a ranking $C_1 >> C_2$, C_1 is always stronger, and a violation of C_2 is always allowed in order to satisfy C_1 (cf. Section 1). The ordinal ranking of standard OT is abandoned in stochastic OT, and replaced by a continuous ranking of constraints. The result is that constraints have overlapping ranges. This is illustrated in Figure 4.

If two constraints C_1 and C_2 have overlapping ranges and there is a certain degree of 'noise' in the system, which slightly perturbs the ranking at every evaluation of an input, the order $C_1 >> C_2$ arises in most cases, but the order reverses to $C_2 >> C_1$ in some cases. This may affect the output, if the optimal candidate under the ranking $C_1 >> C_2$ is

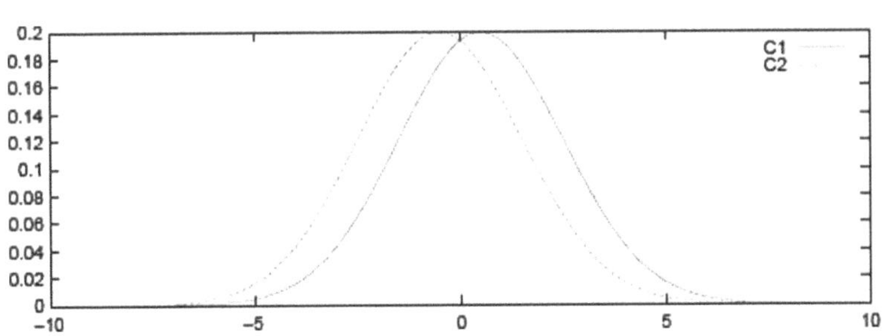

Figure 4 Overlapping constraints (from Jäger 2003)

some candidate A, but the optimal candidate under the ranking $C_2 >> C_1$ is some other candidate B. The degree of overlap between the constraints governs the distribution between candidates A and B in the output. If there is total overlap, A and B will both win in about 50% of the cases. If the degree of overlap is smaller, the distribution between A winners and B winners is different.

Stochastic OT emerged with the work by Boersma (1998) and Boersma and Hayes (2001). In their work, the focus is on acquiring phonological contrasts from phonetic input. Bresnan et al. (2001), Bresnan et al. (2007a, b), and Bresnan and Hay (2008) use stochastic OT to develop 'gradient' grammars, modeling an overlapping range between two possible grammars in neighboring dialects or varieties of English. Koontz-Garboden (2004) exploits stochastic OT to offer a sociolinguistic analysis of the alternation between the imperfective and the periphrastic progressive in varieties of Spanish spoken in Latin America and the United States.

In this book, stochastic OT is used to account for features of negation in natural language that are outside of the scope of the ordinal OT account. In particular, stochastic OT models the evolution of negation in such a way that negation emerges as a universal category of natural language (Chapter 3). Furthermore, stochastic extensions of the standard ordinal OT analysis account for intermediate stages in language typology and language change. Chapter 3 uses stochastic OT to model intermediate phases in the Jespersen cycle. Chapter 5 analyzes negation systems that are in between strict and nonstrict negative concord in stochastic OT. Chapter 6 uses a stochastic OT semantics to model ambiguities between double negation and negative concord readings of a sequence of negative indefinites. Aside from these special cases, the main patterns of negation in natural language are modeled in ordinal OT.

2.6 Conclusion

The key insight explored in this book is that languages make use of the same underlying mechanisms, but exploit the relation between form and meaning in different ways. OT can capture this kind of generalization, because the constraints are universal, but the ranking of the constraints is language specific. The empirical phenomenon of negation in natural language is situated at the syntax–semantics interface, and throughout the book, I will emphasize the need to optimize in two directions (from meaning to form and from form to meaning). In this book, ordinal OT and strong bidirectional OT are used wherever possible. I resort to stochastic OT or weak bidirectionality only in cases where the modeling of specific empirical phenomena requires such extensions.

Chapter 3
Markedness of Negation

Introduction and overview Chapter 1 stated that all languages have ways to express negation, and contain some form that conveys the meaning of the first-order logic connective $\neg$. The markedness of negation with respect to affirmation was empirically established in Chapter 1. This chapter formalizes the basic intuition, and considers its implications for the grammar of natural language.

In Section 1, negation emerges as a universal linguistic category as the result of bidirectional evolutionary learning. I formulate two constraints that are motivated by the asymmetry between affirmation and negation. The constraint ranking is universal, and not subject to typological variation. However, violations of the highest ranked constraint are found in grammars under development (as in first language acquisition) and in certain pathological linguistic systems (as in aphasic language use) (Section 2). These special cases provide independent support in favor of the overall approach.

Negative markers are most widely used to realize negation. Section 3 reflects on the typological variation in the placement of the marker of sentential negation. Section 4 puts this variation in a diachronic perspective and proposes a reinterpretation of the well known Jespersen cycle in OT. This book focuses on main clauses, but Section 5 offers some remarks on subordinate clauses and nonfinite constructions. Section 6 concludes.

This chapter only discusses instances of propositional negation realized by means of a marker of sentential negation. Constructions in which negation is realized by a negative indefinite or a sequence of multiple negative indefinites are the subject of Chapter 4. Chapter 5 integrates the results from both chapters, and investigates the co-occurrence restrictions on the marker of sentential negation and negative indefinites.

Chapter 3 explores the markedness of negation with respect to affirmation. Of course, markedness is a relative issue, and negation constitutes the unmarked member of the pair <single negation, double negation>. Double negation readings are not investigated here, but are discussed in Chapters 4 and 6.

H. de Swart, *Expression and Interpretation of Negation: An OT Typology*, Studies in Natural Language and Linguistic Theory,
DOI 10.1007/978-90-481-3162-4_3,

3.1 Propositional Negation

The aim of this section is to determine how languages express a meaning that could be represented in first-order logic as ¬p, and how they interpret propositional negation. I first examine the generation question, and propose an OT syntax where the input is a meaning (a first-order formula), the set of candidates generated by GEN is a set of possible forms, and a ranked set of violable constraints selects the optimal form for any given meaning (Section 1.1).

The set-up of the system leads to negation as a universal category of natural language. I argue that this is the result of Horn's division of pragmatic labor, according to which unmarked meanings pair up with unmarked forms, and marked meanings pair up with marked forms (Section 1.2). The markedness of negation is modeled as an iconicity effect in evolutionary bidirectional OT.

The OT syntax is combined with an interpretation mechanism in OT semantics, where the input is a form (a well-formed sentence), the set of candidates is a set of possible meanings (first-order formulae), and a ranked set of violable constraints selects the optimal interpretation for the given form (Section 1.3). The result is a strong bidirectional analysis of propositional negation.

Even though negation is a universal category of natural language, it remains useful to maintain a system in terms of soft constraints, in view of the fact that negation is not always expressed in child language, and may break down in sign users who later suffer brain damage (Section 2).

*3.1.1 A Faithfulness and a Markedness Constraint: FNEG and *NEG*

The concept of markedness is defined in different ways in the literature, cf. Haspelmath (2006) for recent discussion and references. As Jacobs (1991) points out, negation is not marked in the sense that it is cross-linguistically rare. On the contrary, negation is a universal category of natural language (Dahl 1979: Chapter 1). That is, every language has sentences conveying truth-conditional negation (¬) as English does using *not* in sentences like (1).

(1) a. It is *not* raining.
 b. John is *not* sick.
 c. Sue did *not* invite Peter.

However, negation is marked in the sense that the expression of negation involves special grammatical means, whereas the expression of affirmation does not. As a result, negative sentences are morphologically or syntactically more complex than their affirmative counterparts. The starting point of the investigation is the observation that negation is formally and interpretationally marked compared to affirmation.

Negation is not a sentential force in the sense described by Portner and Zanuttini (2003) because it can be found in various types of clauses (declarative, interrogative, exclamative). There are strong constraints on the possibility to produce negation in different speech acts, as discussed by Portner and Zanuttini (2000) and Krifka (2001, 2003), but this does not necessarily mean that negation is to be viewed as a sentential force. Nevertheless, there are important similarities.

According to Portner and Zanuttini (2003), all exclamatives share the need to represent two semantic properties in the syntax: namely that they are factive and that they denote a set of alternative propositions. The sentential force of exclamation thus needs to be visible in the form of an exclamative utterance. I extend Portner and Zanuttini's ideas about the need to represent semantic properties in the syntax to negation.

In an OT model, the correspondence between semantics and syntax is modeled by means of faithfulness constraints. Faithfulness constraints aim at a faithful correspondence between input and output (cf. Chapter 2, Section 2). In the process of optimization over forms, a faithfulness constraint governing affirmation and negation requires the syntax to reflect the fact that negative sentences are distinct from affirmatives. The constraint that deals with this is called FNEG (Faith negation):[1]

- FNEG
  Be faithful to negation, i.e. reflect the nonaffirmative nature of the input in the output.

The formulation of FNEG is neutral as to what constitutes the input and what the output. In the course of this chapter, it will become clear that this constraint is relevant to semantics as well as syntax, and can be used in both directions of optimization. For now, the role of FNEG in syntax is the subject of investigation; the semantics of negation will be the topic of Section 1.3.

Within the generation perspective (OT syntax), FNEG requires negation in the meaning (input) to be reflected in the output (form). Thus, the expression of negation satisfies FNEG if there is a formally visible reflection of negation. In OT, faithfulness constraints are balanced by markedness constraints. Markedness constraints are output oriented and typically aim at the reduction of structure in the output. The markedness constraint that plays a role in negative statements is *NEG:

[1] Readers familiar with the OT distinction between Max and Dep constraints can read FNEG as a Max constraint, which preserves information from the input in the output. The reason I do not use the term Max constraint is that this would give rise to confusion in Chapter 4, where the constraint MAXNEG is introduced, with a different meaning, namely maximizing the use of negative indefinites. There is no need for a Dep constraint for negation (blocking insertion of negation in the output when it does not occur in the input), as the economy constraint *NEG (to be introduced shortly) takes care of this.

♦ *NEG
Avoid negation in the output

Again, *NEG is neutral as to what constitutes the output, and in the course of this chapter, I will apply this constraint both in OT syntax and in OT semantics. The intuition behind *NEG is that negation is marked, both in form and in meaning. Marked forms and meanings should be avoided, so negation should be avoided both in the syntax and the semantics. Section 1.2 embeds the markedness of negation in an evolutionary bidirectional OT model.

The markedness constraint *NEG is obviously in conflict with the faithfulness constraint FNEG. FNEG requires a reflection in the output of negative features found in the input, whereas *NEG blocks negation in the output. The two requirements cannot be satisfied at the same time. Such conflicting constraints are characteristic of OT style analyses (cf. Chapter 2).

Both FNEG and *NEG are violable constraints, and the conflict between them is resolved by the ranking of constraints in terms of strength. Ranking FNEG higher than *NEG, and making it a stronger, more important constraint derives the formal expression of negative meanings, as illustrated in Tableau 1.

The input shown in Tableau 1 represents a particular meaning, and the output candidates for evaluation by the grammar are the candidate forms. All the generation tableaux in this book will be constructed in this way. The ranking FNEG >> *NEG reflects the generally accepted view that negative statements are cross linguistically more marked in form than their affirmative counterparts (Payne 1985, Horn 1989, Haspelmath 1997).

Which item functions as the marker of sentential negation in a language, and satisfies FNEG is a lexical matter. In English, this is *not* (cf. the examples in (1)). In other languages, sentential negation is lexicalized by some other lexical item. All the sentences in (2) express a negative proposition, and contain a linguistic marker of sentential negation (in italics), which is glossed as SN:[2]

(2) a. *Ou* petetai Sokrates. [Ancient Greek]
SN flies Sokrates.
'Socrates doesn't fly'

Tableau 1 Generation of negative sentences

Meaning ¬p	Form	FNEG	*NEG
	S	*	
☞	not S		*

[2] Examples are from Payne (1985) (1b), Borsley and Jones (2005) (1c), de Groot (1993) (1d), and Sells (2001) (1e).

b. On *ne* igraet. [Russian]
he SN plays.
'He doesn't play.'

c. *Nid* oedd Sioned yn gweithio. [formal Welsh]
SN be.IMPF.3SG Sioned PROG work
'Sioned was not working.'

d. János *nem* dohányz-ik. [Hungarian]
János SN smoke.3SG
'János doesn't smoke.'

e. Jag kisste *inte* Anna. [Swedish]
I kissed SN Anna
'I didn't kiss Anna.'

The two constraints in their order FNEG >> *NEG guarantee the introduction of a negative expression in sentences that describe a negative proposition. According to Dahl (1979), negation is a universal category of natural language. As far as I have been able to determine, there are no languages in which *NEG outranks FNEG. The ranking is universal, because FNEG >> *NEG is an evolutionary stable equilibrium of the linguistic system. This result is derived in Section 1.2.

However, the two constraints, *NEG and FNEG, do not say anything about the way a negative expression is realized in natural language. So far, the form taken to express clausal negation (¬p) is 'not S' as seen in Tableau 1. But the expression of clausal negation takes various forms across languages, (cf. Jespersen 1917, 1933, Dahl 1979, Payne 1985, Horn 1989, Ladusaw 1996, Bernini and Ramat 1996, and Haspelmath 1997, and Chapter 1, Section 2 for overviews of the facts). This chapter focuses on the most frequent type, which involves the use of a negation marker.

Two issues emerge from the reported observations in the literature. The first concerns the position of the marker of sentential negation in the sentence, and the second the relation between the expression of clausal negation and the marking of propositional negation on an argument of the verb. The first issue will be addressed in Sections 3 and 4. The second will be deferred until Chapter 4. Before I work out these issues, the universal nature of the ranking FNEG >> *NEG will be grounded in evolutionary bidirectional OT.

3.1.2 Negation as a Universal Category of Natural Language

The empirical data discussed in Section 1.1 strongly suggest that there are no languages in which *NEG outranks FNEG. In the course of this book, more constraints will be added to the OT grammar, but FNEG is always at the top of the constraint hierarchy. The constraints postulated so far raise the question why it is negation that is marked in natural language and not affirmation. Why are the relevant constraints FNEG and *NEG, rather than, say FAFF and *AFF (for Faith Affirmation and Avoid Affirmation, respectively)?

Haspelmath (2006) argues that frequency asymmetries lead to a direct explanation of the observed structural asymmetries. The frequency argument works well for the empirical phenomenon of negation, because affirmative meanings are more frequently in need of expression in natural language than negative meanings. This section argues that the relative rarity of negative meanings as compared to affirmative meanings makes FNEG >> *NEG the universally preferred ranking. The iconicity view is motivated by means of the evolutionary bidirectional learning algorithm developed by Zeevat and Jäger (2002), Jäger (2003) and Mattausch (2005, 2007). This approach is in line with Haspelmath's (2006) view. The additional value of the evolutionary bidirectional OT learning algorithm resides in a precise modeling of the step from frequency to the distribution of marked and unmarked forms.[3]

The idea of bidirectional evolutionary OT hinges on three concepts: stochastic ranking of OT constraints, gradual bidirectional learning, and iterated learning over different generations. Standard optimality theory has an ordinal ranking. That is, in a ranking $C_1 >> C_2$, C_1 is always stronger, and a violation of C_2 is always allowed in order to satisfy C_1. The ordinal ranking of standard OT is abandoned in stochastic OT, and replaced by a continuous ranking of constraints (Boersma 1998). The result is that constraints have overlapping ranges. This is illustrated in Figure 1. If two constraints C_1 and C_2 have overlapping ranges and there is a certain degree of 'noise' in the system, which slightly perturbs the ranking at every evaluation of an input, the usual order is $C_1 >> C_2$ but the order $C_2 >> C_1$ arises in some cases (cf. Chapter 2, Section 5).

Boersma (1998) and Boersma and Hayes (2001) combine stochastic OT with learning theory and develop the so-called gradual learning algorithm. This algorithm

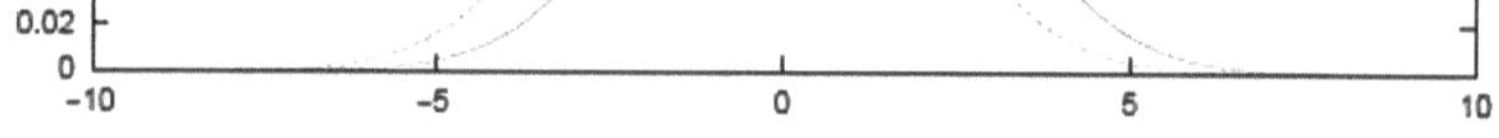

Figure 1 Overlapping constraints (from Jäger 2003)

[3] A closely related game-theoretical version of the same idea is explored by Dekker and van Rooy (2000), van Rooy (2004), and Jäger and van Rooy (2007), but will not be discussed here.

allows the learner to develop a stochastic OT grammar based on observed linguistic behavior. Jäger (2003) proposes a bidirectional version of the gradual learning algorithm by stipulating a recoverability restriction for optimality. Forms compete in the optimal recoverability of the intended meaning. The bidirectional optimization process is asymmetric, for there is no similar competition among meanings.

Formally, asymmetric bidirectional optimality is defined in (3) (Jäger 2003):

(3) Asymmetric bidirectional optimality
   a. A form-meaning pair ⟨f,m⟩ is hearer optimal iff there is no pair ⟨f,m'⟩ such that ⟨f,m'⟩ < ⟨f,m⟩, where < means 'better', 'more harmonic' or 'less marked'.
   b. A form-meaning pair ⟨f,m⟩ is optimal iff:
      – either ⟨f,m⟩ is hearer optimal, and there is no distinct pair ⟨f',m⟩ such that ⟨f',m⟩ < ⟨f,m⟩ and ⟨f',m⟩ is hearer optimal, or
      – no pair is hearer optimal, and there is no distinct pair ⟨f',m⟩ such that ⟨f',m⟩ < ⟨f,m⟩.

In words, a form–meaning pair is *hearer-optimal* if and only if there is no better meaning for the same form. A form–meaning pair is optimal if and only if the pair is hearer-optimal, and there is no better form for the same meaning or, in case there is no hearer-optimal form, there is no better form for the same meaning.

The introduction of hearer optimality means that the OT syntax has to take the interpretation into account while evaluating forms. Crucially, learning is also asymmetric, and has both a speaker perspective (comparison of forms) and a hearer perspective (comparison of meanings). The combination of speaker and hearer perspectives with stochastic OT leads to an adjustment of the constraint values. Jäger (2003, 2007) shows how this explains the correlation between animacy, subject/object position, and case-marking patterns observed by Aissen (1999, 2003). Mattausch (2005, 2007) discusses the ideas underlying the evolutionary bidirectional approach in more abstract terms. I will first present his general view, and then apply it to the case of negation.

Suppose there are two forms, one marked (m) and one unmarked (u), and suppose there are two meanings, a more common meaning α and a less frequent meaning β. Their combination leads to four possible form-meaning pairs: ⟨u,α⟩, ⟨m,α⟩, ⟨u,β⟩, ⟨m, β⟩. The question is which pairs are the optimal, most harmonic pairings of form and meaning. In order to model this situation, Mattausch proposes four bias constraints on the relation between form and meaning:

(4) Bias constraints
   *m,α: the (marked) form m is not related to the (frequent) meaning α.
   *m,β: the (marked) form m is not related to the (infrequent) meaning β.
   *u,α: the (unmarked) form u is not related to the (frequent) meaning α.
   *u,β: the (unmarked) form u is not related to the (infrequent) meaning β.

The bias constraints in (4) mitigate against all possible form–meaning combinations. Obviously, they must be modeled as soft constraints.

Furthermore, there is a general markedness constraint *MARK on forms, which avoids the use of the marked form. *MARK models a notion of economy, a preference for simpler forms over more complex ones. The ranking of *MARK with respect to the bias constraints reflects the balance of economy considerations with faithful correspondence relations between forms and meanings. Mattausch derives the constraint ranking from iterated learning over several generations in a computational model. In the initial training corpus, a grammar is used in which all four bias constraints are ranked equally high. Due to learning effects, the rankings can shift, and the association between forms and meanings shifts along with the rankings. Learning is driven by frequency distributions in a corpus.

It is crucial to start out with a frequency asymmetry between meanings α and β, α occurring say 90% of the time, and β 10% of the time. There is an even distribution of marked and unmarked forms over the meanings α and β, so that 45% of the unmarked forms are associated with meaning α, and 5% of the unmarked forms with meaning β, and the same distribution of meanings α and β (45–5%) holds for the marked forms. Adjustment of constraint values is dependent on comparisons of forms (by the speaker) and comparisons of meanings (by the hearer).

When it comes to the association of meanings to a particular form (hearer mode), a bias constraint *u,α will be promoted when the learner observes the use of an unmarked form to express the meaning β, and demoted when an unmarked form is found with the meaning α. A bias constraint *m,α is promoted when the learner observes the use of a marked form expressing the meaning β, and demoted when the marked form is found with the meaning α.

When it comes to the comparison of forms (speaker-mode), a bias constraint *u,α is promoted when the learner observes the use of a marked form to express the meaning α, and demoted when the learner observes the use of an unmarked form to express that meaning. A bias constraint *u,β is promoted when the learner observes the use of a marked form used to express the meaning β, and demoted when the learner observes the use of an unmarked form to express that meaning. The constraint *MARK is promoted when the learner observes the use of an unmarked form, and demoted when the learner observes the use of a marked form.

Because of the discrepancy between the number of α meanings and the number of β meanings in the training corpus, any form is much more likely to occur with meaning α than with β. In the interpretive dimension, this demotes *u,α, and *m,α and promotes *u,β and *m,β, roughly in accordance with the frequencies of the input training corpus. But in speaker mode, the constraint *MARK interacts with the bias constraints. *MARK favors the use of unmarked form for the frequent meaning α as well as the infrequent meaning β. But given that meaning α is more frequent than meaning β, the compromise between optimization over forms and optimization over meanings is to promote *m, α more than *m, β. Balancing the constraints leads to a new frequency distribution in the corpus, in which more than half of the infrequent meanings β are expressed by marked forms. Such statistical tendencies become visible after one generation of training the network.

The evolutionary perspective comes in when the bidirectional learning algorithm is integrated with the iterated learning model of Kirby and Hurford (1997).

The iterated learning model of Kirby and Hurford (1997) is based on the idea that learners produce language according to their acquired grammar. This means that the corpus frequencies produced by the first generation, after adjusting the OT grammar, are slightly different from the frequencies of the original training corpus.

The second-generation learner is exposed to the set of frequencies in the speech of the first-generation learner. Instead of having an equal distribution of marked and unmarked forms for the meanings α and β, the corpus will now contain more instances of the rare meaning β expressed by marked forms. This distribution reinforces the tendency to avoid marked forms for the more frequent meaning. The frequencies produced by the second-generation learner will be slightly different from those of the first-generation learner, and will constitute the input for the third-generation learner. The iteration of this process over multiple generations leads to a stable ranking of {*u,β; *m,α} >> *MARK >> {*u,α; *m, β}, as in Figure 2.

The vertical axis of Figure 2 indicates the relative strength of the constraints with respect to each other. The learning curve shows how an original training corpus with 50–50% frequencies for marked and unmarked forms, but an asymmetric distribution of meanings, develops into a grammar in which the unmarked meaning is 100% associated with the unmarked form, and the marked meaning is 100% associated with the marked form.

In this way, the bidirectional iterated learning system models the emergence of Horn's division of pragmatic labor as the optimal communication strategy that arises under evolutionary pressure. Given that the system is driven by frequency distributions, the emerging notion of markedness fits in with Haspelmath's (2006) views that frequency is the driving force behind structural asymmetries.

The model presented by Mattausch is fairly abstract, and can be applied to a range of different linguistic phenomena. An adjustment of the relation between form and

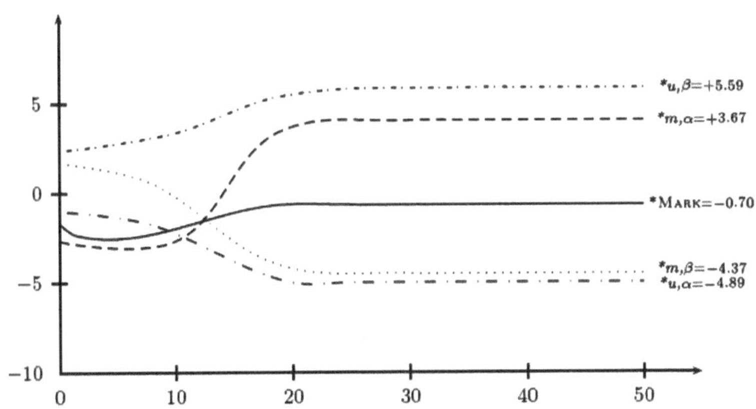

Figure 2 Bidirectional iterated learning (generations 1–50) (from Mattausch 2007)

meaning plays a role in diachronic change and grammaticalization processes, as illustrated for the expression of the relation of possession in Jäger and Rosenbach (2006). Applications to binding theory are developed in Mattausch (2005, 2007) and Hendriks and Spenader (2005, 2006). Farkas and de Swart (2009) exploit the interaction of bias constraints and *MARK to model the semantics of singular and plural nominals. Here, I present the distribution of meanings of affirmation and negation across 'zero' marked sentences and sentences with an overt marker as yet another example of Horn's division of pragmatic labor, which can be modeled by means of the interaction of a set of bias constraints with a markedness constraint.

An application of the general, abstract pattern presented so far to the empirical phenomenon of negation requires an instantiation of the meanings α and β, as well as the forms u and m. The frequent meaning α is affirmation, and the infrequent meaning β is negation. The semantic markedness of negation is thus directly related to its infrequency. The unmarked form u is a sentence with 'zero' marking. The marked form m is a sentence with an explicit marker. The iterated learning algorithm models the development of this marker into the negation marker *not* in (1), and its counterparts in other languages in (2).

Under the abstract model, the two meanings can pair up in various ways with the two possible forms. In order to model the most harmonic pairing of forms and meanings, the following instantiations of the bias constraints are called for:

(5) Bias constraints involving negation

*m,aff: the (marked) form m is not related to the meaning affirmation.
*m,neg: the (marked) form m is not related to the meaning negation.
*u,aff: the (unmarked) 'zero' form u is not related to the meaning affirmation.
*u,neg: the (unmarked) 'zero' form u is not related to the meaning negation.

The bias constraints interact with the markedness constraint *MARK, which penalizes marked forms. The initial corpus ranks all four bias constraints as equally strong, but the negation meaning is less frequent in the corpus than the affirmation meaning.

In the hearer-mode, the frequency distribution leads to a promotion of constraints related to the expression of negation (*m,neg *u,neg), and a demotion of the constraints related to the expression of affirmation (*m,aff and *u,aff). In the speaker-mode, the markedness constraint *MARK favors the use of unmarked forms for affirmation as well as negation. However, because of the infrequency of the meaning of negation, *m,aff is promoted more than *m,neg. This leads to an adjustment of constraint values, accompanied by a slightly higher production of marked forms for negation after one generation.

The application of the bidirectional OT learning algorithm in combination with iterated learning in an evolutionary setting reinforces the asymmetry. As a result, the system eventually stabilizes on the ranking {*u,neg, *m,aff} >> *MARK >> {*u,aff, *m,neg}. Under this ranking, affirmation is expressed by unmarked forms, and sentences containing 'zero' marking ('John is sick') will be interpreted as affirmative. The expression of a negative meaning requires the insertion of a special marker, and negative sentences containing a marker of negation ('John is not sick') are interpreted as negative. Mattausch's model thus permits the grounding of the markedness of negation with respect to affirmation in evolutionary principles that derive Horn's division of pragmatic labor.

The bias constraints that Mattausch uses are not very commonly used in OT more generally. Therefore, I rewrite the results for negation in terms of the faithfulness and markedness constraints used in this book so far. The bias constraints *u,neg and *m,aff require the unmarked form not to be related to negation, and the marked form not to be related to affirmation. These constraints are captured by the faithfulness constraint FNEG, introduced in Section 1.1, and repeated here:

- FNEG: Be faithful to negation, i.e. reflect the nonaffirmative nature of the input in the output.

Within the generation perspective (OT syntax), FNEG requires negation in the meaning (input) to be reflected in the output (form) (*u,neg). In the interpretation perspective (OT semantics), FNEG requires a formal expression of negation to be interpreted as contributing a semantic negation (*m,aff).

The bias constraints *u,aff and *m,neg require the unmarked form not to be related to affirmation, and the marked form not to be related to negation. These constraints can be captured by the faithfulness constraint FAFF (Faith Affirmation).

- FAFF: Be faithful to affirmation, i.e. reflect the affirmative nature of the input in the output.

Within the generation perspective (OT syntax), FAFF requires affirmation in the meaning (input) to be reflected in the output (form) (*u,aff). In the interpretation perspective (OT semantics), FAFF requires a formal expression of affirmation to be interpreted as contributing a semantic affirmation (*m,neg).

Given the results of the evolutionary learning process, the ranking FNEG >> *MARK >> FAFF instantiates a stable communicative pattern for all languages, which relates negation to an overt marker, whereas zero marking conveys affirmation. However, if FAFF is always ranked below *MARK, the constraint is in fact inoperative. Thus, it might as well be left out of the system. The contrast between FNEG and FAFF illustrates that faithfulness constraints in linguistic applications of OT target marked, rather than unmarked expressions and meanings. In the remainder of this book, I will use FNEG, but ignore FAFF, because it does not do any work for the analysis.[4]

[4] Note that emphatic affirmation may require a special expression, for instance, in the context of affirmative answers to questions presupposing a negative answer. Obviously, emphasis adds a new dimension to the markedness discussion, and the claims made here concern nonemphatic affirmation. Compare Martins (2006) for a recent study of strategies for marking emphatic affirmation in various Romance languages. See Section 4.4 and Chapter 5 (Sections 9 and 10) for some remarks on emphatic negation.

As a result of the analysis, the 'zero' marker is associated with affirmation, and the overt marker with negation. This derives the interpretation of the marker *not* as a negation marker (rather than a marker of affirmation). The markedness constraint *NEG, introduced in Section 1.1 can be viewed as a sub-constraint of *MARK, aimed at the avoidance of the more complex negative forms in the OT syntax. In the end, the universal ranking of FNEG >> *NEG is nothing other than an iconicity pattern, that is, an instance of Horn's (1984, 2001) division of pragmatic labor or Levinson's (2000) M-heuristics. I conclude that negation constitutes a universal category of natural language, because negation is a meaning that humans need to express in their languages (de Swart 2009 for an evolutionary argument based on L2 acquisition), but less frequently than affirmation.

3.1.3 Negation in OT Semantics

Jäger (2003), Jäger and Rosenbach (2006), and Mattausch (2005, 2007) use an asymmetric version of bidirectional optimization, in which forms are disqualified as candidates if the intended meaning is not optimally recoverable. Such a model is stricter than a symmetric version of bidirectional optimization, which adds up the directions of optimization over meanings and optimization over forms (cf. Beaver and Lee 2004 and Chapter 2, Section 5 for discussion of different models of bidirectional optimization). For the expression of sentential negation (¬p), the stricter asymmetric model would be sufficient. After all, the negative (marked) meaning is directly recoverable from the negative (marked) form.

However, in Chapters 4 and 6, I need to build an OT semantics that mirrors the OT syntax, because I want to model how negative meanings arise as the optimal interpretation of negative sentences in constructions in which the meaning is not directly recoverable. That requires me to zoom in on the competition between different meanings as possible candidate meanings for a given particular form. This requires a symmetric model of bidirectional optimization. In a symmetric model of bidirectional optimization, the interpretation of utterances that contain a marker of sentential negation is straightforward.

In Section 1.1, I already mentioned that the two constraints FNEG and *NEG are neutral as to what they take to be the input and what the output. They were deliberately phrased in this way, so that they could be used in both OT syntax and OT semantics. FNEG is satisfied in OT semantics if a form marked as negative is mapped onto a negative meaning. *NEG is satisfied in OT semantics if the meaning representation does not involve a negation. The same constraint ranking FNEG >> *NEG adopted in OT syntax provides the desired interpretation of negative sentences.

The input in Tableau 2 is a form (*not* S), and the output candidates evaluated by the grammar are meanings. All interpretation tableaux in this book are set up in this way. The choice is between an affirmative and a negative interpretation of the sentence.

If FNEG outranks *NEG, the negative meaning (¬p) comes out as the optimal interpretation of a negative sentence like those in (1) and (2) in Section 1.1.

The syntax and semantics of propositional negation are connected by means of strong bidirectional OT, which adds up the two directions of optimization over forms and over meanings. Tableau 3 illustrates. The format of the bidirectional tableau is slightly different from that of the unidirectional ones. Recall from Chapter 2 (Section 4) that strong bidirectional OT optimizes over form–meaning pairs. Accordingly, the input consists of form–meaning pairs [f,m], with different possibilities for the values of *f* and *m*.

Tableau 3 shows that two bidirectionally strong pairs (indicated by the victory hand ✌) emerge out of the comparison between marked and unmarked sentences under the universal constraint ranking FNEG >> *NEG. The 'zero' form S pairs up with the affirmative meaning p, because this pair does not violate any constraints. The marked form *not* S pairs up with ¬p. Although this pair violates *NEG twice, it is better than the two alternatives, each of which incur one violation of the higher ranked constraint FNEG.

3.2 Negation in Users of Sign Language Who have Suffered Brain Damage

In Section 1.1, the ranking FNEG >> *NEG was posited as part of the grammar of all natural languages. The universal category of negation is thereby rooted in the OT constraint ranking. The universality of this pattern was motivated as an evolu-

Tableau 2 Interpretation of propositional negation

Form not S	Meaning	FNEG	*NEG
	p	*	
☞	¬p		*

Tableau 3 Propositional negation in strong bi-directional OT

input [f,m] f_1: S; f_2: not S m_1: p; m_2: ¬p	FNEG	*NEG
[S, p] ✌		
[S, ¬p]	*	*
[not S, p]	*	*
[not S, ¬p] ✌		**

tionarily stable result that fits into Horn's division of pragmatic labor (Section 1.2). This outcome is potentially problematic for the system, as OT is founded on the notion of soft constraint. This means that for any constraint, it should be possible to find cases where the constraint is violated. The status of FNEG as a soft constraint cannot be tested if it is never violated.

This section shows that in certain situations where language is developing (L1 acquisition by children) or has broken down (aphasia due to brain damage), there may be systems in which FNEG loses against *NEG. Such restricted linguistic systems support the view that FNEG is a soft constraint. These observations do not invalidate the argumentation set up in Section 1.2, because young children and people who suffer from brain damage encounter communicative problems that may affect the Horn pattern of optimization. However, a restricted linguistic system in which negation is not expressible and not interpretable indirectly supports the treatment of negation advanced here, because it shows that in extreme cases FNEG can behave like a soft constraint.

Children sometimes convey a negative meaning without using a negative form, as in the Dutch utterance *Aankomen!* ('touch!'), when they mean *Niet aankomen*! ('don't touch'). The reason for omitting the adverb might be the ranking of the constraint *NEG above FNEG. The meaning the child intends to convey is clearly negative: *Niet aankomen*! ('don't touch!') is a common command in the presence of children, and its positive counterpart *Aankomen!* ('touch!') is never used by adult speakers.[5]

The production patterns must be extended to comprehension with caution. Comprehension of negation is difficult to test with young children, because their linguistic and cognitive development proceeds in tandem. Given that this book focuses on truth-functional negation, and early negation in child language might have pre-logical meanings (cf. Horn 1989), I do not intend to develop a full theory of the L1 acquisition of negation, but cf. de Swart (2009) for an extension of the model to L2 acquisition, with implications for language genesis.

Atkinson et al. (2004) offers a study on the understanding of negation by users of British sign language (BSL) with unilateral left and right hemisphere lesions (LH and RH). The remainder of this section heavily relies on their insights.

People with normal hearing use a variety of linguistic and paralinguistic ways to express negation. The lateral head-shake conveys negation in many cultures, and a furrowed brow is a universal feature of a communicative display suggesting negativity in intention or emotion. Languages allow more focused aspects of negation to be expressed through lexical forms (*not*, *never*, …) or affixes (*in* + *exact*, …), sometimes accompanied by special morpho-syntactic structures (*he left*/*he didn't leave*). Both speakers and signers make use of gestures. Users of spoken language, may use facial, manual or vocal gestures. Manual gestures can occur

[5]Thanks to an anonymous reviewer for drawing my attention to the child language data.

alongside linguistic elements. Manual gestures may resemble the signs of sign language but they are processed independently of signs.

British Sign Language (BSL) is conveyed using both hands that function relatively independently, and this process is supported by further articulators in the face and head. The two hands generate simultaneous syntactic structure in a specific way that is not available to the users of spoken language. The face can convey important phonological, morphological and syntactic information at the same time.

Negation in signed languages is achieved by a combination of manual and nonmanual elements. Sign languages for which the expression of negation has been investigated are strikingly similar in this respect: a manual negation element is combined with nonmanual elements and the manual sign is usually optional. In BSL, negation is indicated by a variety of nonmanual negation elements. These include one or more short lateral head shakes, a furrowed brow, narrowed eyes, and down-turned mouth, either alone or in combination. There are also lexical and affixal forms of negation, which use manual actions as well as the mandatory facial and head movements. However, facial negation is an *obligatory* feature of negation in sign language, whereas manual negation is *optional*. In this respect, BSL contrasts with spoken English, where a negative statement is always marked by a lexical or morphological feature.

In sign languages, negation can occur without lexical or morphological marking, using face–head actions only. Atkinson et al. are interested in the status of face–head actions: are they part of syntactic structure, or do they involve prosody? Prosodic expression of negation is rare in spoken languages, but not impossible: in a small number of African and Austronesian languages, negation is realized through prosodic change only, primarily by a change of tone or lengthening of vowels (Dahl 1979). Given that both syntactic and phonological information can be conveyed by the face in BSL, the two options are a real issue in sign language. The study by Atkinson et al. was set up to decide between these two options.

The purpose of the study was to explore the extent to which adults who are either native users of BSL or acquired it at an early age, and who then suffered unilateral brain lesions, show specific anomalies in processing negation. Producing facial negation is a problem for the BSL users with RH lesions (Bencie Woll, 2005, personal communication). This is caused by the neurological damage related to RH brain damage which means that these patients have a 'mask-like' face that does not express emotion. As a result, these BSL users do not have access to the articulator in the face in their sign production. Therefore, the main accent of the study is on processing negation.

Atkinson starts from the assumption that language perception and production is localized in the same way irrespective of whether people speak or sign. This raises the question whether BSL users who have suffered a unilateral lesion show dissociations in their understanding of negative statements and, if so, whether this varies according to the type of utterance. Atkinson's study looks at how well these patients comprehend negation expressed through face and head actions alone as compared to these actions together with manual elements (lexical or morphological).

If facial negation is a direct surface realization of syntax, then users with lesions in their left-hemisphere (LH) should show impaired negation processing along with other language processing difficulties. However, there should be no distinction between the different realizations of negation: both face-only and face-and-hands (lexical) displays would be difficult for LH lesion patients to understand because in this case the facial and manual signs are both related to syntax. Even if they are unable to understand emotive facial expressions, BSL users with right hemisphere (RH) lesions should be able to understand combinations of facial expression and hand movements as expressions of negation, since for sign users these movements form part of their language system and the language processing centre is located in the (undamaged) left hemisphere (LH).

If facial negation maps to surface prosodic rather than to syntactic structures, the prediction is quite different. In this case, patients with RH lesions would be expected to show a spared understanding of negation that is conveyed by means of a manual (lexical) element, since this can be processed by the unimpaired language module in the LH. However, there should be impaired processing of facial negation in line with other impairments in prosody or facial expression processing in this group. For LH lesion patients, negation should be understood relatively well in comparison to their other linguistic problems since all negation in BSL, whether or not it includes a manual gesture, includes facial and head gestures. This means there should be no dissociation between manual and facial negation in BSL users who have LH lesions.

Atkinson et al. carried out a number of tests to investigate and compare language comprehension in the two groups of patients. In general, the RH lesion group scored within normal limits on the comprehension test of BSL. None of the individuals with RH damage displayed aphasia in conversations. However, they had problems with the pragmatic and discourse aspects of language. The individuals with LH damage scored outside normal limits on all the language tests. According to Atkinson et al., these findings are consistent with studies on deaf American sign language users who have brain lesions: people with LH damage display sign aphasia, while those with RH damage do not.

The negation experiment that Atkinson et al. carried out was a comprehension test. In this task, patients with LH or RH lesions were presented with two pictures: an image and its opposite. Investigators used BSL to communicate either a positive or a negative statement to the patients and asked them to match it to the appropriate picture. Different word classes were tested: nouns, adjectives, and verbs. Negative statements comprised a single noun, adjective, or verb together with a negative face and head marker. Half of the negative items, additionally, used a lexical/manual marker. The test, therefore, compared comprehension of lexically marked (manual) and unmarked (facial) negation as a function of word class. Atkinson et al. found that RH lesions impaired performance more than LH lesions. Problems found in the RH group were almost entirely to do with items requiring comprehension of facially marked negation in the absence of a lexical/manual marker. All patients with RH damage were significantly worse at comprehending negation without the lexical/manual marker.

There was one person with RH damage who had normal hearing, but who had learnt sign language at a young age because both parents were deaf. This provided an opportunity to test the comprehension of spoken English negation and compare it to comprehension in sign language. The test Atkinson et al. used for spoken language mirrored the format of the BSL task. This particular patient performed perfectly on the spoken English test, but displayed problems characteristic of RH damage in the sign language test.

In general, Atkinson et al. found that BSL users with RH lesions are impaired with respect to reading facial negation in comparison to manual (lexical and morphological) negation. This dissociation was not detected in the group who had LH lesions and various language difficulties. Atkinson et al. conclude that nonmanual negation in sign language may not be a direct surface realization of syntax. They propose that some aspects of the linguistic analysis of sign language are achieved by prosodic analysis systems (analysis of face and head gestures) that are lateralized to the right hemisphere.

These results are of great relevance to the study of the universal order FNEG >> *NEG. BSL is like the spoken languages investigated so far in that negative sentences are marked in ways that affirmative sentences are not. Thus, the grammar of BSL instantiates the ranking FNEG >> *NEG. Prosodic knowledge of sign language users is assumed to be located in the right hemisphere. If the right hemisphere is damaged, production and comprehension in relation to the prosodic system should be entirely blocked. In OT terms, the brain damage results in a general ranking of *MARK >> FAITH in the prosodic domain, where *MARK is an overall markedness constraint (blocking all structure and all interpretation), and FAITH is an overall faithfulness constraint (requiring a correlation between input and output). Under the overall ranking *MARK >> FAITH, no prosodic signs are produced, and none are understood: the output from the prosodic part of the system is always unmarked, independently of the input.

If *NEG constitutes a subconstraint of *MARK (cf. Section 1.2), and FNEG a subconstraint of FAITH, the ranking *NEG >> FNEG can be viewed as a particular instance of the general constraint ranking *MARK >> FAITH. If negation is part of the prosodic realization of the utterance, it will not be processed as semantic negation by the hearer. No prosodic negation is produced under this ranking. The two directions of optimization block the expression and the comprehension of negation in the prosodic domain.

The L1 acquisition and the aphasic system of sign language users provide the two situations I found in which the universality of the ranking FNEG >> *NEG is violated. Of course, young children have not yet developed the full communicative skills of adults (cf. Hendriks et al. 2009, Chapters 4 and 5), and brain damage may very well affect general human cognition. So the linguistic systems of BSL users with LH- or RH lesions may be subject to very different constraints from those of full linguistic systems used by unimpaired adults. In particular, the fact that children and aphasics may not be able to follow the Horn patterns of division of pragmatic labor because of immature cognition or damage to certain parts of the brain does not necessarily lead to similar patterns in the languages of the world.

The important insight is that first language acquisition and aphasia offer instances of the grammar in which FNEG is violated, and must be characterized as a soft constraint that can be demoted in the constraint hierarchy. The OT system permits a description of the normal adult stage as well as the acquisition pattern and the pathological case in terms of the same constraints. The data from L1 acquisition and the aphasic pattern then support the treatment of FNEG as a soft constraint.

3.3 Typological Variation in the Placement of Negation

There is a large body of literature about the syntax of negation, in all kinds of theoretical frameworks, and in synchronic research as well as diachronic work. It is impossible for me to review this literature and do justice to it here. It would also lead me too far away from my concerns with the syntax–semantics interface. Instead, what I will do is rephrase some of the key insights from the descriptive, typological and theoretical literature in OT terms, thereby shedding light on the underlying similarities of natural languages, as well as respecting the complex patterns of differences between them. Thus the focus is on the range and limits of cross-linguistic variation.

Section 3.1 discusses preverbal and postverbal negation. Section 3.2 adds discontinuous negation. Section 3.3 describes Jespersen's typology of the placement of negation in natural language in terms of the joint maximization of two constraints, as opposed to a third, weaker constraint.

This section is restricted to the placement of the negation marker in finite main clauses. In many languages, the position of negation in subordinate clauses and nonfinite constructions raises a separate set of complex syntactic questions. Some of the relevant issues will be sketched in Section 5, but they will not receive a full analysis in this book. Note also that this chapter is concerned with the expression of propositional negation by means of a marker. Some restrictions on the position of negative indefinites will be addressed in Chapter 4 (Section 1).

3.3.1 Preverbal and Postverbal Negation

The fact that all (full) natural languages have ways to express propositional negation is reflected in the universal ranking FNEG >> *NEG. But these two constraints do not say anything about the way negation is realized in natural language. The most common realization of negation involves the use of a marker of sentential negation. This marker, glossed as SN, realizes the propositional negation $\neg$ in sentences (1) and (2) in Section 1.

Once it has been established that there is a special negative marker, the question arises where this marker is placed in the sentence. This section investigates the position of negation in the sentence across languages. The focus is on the placement of negation relative to the verb.[6] In (6) and (7), I provide examples of negation in preverbal and postverbal positions, respectively:[7]

(6) Preverbal negation

a. Maria *non* parla molto. [Italian]
Maria SN talks much.
'Maria doesn't talk much.'

b. Juan *no* ha llamado a su madre. [Spanish]
Juan SN has called to his mother.
'Juan hasn't called his mother.'

c. ʔəli *ma:* ra:ħ lidda: ˀirə [Baghdad Arabic]
Ali SN went to the office '
Ali didn't go to the office.'

d. A vaga koŋ *ba* bɛnɛ. [Koromfe]
ART dog.SG DET.NONHUMAN.SG SN come.PASt
'The dog did not come.'

e. tā *bu* sǐ [Chinese]
3SG SN die
'S/he refuses to die/won't die.'

f. János *nem* dohányz-ik. [Hungarian]
János SN smoke.3SG
'János doesn't smoke.'

g. On *ne* igraet. [Russian]
he SN plays.
'He doesn't play.'

[6] Several attempts have been made to relate the position of the marker of negation to the basic SVO order of the language. Although there are some general tendencies, Dahl (1979) and Dryer (1988) find exceptions to any strict correlation. The placement of negation with respect to the verb seems to have the most important implications for the syntax–semantics interface, so I focus on preverbal and postverbal positions. Compare also Chapter 1, Section 2 for remarks on this issue.

[7] The Romance examples are from Zanuttini (1991, 1996). The Baghdad Arabic example is from Payne (1985). The Koromfe example and the Gbaya Kaka example are from Dryer (2007). Koromfe is a Niger-Congo language spoken in Burkina-Fasso and Mali; Gbaya Kaka is a Niger-Congo language spoken in Cameroon. The Chinese example and the Tamil example are from Croft (1991). The Turkish and Japanese examples are from Morimoto (2001). For the examples quoted, I copy the glosses from the source, except for the marker of sentential negation, which I consistently gloss as SN, even if not in the original, in order to maintain uniformity.

(7) Postverbal negation

a. Maria a parla *nen* tant. [Piedmontese]
Maria CL talks SN much.
'Maria doesn't talk much.'

b. Maria spricht *nicht* viel. [German]
Maria talks SN much.
'Maria doesn't talk much.'

c. Jag kisste *inte* Anna. [Swedish]
I kissed SN Anna
'I didn't kiss Anna.'

d. Mi-zɔk wi ndɔng *na* [Gbaya Kaka]
IsG-see person that SN
'I do not see those people.'

e. naan pooka-*le* [Tamil]
I go-SN
'I didn't go.' / 'I am not going.'

f. John elmalar-i ser-*me*-di-Ø [Turkish]
John apples-ACC like-SN-PAST3SG
'John didn't like apples.'

g. Taroo-wa asagohan-o tabe-*na*-katta. [Japanese]
Taroo.TOP breakfast.ACC ate.SN.PAST
'Taroo didn't eat breakfast.'

The marker of sentential negation need not be an independent word; it can be an affix as in (7f, g). Preverbal negation frequently cliticizes onto the verb (as French *ne*, cf. example 9b), and can even be incorporated in the verb. Mazzon (2004: 29) reports that Old English had many commonly used verbs in which negation *ne* was incorporated: forms of *wesan* ('to be') (*nis/nys* beside *is*, *nere* beside *wære*, etc.), and all the forms of *nillan* (beside *willan > will*), *nabban* (from *habban > have)*, *nagan* (from *agan*), and *nytan* beside *witan*. Most of these forms disappeared from later stages of the language.

Borsley and Jones (2005: 49) report distinct negative forms with initial *d/t* for a number of frequent verbs in modern colloquial Welsh, and provide pairs of sentences like the following:

(8) a. Oedd Sioned yn gweithio. [colloquial Welsh]
be.IMPF.3SG Sioned PROG work.
'Sioned was working.'

b. Doedd Sioned ddim yn gweithio.
SN.be.IMPF.3SG Sioned SN PROG work
'Sioned was not working.'

The preverbal particle *ni(d)* that characterizes formal Welsh has disappeared from colloquial Welsh, but survives as a special negative form of the verb, at least for certain verbs. Incorporation of negation is not only seen in verbs, but also in pronouns and adverbs (e.g. English *ever-never*). Negative incorporation into indefinites will be treated in Chapter 4.

I will not distinguish between negative particles and negative clitics/affixes, but gloss them all as SN. In so far as the syntax interacts with the morphophonological system of a language, the effects of this interaction on the marker of sentential negation are outside the scope of this book.

Even though the examples in (1), (2), (6)–(9) do not reflect the full range of typological variation, they indicate two important patterns. In Italian, Spanish, formal Welsh, Chinese, Russian, Hungarian, and many other languages, sentential negation directly precedes the verb (6), (2a-e). In Piedmontese, a dialect spoken in northern Italy, the sentential marker *nen* follows the finite verb (7a). The same situation is found in Germanic languages such as German (7b) and Swedish (7c). In English, negation follows the auxiliary (1a, b), but precedes the main verb (1a, c, cf. Section 3.4).

These two main tendencies have already been described by Jespersen (1917). On the one hand, there is a strong tendency "to place the negative first, or at any rate as soon as possible, very often immediately before the particular word to be negated (generally the verb)" (Jespersen 1917: 4). Horn (1989: 292–293) uses the term NegFirst for this tendency. NegFirst is motivated by communicative efficiency, i.e. to "put the negative word or element as early as possible, so as to leave no doubt in the mind of the hearer as to the purport of what is said" (Jespersen 1924: 297), quoted by Horn (1989: 293). The close connection between the verb and sentence negation is expected if Aristotle's and Jesperson's view of negation as predicate denial is adopted, as argued extensively in Horn (1989). If (propositional) negation typically bears on the verb, an immediately preverbal slot is the natural position for the marker of sentential negation. Evidence from early L2 acquisition supports this claim (cf. de Swart 2009).

Although many languages have a preverbal marker of sentential negation, the examples of postverbal negation in (7) indicate that NegFirst is not an absolute rule. NegFirst is opposed by another strong tendency, which I label as FocusLast. FocusLast reflects that given information comes early in the sentence, and new or significant information comes last in the sentence. FocusLast is not specific to negation, but is a pragmatic strategy or an instance of information structure that mainly operates at the discourse level. In languages in which word order is not strict, principles of information structure often interact with linear order. If negation is part of the new information expressed by the sentence, it is expected to show up late, rather than early in the sentence. FocusLast for negation is then motivated by the idea that the negative force is stronger if the negator comes later in the linear order (Mazzon 2004: 97).

In languages like German and Swedish, the postverbal position of the marker of negation is the result of a diachronic development where postverbal adverbials that originally served as emphasis for the negation gradually took over the negative force of the sentence, while maintaining their postverbal position (7b, c). The outcome of this diachronic process is a grammaticalization of the postverbal position of negation (at least in main clauses). In these languages, FOCUSLAST as applied to negation does not operate as a purely pragmatic constraint, but functions in the syntax. I will come back to the diachronic development commonly referred to as the 'Jespersen cycle' in Section 4. For now, I develop a typological approach in a synchronic perspective.

Although NegFirst and FocusLast are grounded in information structuring principles, I treat them here as grammatical constraints. Given that NegFirst and FocusLast are both strong tendencies, but not hard rules, they work best as violable constraints that can be ranked with respect to each other and to other constraints. I propose the following formulations for the two constraints in OT:

- NegFirst
  Negation precedes the finite verb
- FocusLast as applied to negation
  Negation as new information comes last in the sentence

In this book, I am only concerned with the role of FocusLast in the placement of negation.[8]

Languages with preverbal negation then have grammars in which NegFirst is strong, and languages that do not have preverbal negation have grammars in which NegFirst is ranked low. Similarly, if a language has a grammar with a high ranked position for FocusLast as applied to negation, it can push the marker of sentential negation to a later position in the sentence.

The assumption that both rankings NegFirst >> FocusLast and FocusLast >> NegFirst are available in the grammar gives rise to tableaux 4 and 5 for the two general positions of the marker of sentential negation with respect to the verb.[9]

Tableau 4 Preverbal negation (Italian, Spanish, formal Welsh,...) (first version)

Meaning ¬p	Form	NegFirst	FocusLast
☞	(S) sn V (O)		*
	(S) V sn (O)	*	

Tableau 5 Postverbal negation (Piedmontese, German, Dutch, ...) (first version)

Meaning ¬p	Form	FocusLast	NegFirst
	(S) sn V (O)	*	
☞	(S) V sn (O)		*

[8]Aissen (1999, 2003) uses a general markedness constraint *Case, which can be split up into different subconstraints. In a similar way, FocusLast can work out in different ways for negation and other constructions sensitive to focus within a language (cf. Rooth 1985).

[9]FNeg is always the highest ranked constraint, and it is left out here, so the tableaux concentrate on candidates that realize negation at least once. In order to avoid the discussion about word order, I give the tableaux in SVO form, but with S and O between brackets, to indicate that these could also get a different position. An anonymous reviewer correctly points out that NegFirst does not distinguish between S sn V (O) and sn S V (O). According to Dryer (1988), SVO languages are most commonly S snVO, and the examples in (6) above illustrate this preference. I am only concerned with the position of negation with respect to the verb (cf. footnote 6), so I leave out the competition between S sn V (O) and sn S V (O).

The occurrence of preverbal or postverbal negation in a particular language is then governed by the ranking of the two constraints FocusLast and NegFirst.

3.3.2 Adding Discontinuous Negation

In Tableaux 4 and 5, the competition is limited to NegFirst and FocusLast. This set-up suggests that there is a forced choice between the preverbal and the post-verbal position. However, languages can also display discontinuous negation. This configuration combines two negation markers, one of which usually occurs in a preverbal position, and the other one in a postverbal position. The examples in (9) illustrate.[10]

(9) Discontinuous negation

a. *Ne* bið he *na* geriht. [Old English]
SN is he SN righted
'He is not/never set right (=forgiven)'

b. Elle *ne* vient *pas*. [written French]
She SN comes SN.

c. U *n* li sent *nent*. [Cairese Piedmontese]
3.CL SN him hears SN.
'He can't hear him.'

d. Igl bab *na* lavoura *betg*. [Surmeiran]
the father SN works SN
'The father doesn't work.'

e. *Ni* soniodd Sioned *ddim* am y digwyddiad. [formal Welsh]
SN mention.PAST.3SG Sioned NEG about the event
'Sioned did not talk about the event.'

f. *Doedd* Gwyn *ddim* yn cysgu. [informal Welsh]
SN.be.IMPF.3SG Gwyn NEG PROG sleep
'Gwyn was not sleeping.'

g. baba *wo*-shìi nai tapa *u*. [Kanakuru]
father SN-he drink tobacco SN
'My father does not smoke tobacco.'

[10] The Old English example (9a) is from Mazzon (2004). The Romance examples are from Zanuttini (1996: 5). Kanakura is a West Chadic language spoken in Nigeria. The Welsh examples are from Borsley and Jones (2005: 22, 26). The Kanakura example (9g) is from Dryer (2007). The Afrikaans example (9h) is from K. van Gass (2007, personal communication), and the Breton example in (9i) from Stump (1989), quoted in Legendre (2001). The Moroccan Arabic example (9j) is from Benmamoun (2000).

h. Haar suster het *nie* haar verjaarsdag vergeet *nie*. [Afrikaans]
Her sister has SN her birthday forgotten SN
'Her sister didn't forget her birthday.'

i. *Ne* lenn *ket* Anna al levr. [Breton]
SN read-3 SN Anna the book
'Anna does not read the book.'

j. Omar *ma*-taykteb-*sh* [Moroccan Arabic]
Omar SN-ASP-3M-write-SN
'Omar doesn't write.'

I call this phenomenon 'discontinuous negation', because there is only one negation in the semantics. That is, all the sentences in (9) express a proposition of the form ¬p, with p an atomic proposition. However, negation is expressed by two 'bits' of form, one preceding the verb, the other following it. Syntactically then, we have double negation, but semantically, the sentence conveys just a single negation.

Mazzon (2004: 27) indicates that the discontinuous negation in (9a) was a rather unstable phenomenon in the late Old English and Early Middle English period. The written French example in (9b) illustrates the bleaching of preverbal *ne* to a syntactic negator, whereas the expressive force of negation is borne by the postverbal negator *pas* (cf. Godard 2004 and references therein). In spoken French, *ne* is on its way out: *ne* is dropped in many cases, in favor of a system which only uses the postverbal negator *pas*. However, presumably under the pressure of prescriptive grammar, *ne* is maintained in the written language.

In formal Welsh, which reflects an older stage of the language, the postverbal *ddim* is optional (9e). In informal Welsh (9f), the preverbal particle has disappeared, but it survives in incorporated form on some verbs, such as *oedd-doedd*, as already pointed out in (8). Although the verb appears in a negative form, it is unable to express semantic negation, and the presence of the postverbal adverb *ddim* is obligatory. Negation on the verb in (9f) is then semantically bleached, in the same way as preverbal *ne* is in written French (9b).

Discontinuous negation is crosslinguistically relatively rare; it is usually not very stable in a diachronic sense, and one of the markers may undergo semantic bleaching. The English, French, and Welsh data shown in (9) support this view. Reasons of economy might explain the rarity of discontinuous negation: syntactically, the double marking found in discontinuous negation is of course rather costly.

In terms of the OT system, discontinuous negation arises when both NEGFIRST and FOCUSLAST are satisfied, whereas in the examples in (6) and (7) either one or the other was violated. Discontinuous negation then involves a grammar in which both NEGFIRST and FOCUSLAST are ranked above *NEG (Tableau 8). Systems with preverbal or postverbal negation are then properly captured by the insertion of *NEG in between the constraints NEGFIRST and FOCUSLAST (Tableaux 6 and 7).

Tableau 6 Preverbal negation (Italian, Spanish, formal Welsh,...) (final version)

Meaning ¬p	Form	NegFirst	*Neg	FocusLast
☞	(S) sn V (O)		*	*
	(S) V sn (O)	*	*	
	(S) sn V sn (O)		**	

Tableau 7 Postverbal negation (Piedmontese, German, Dutch, ...) (final version)

Meaning ¬p	Form	FocusLast	*Neg	NegFirst
	(S) sn V (O)	*	*	
☞	(S) V sn (O)		*	*
	(S) sn V sn (O)		**	

Tableau 8 Discontinuous negation (Old English, written French, colloquial Welsh, Kanakuru, ...)

Meaning ¬p	Form	NegFirst	FocusLast	*Neg
	(S) sn V (O)		*	*
	(S) V sn (O)	*		*
☞	(S) sn V sn (O)			**

Tableaux 6–8 indicate that *Neg is a gradable constraint that incurs one violation for every instance of a negative form. In the case of discontinuous negation, *Neg is violated twice in order to satisfy the two faithfulness constraints ranked above it.

Grammars with a low ranking of *Neg in the OT syntax are the hallmark of negative concord systems, as Chapter 4 will show. Discontinuous negation is then part of our understanding of the syntax and semantics of negative concord and double negation.

3.3.3 A Typology of the Placement of Negation in Natural Language

In OT, typological variation is accounted for in terms of reranking. A full factorial typology would lead to six possible constraint rankings for the three constraints *Neg, NegFirst, and FocusLast. However, natural languages display only three main patterns, namely preverbal negation, postverbal negation, and discontinuous negation. A closer look at the constraint rankings reveals that each case involves the joint ranking of two constraints as higher than the third one.

Table 1 Typology of placement of negation w.r.t. the verb

Preverbal negation	{NEGFIRST, *NEG} >> FOCUSLAST
Discontinuous negation	{NEGFIRST, FOCUSLAST} >> *NEG
Postverbal negation	{FOCUSLAST, *NEG} >> NEGFIRST

Note that the two highest constraints in Tableaux 6–8 are connected by a dotted line, rather than a straight line. The dotted line indicates that the relative ranking of the two constraints cannot be decided on the basis of the candidates displayed, so they are ranked equally high. The three grammars are summarized in Table 1, where two constraints ranked equally high are enclosed in curly brackets.

Given that two different constraints are paired up in each case, the three constraints are not harmonically bound, and no reduction in the number of constraints is possible. However, the result of the pairing up of constraints is that the factorial typology leads to three main rankings, rather than six.

The idea behind the three-way partition in the position of negation is that all three constraints capture a fundamental and highly valued aspect of the expression of negation, namely the markedness of negation, and its preference for either a preverbal position or a focus position late in the sentence. It is impossible to satisfy all three constraints at the same time, because they are partially conflicting. However, it is possible to maximize the satisfaction of two constraints by accepting the violation of the third one. This provides a better optimization strategy than a full factorial typology.

The joint maximization of two constraints as opposed to a third, weaker constraint leads to three possible rankings, which correspond to the three main patterns found in natural language. In this way, the OT analysis models Jespersen's (1917) findings in a principled way.

3.3.4 Refinements in the Postverbal Domain

The basic typology corresponding with Jespersen's (1917) findings is worked out in Section 3.3. This section considers some more complex cases that rely on variants of the constraint NEGFIRST. English *do*-support and postverbal discontinuous negation in Afrikaans are central to the discussion.

It is generally admitted that negation in modern English is difficult to characterize in terms of NEGFIRST and FOCUSLAST, because negation follows the auxiliary verb (10a, b), and triggers *do*-support with lexical verbs (10c, d):

(10) a. I am *not* sick.
b. He may *not* be available.
c. I do *not* sing.
d. Mary does *not* talk much.

Dryer (1988: 93) and Horn (1989: 456) suggest that the construction of *do*-support finds its motivation in NegFirst. The periphrastic *do*-support construction is known to have already existed in the 15th century and became standard by the 17th century. As pointed out by Horn (1989: 435, 436), Shakespeare uses both the forms 'I love you not' and 'You do not love me'. In diachronic terms, *do*-support involves restoration of a version of NegFirst (not preceding the finite verb, but preceding the main verb) in the presence of a postverbal adverbial *not.*

Not, which originates from *nawiht/nogh/nahtet* 'nothing', has taken over the negative force in modern English. Mazzon (2004: 75) argues that English verbs come in two classes, operators (modals and auxiliaries as in 10a, b) and nonoperators (lexical verbs as in 10c, d). This suggests that the version of NegFirst operative in modern English involves precedence with respect to a nonoperator, i.e. the (main) lexical verb.[11]

- ♦ NegFirst (lexical verb)
  Negation precedes the (main) lexical verb

The version of NegFirst used so far in the characterization of Italian, formal Welsh, and other languages that have a marker of sentential negation that precedes the finite verb does not apply to modern English. However, the replacement of NegFirst by NegFirst (lv) brings English within the range of languages that express negation 'early' in the sentence. English negation is captured by the grammar {NegFirst (lexical verb), FocusLast} >> *Neg, as illustrated in Tableau 9.

Tableau 9 illustrates that a grammar in which both (a version of) NegFirst and FocusLast rank higher than *Neg does not necessarily need discontinuous negation. The auxiliary *do* is a 'expletive' verb that does not carry semantic meaning,

Tableau 9 *Do*-support in modern English negation

Meaning ¬Talk(m)	Form	NegFirst (lv)	FocusLast	*Neg
	Mary not talks much		*	*
	Mary talks not much	*		*
	Mary not talks not much			**
☞	Mary does not talk much			*

[11] Neg-Raising seems to be related to NegFirst in a more general sense, cf. also Horn (1989) and Mazzon (2004: 97). NegRaising is the phenomenon that raises the sentential negation from the subordinate clause to the main clause with certain verbs of communication, cf. (ii) as opposed to (i):

i. I think you will not find him.
ii. I don't think you will find him.

NegRaising leads to the expression of negation earlier on in the complex sentence. It carries special meaning effects (cf. Horn 1989: 321–359). Given that this study is restricted to simple clauses, I will not attempt to formulate rules for NegRaising in the OT system.

but is inserted in order to satisfy FocusLast (postverbal position) as well as NegFirst (in the version in which negation precedes the lexical verb).[12]

An important conclusion to draw from the discussion of modern English is that NegFirst is possibly not a single constraint. Rather, it is a particular pattern that supports the occurrence of negation 'early' in the sentence or 'high' in the hierarchical structure. Accordingly, it should probably be seen as a cluster of tightly related constraints that interact with the rest of the syntax. Usually, one particular version of NegFirst will do for the grammar of a particular language. 'Negation precedes the finite verb' seems to be the most common version, but in modern English another version of NegFirst is operative.

Further support for a reformulation of NegFirst comes from a language like Afrikaans. Section 3.2 analyzes patterns of discontinuous negation in which one bit occurs to the left of the verb, and another bit to its right. However, Afrikaans displays a pattern of postverbal discontinuous negation in which both composing parts follow the finite verb in main clauses, as illustrated in (11):

(11) a. Ik het hom *nie* gesien *nie*. [Afrikaans]
I have him SN seen SN
'I have not seen him.'
b. Ons maak *nie* 'n gemors *nie*.
We make SN a mess SN
'We're not making a mess.'

The pattern in (11) suggests that the first occurrence of *nie* should immediately follow the finite verb (except for clitics), and the second occurrence of *nie* should be sentence final. The sentence-final position of *nie* corresponds with a high ranking of FocusLast. The first occurrence of *nie* can be licensed by a revised version of NegFirst, which requires the negation marker to precede all other postverbal material.[13] Both constraints outrank *Neg, in line with the analysis of discontinuous negation developed in Section 3.2.

♦ NegFirst (Afrikaans)
Negation must precede all material following the finite verb

If NegFirst is replaced with NegFirst (Afrikaans) in Tableau 8, the discontinuous negation pattern illustrated in (9h) and (11) falls out immediately.

Chapter 5 offers versions of NegFirst that are operative in varieties of colloquial Welsh that use the postverbal marker of negation *ddim*. This discussion is

[12] As a finite verb, the auxiliary *do* of course carries inflection for person and tense. However, it does not contribute the kind of semantics found with modal auxiliaries or *have/be* in the construction of perfect tenses. This is the reason I label it an 'expletive' verb along the lines of the expletive subjects discussed in Chapter 2 (Section 2).

[13] The constraint does not handle the exceptional case of clitics. Within OT, it is possible to account for examples like (11a) as opposed to (9h, 11b) by imposing a stronger constraint governing the position of clitics. I will not spell out such a constraint here, but the gist of it will be clear.

postponed until Chapter 5, because the Welsh data imply an interaction between the marker of sentential negation and negative indefinites.

What emerges from a range of typological studies is that natural languages have a strong desire to express negation early in the sentence, even those that locate the marker of negation in the postverbal position (Dryer 1988, 2007). These findings support the view that preverbal negation in early L2 acquisition and, possibly, language genesis constitutes an instance of the 'emergence of the unmarked' (de Swart 2009). Accordingly, the constraint NegFirst plays an important role in the grammar of many languages, as will be underscored in Chapters 4 and 5.

The examples of preverbal, postverbal, and discontinuous negation in (6)–(11) indicate that grammars strike a balance between different desirable properties, which cannot all be satisfied simultaneously. What constitutes the optimal position for negation depends on the strength of the three constraints, NegFirst, FocusLast, and *Neg, and may vary from one language to the next. Of course it is not a coincidence that the three basic constraint settings illustrated here correlate with the three main stages of the Jespersen cycle. The patterns of diachronic change are worked out in Section 4.

3.4 A Dynamic Analysis of the Jespersen Cycle

Section 3 discussed the position of the marker of sentential negation in the sentence in terms of two opposing tendencies: NegFirst and FocusLast. I argued that the contrast between preverbal expression of negation in languages like Italian, formal Welsh, etc. is the result of the dominance of NegFirst, whereas the postverbal expression of negation in languages like Piedmontese, Dutch, and German illustrates the effect of FocusLast. Discontinuous negation as in Old English, written French, colloquial Welsh, and Kanakura arises when both constraints are satisfied, and *Neg is ranked low.

The typological observations made in Section 3.3 are here connected to the pattern of diachronic change commonly referred to as the 'Jespersen cycle'. Section 4.1 describes the empirical patterns, and Section 4.2 provides the OT analysis. Sections 4.3–4.4 provide reflections on intermediate stages. Section 4.5 relates the developments to the pragmatics of negation, and Section 4.6 discusses the advantages of a dynamic versus a static approach.

3.4.1 Patterns of Diachronic Change

Jespersen formulates the diachronic pattern as follows: 'The history of negative expressions in various languages makes us witness the following curious fluctuation: the original negative adverb is first weakened, then found insufficient and therefore strengthened, generally through some additional word, and this in turn may be felt as the negative proper and may then in course of time be subject to the same development as the original word' (Jespersen 1917: 4), quoted by Horn (1989: 452).

A few pages later, Jespersen adds: "Now, when the negative begins a sentence, it is on account of that very position more liable than elsewhere to fall out, by the phenomenon for which I venture to coin the term of *prosiopesis* (the opposite of what has been termed of old aposiopesis): the speaker begins to articulate, or thinks he begins to articulate, but produces no audible sound (either for want of expiration, or because he does not put his vocal chords in the proper position) till one or two syllables after the beginning of what he intended to say. (...) The interplay of these tendencies – weakening and strengthening and protraction – will be seen to lead to curiously similar, though in some respects different developments in Latin with its continuation in French, in Scandinavian and in English." (Jespersen 1917: 6).

The trajectory of the Jespersen cycle is well documented for English (Jespersen 1917, 1924, 1933, Horn 1989, Mazzon 2004, Wallage 2005, 2008), French (Bréal 1897, 1900, Jespersen 1917, Horn 1989, Rowlett 1998, Godard 2004), German (Jäger 2005, 2008), and Dutch/Flemish (Hoeksema 1997, Postma 2002, Zeijlstra 2004, Van der Auwera and Neuckermans 2004, Breitbarth and Haegeman 2008). Although Borsley and Jones do not describe it in these terms, it is traceable for Welsh in their (2005) book (cf. Section 4.3). Horn's (1989: 455) view of the English and French development is summarized in Table 2. Mazzon (2004) qualifies the development by indicating that there are overlapping patterns in different stages of English, but agrees with the overall picture.

Zeijlstra's (2004: Chapter 4) summary of the diachronic development in Dutch is in Table 3. In modern Dutch spoken in the Netherlands, the cycle has been completed. In other Dutch dialects (mostly spoken in the south of the Netherlands and in Flanders, Belgium), discontinuous negation is still extensively used, although the preverbal enclitic *en* is typically optional (cf. example 20, Haegeman and Zanuttini 1996, Breitbarth and Haegeman 2008, Van der Auwera and De Vogelaer 2008, and Section 4.5 for discussion).

Table 2 Jespersen cycle in English and French

Old French	Jeo *ne* dis I SN say	Old English	Ic *ne* secge I SN say
Modern French (written/ standard)	Je *ne* dis *pas* I SN say SN	Middle English	Ic *ne* seye *not* I SN say SN
Modern French (colloqu ial)	Je dis *pas* I say SN	Early Modern English	I say *not* I say SN
		Modern English	I do*n't* say I do SN say

Table 3 Jespersen cycle in Dutch

Old Dutch	Inde in uuege sundigero *ne* stûnt And in way sinners.GEN SN stood.3SG 'And didn't stand in the way of sinners'
Early Middle Dutch (13th century)	*En* laettine mi spreke *niet* SN let.he me speak SN 'If he doesn't let me speak'
Late Middle Dutch (16th century, Holland)	Mine herberge ontseggic u *niet* My tavern take.away.I you SN 'My tavern I won't take away from you'
Modern Dutch (Netherlands)	Jan loopt niet. Jan walks SN 'Jan doesn't walk.'

Table 4 Jespersen cycle in German

Old High German (AD 750–1050)	thaz thu irrímen *ni* math. that you name SN can 'that you cannot name it.'
Middle High German (AD 1050–1350)	daz ich drîzic pfunt *niht en*naeme. that I thirty pound SN SN-take 'that I would not take thirty pounds.'
Early New High German (AD 1350–1650)	Do wolt er *nicht* gen. then wanted he SN go 'He did not want to go then.'
Modern German	wenn sie *nicht* nach Hause kommt. if she SN to home comes 'if she does not come home.'

Jespersen (1917) discusses similar diachronic patterns for Norse and German. A. Jäger's (2008) description of the German development is summarized in Table 4. Old High German has a preverbal negation *ni*. Toward the end of the Old High German period, this preverbal negation is strengthened with the postverbal *niht*. Toward the end of the Middle High German period, the enclitic *en* disappears, and the final stage is the postverbal negation *nicht* familiar from Modern German.

Jespersen describes the diachronic development as a cycle, so in principle the postverbal negator present in the last stages of Tables 2–4 is expected to give away to a new preverbal marker of negation, under the pressure of NEGFIRST. The *do*-support construction found in Modern English signals a return to the preverbal position of negation (cf. Sections 3.4 and 4.3).

Further evidence for the cyclic nature of the development is provided by data from French Creoles, spoken in Haiti, Guadeloupe, and the Seychelles, which have reanalyzed the postverbal marker *pas* from standard French as a preverbal negator. The examples in (12) illustrate (from Posner 1985).[14] [15]

(12) a. li *pas* t a ap vi_ni_ [Haitian Creole]
him SN PAST FUT PROG come
'He wouldn't be coming.'
b. person *pa* pu pik u. [Seychelles]
nobody SN FUT prick you
'Nobody is going to prick you.'

Obviously, Creole languages do not constitute a straightforward historical development of the standard language, and creolization is a special process, subject to extensive discussion in the literature, so the sentences in (12) do not sketch the next step in the development of modern French. In fact, there may be sufficient pressure from other word order principles to block the development of a preverbal marker in standard modern French. However, the example in (12) illustrates that it is possible in principle to reanalyze *pas*, and shift back from a postverbal to a preverbal marker of negation.

In the literature, it has been observed that the Jespersen cycle involves changes in negation being a syntactic head (a clitic, an affix) or a maximal projection (an adverb), cf. Zanuttini (1991) and Haegeman (1995) for discussion along these lines. It is not entirely clear whether these distinctions provide the key to an analysis of the Jespersen cycle (cf. Déprez 1997a, b for critical discussion). The OT analysis of the Jespersen cycle does not rely on the head/maximal projection distinction for the marker of sentential negation, but makes crucial use of the conflicting preferences for an early and a late realization of negation in the sentence, and balances these against the markedness constraint *NEG.

The OT system developed so far allows the following informal description of the Jespersen cycle as an instance of language change. NEGFIRST is the driving force behind the expression of negation 'early' in the sentence. At the same time, this position is subject to erosion, which leads to a reinforcement of negation. Because the reinforcement of negation is emphatic, and bears focus, it occurs later in the sentence under the influence of FOCUSLAST. This leads to negation occurring in postverbal position.

---

[14] DeGraff (1997) proposes that the effect observed in (12) is that in Haitian Creole, the lexical verb no longer moves past the negative marker, this in contrast with French. In the OT analysis, which does not rely on movement operations, this corresponds to a reordering of the constraints NEGFIRST and FOCUSLAST, which govern the placement of the negation marker in the sentence.

[15] Note also that there is a negative concord reading for the combination of *pa* and *person* in (12b), whereas similar combinations in Standard French normally lead to a double negation reading. Compare Chapter 5 (Section 6) and Chapter 6 (Section 5) for an elaboration.

In his sketch of two opposing tendencies, Jespersen describes the two roles of negation as closely connected to the verb, and as an element carrying important new (i.e. focused) information. By relating these opposing tendencies to the dynamics of language change, Jespersen sketches a pattern of diachronic change in which preverbal negation is doubled with a postverbal emphatic expression that reinforces negation. The new postverbal negator gradually takes over the negative force of the original negator, eventually leading to the disappearance of the preverbal marker of sentential negation. Furthermore, this process is subject to iteration. Section 4.2 works out these ideas more precisely.

3.4.2 Modeling the Jespersen Cycle in OT

This section links the synchronic typological variation from Section 3 to the diachronic change known as the Jespersen cycle. Across a wide range of languages, there are three positions for a marker of sentential negation with respect to the verb: a single marker of sentential negation in preverbal position, a discontinuous negation surrounding the verb, or a marker of sentential negation in postverbal position. The grammars proposed for these three positions of negation (Section 3.3) are summarized in Table 1.

The three positions of the marker of sentential negation are easily linked to the three main stages of the Jespersen cycle. In line with Chapter 1 (Section 5), grammatical change is taken to be gradual. Accordingly, I propose a step-wise change in the grammar, whereby one constraint changes position in the ranking at every stage. This leads to the modeling of the Jespersen cycle in Table 5.

Given that the joint ranking of two constraints as opposed to the weaker position of the three constraints captures each of the three stages (Table 1), a complete ranking for each stage always involves two possible fully ordinal rankings (Table 5). The distinction between the two rankings posited for each stage is invisible in the language production. For example, the rankings in stages 1.1 and 1.2 both lead to the expression of negation in a preverbal position.

Table 5 Jespersen cycle in OT

Stage 1 (preverbal negation)	1.1	*NEG >> NEGFIRST >> FOCUSLAST
	1.2	NEGFIRST >> *NEG >> FOCUSLAST
Stage 2 (discontinuous negation)	2.1	NEGFIRST >> FOCUSLAST >> *NEG
	2.1	FOCUSLAST >> NEGFIRST >> *NEG
Stage 3 (postverbal negation)	3.1	FOCUSLAST >> *NEG >> NEGFIRST
	3.2	*NEG >> FOCUSLAST >> NEGFIRST

However, at each stage, the ranking has to have shifted to the second possible ranking in order to allow the transition to the next stage by means of the shift of a single constraint. For example, the ranking in 1.2 allows the transition to the ranking in 2.1 by means of the raising of FocusLast above *Neg. In order to allow for gradual change in which the transition between two stages does not involve a revised ranking for more than one constraint, it is useful to spell out the first and second (full) rankings for each stage, as in Table 5.

Note that one change in the ranking of FocusLast and NegFirst leads back from stage 3.2 to stage 1.1. This completes the cycle, and allows the diachronic process to repeat itself, as Jespersen suggested. In general, a change in stage occurs when re-ranking affects the lower two constraints in the ranking. Re-ranking of the highest two constraints in the ranking does not affect the stage the grammar is in, because these constraints pair up in a battle of two against one, as argued in Section 3.3.

3.4.3 Intermediate Stages: Between Preverbal and Discontinuous Negation

With full ordinal OT, the constraint rankings give rise to three typologically well established cases of preverbal, discontinuous, and postverbal negation. Intermediate stages can be modeled in an extension toward stochastic OT. Several languages have been argued to exemplify negation patterns in between stages 1 and 2, or 2 and 3.

A language that displays a negation system in between stage 1 and stage 2 has an obligatory preverbal marker of negation that is optionally reinforced by postverbal emphatic negation. In a diachronic development, such a language might be moving away from a preverbal negation toward a system with a discontinuous negation. This situation is illustrated by formal Welsh, as illustrated in (13) (data from Borsley and Jones 2005):

(13) a. *Nid* oedd Sioned yn gweithio [formal Welsh]
SN be.IMPF.3SG Sioned PROG work
'Sioned was not working.'

b. *Ni* soniodd Sioned *ddim* am y digwyddiad. [formal Welsh]
SN mention.PAST.3SG Sioned SN about the event
'Sioned did not talk about the event.'

c. *Doedd* Gwyn *ddim* yn cysgu. [informal Welsh]
SN.be.IMPF.3SG Gwyn SN PROG sleep
'Gwyn was not sleeping.'

d. Na'th Emrys *ddim* gweld dim byd. [informal Welsh]
do.PAST.3SG Emrys SN see SN world
'Emrys didn't see anything.'

In (13a), the preverbal particle *nid* alone carries the negative force of the sentence. In (13b), *ni* is optionally reinforced by the postverbal adverb *ddim.* The contrast between (13a) and (b) indicates that formal Welsh is moving from a preverbal negation to a discontinuous negation pattern. Discontinuous negation is well established in colloquial Welsh, with verbs that have a distinctive negative form as in (13c). In sentences that involve verbs which do not have a distinctive negative form, such as *na'th* in (13d), the presence of postverbal *ddim* is the only indication of negation. This situation reflects a stage 3 grammar. Note that in both (13c, d), the presence of *ddim* is mandatory. Without *ddim* (13c) is ill formed, and (13d) expresses an affirmative sentence. Under the assumption that formal Welsh reflects an earlier stage of the language, the differences between formal and informal varieties of the language in (13) show the Jespersen cycle at work.

The OT system can model such an intermediate situation if *NEG and FOCUSLAST have an overlapping range in the stage 1.2 ranking. Overlap between two constraints C_1 and C_2 is represented as C_1 0 C_2. The ranking NEGFIRST >> *NEG 0 FOCUSLAST can then be postulated as the grammar of formal Welsh. The high ranking of NEGFIRST guarantees an obligatory preverbal marker of negation (*nid*). The overlap between *NEG and FOCUSLAST means that in some cases *NEG will win (and there is only a preverbal marker of negation), whereas in other cases FOCUSLAST will win (and the preverbal marker is doubled by postverbal emphatic negation *ddim*).

Other examples of languages that are possibly in a transitional phase between a stage 1 and a stage 2 language are the Tamazight and Taqbaylit varieties of Berber (Ouali 2003, 2005), and Hausa, a West Chadic language spoken in Nigeria (Dryer 2007):

(14) a. *Ur* ssex (*sha*) [Tamazight Berber]
SN drink.PERF.1SG (SN)
'I don't drink.'
b. *Ur* kshimegh (*ara*) [Taqbaylit Berber]
SN entered.PAST.1SG (SN)
'I didn't enter.'

(15) a. *bàn* san sũna-n-sà *ba*. [Hausa]
SN:1SG know name-LINK-3SG SN
'I don't know his name.'
b. *ba* nà zuwà dà kai
SN CONT come:NOMIN with 2SG
'I am not going with you.'

In all these languages, the preverbal marker of negation is mandatory, whereas the postverbal marker is optional.

Catalan displays regional variation, which corresponds with different stages of the Jespersen cycle. Wheeler et al. (1999: 481) show that *pas* is a postverbal negator that optionally doubles the preverbal *no* in central Catalan (16a). In North Catalan *pas* can be the only element in a negative sentence (16b).

(16) a. No, senyor: *no* sóc *pas* l'home que vostè busca. [Central Catalan]
'No, sir: I am not the man you are looking for.'
b. Cantis *pas* aquesta cançο. [Northern Catalan]
'Don't sing that song.'

According to Wheeler et al. (1999: 482), the doubling *pas* is emphatic in (16a), and denies possible implications of what is being said. The optional status and emphatic nature of *pas* suggests the beginnings of a transition from stage 1.2 to stage 2.1 of the Jespersen cycle. A high ranking of NEGFIRST guarantees an obligatory preverbal marker of negation (*no*) in (16a), but overlapping constraints *NEG and FOCUSLAST allow an optional realization of the emphatic negator *pas*. In the North Catalan variety, *pas* has taken on the role of the simple marker of negation, as in Occitan and spoken French (16b), so North Catalan exemplifies stage 3 of the Jespersen cycle. In the grammar of this dialect, FOCUSLAST has risen above *NEG at the expense of NEGFIRST, which has been lowered in the ranking.

The data in (16) suggest that the negation system of Catalan is currently going through a process of diachronic change that is more advanced in certain varieties than in others. Instability in the co-occurrence restrictions on the preverbal negator *no* with negative indefinites support this view (cf. Chapter 5, Section 5).

3.4.4 Intermediate Stages: Between Discontinuous and Postverbal Negation

Stochastic OT can also describe transitions from stage 2 to stage 3. An overlapping range between NEGFIRST and *NEG in the 2.2 ranking underlies the ranking FOCUSLAST >> NEGFIRST 0 *NEG for informal Welsh. Under this ranking, verbs that have a special negative form like *doedd* (in 13c) can coexist with verbs that do not such as *na'th* (in 13d). The overlapping range of constraints allows for an intermediate stage in the Jespersen cycle in which the preverbal marker of negation becomes optional, but the postverbal marker is obligatory.

Dryer (2007) cites the West Chadic language Mupun and the Bongo-Bagirmi language Bongo as languages in which the preverbal marker of negation is optional, and the postverbal marker is obligatory:

(17) (*ba*) kə̀ n=se lua nyer *kas* [Mupun]
(SN) PERF 1SG=eat meat bird SN
'I didn't eat the bird meat.'

(18) a. ma *nja* ami a'ji *wa* [Bongo]
1SG SN make thing SN
'I am not doing anything.'

b. m-u-yɛ le'ji *wa*
1SG-PAST-drink beer SN
'I did not drink beer.'

We do not have diachronic data on Mupun or Bong-Bagirmi, but it is not unlikely that a similar development is taking place in this language.

It is easier to find evidence in favor of overlapping constraints in languages that we have historical data of. The ranking FOCUSLAST >> NEGFIRST 0 *NEG is strongly supported by the situation in modern French, where the formal version of the language requires the presence of preverbal *ne* as illustrated in (19a). In colloquial French, it is quite common to find sentences like (19b), in which negation is exclusively expressed by means of the postverbal adverb *pas*:

(19) a. Je n' ai pas vu Sophie. [formal/written French]
I SN have SN seen Sophie.
'I have not seen Sophie.'
b. J'ai pas vu Sophie. [colloquial French]
I have SN seen Sophie.

Even though discontinuous negation is the norm since Classical French, occasional instances of *ne* drop date back to the 17th century. These establish *pas* as the bearer of semantic negation. Although the presence of *ne* is required by prescriptive grammars *ne* drop is frequent (up to 80% of the time), even in the higher registers of spoken French (Ashby 1981, 2001, Christensen 2003). In the variety of French spoken in Montréal, *ne* has practically disappeared as a productive marker of negativity (Sankoff and Vincent 1977).

These findings indicate that French is going through a transitional phase in which the discontinuous negation is losing against the postverbal marker of negation. The preverbal clitic *ne* has lost its semantic force as a negator, as pointed out already by Bréal (1897, 1900). Formal, written French is still a stage 2 language with obligatory discontinuous negation, supported by prescriptive grammars and the highly influential Académie Française. Colloquial French is a stage 3 language, with a single, postverbal marker of sentential negation. Spoken French in the higher registers is in the intermediate phase between a stage 2 and a stage 3 language: the use of *ne* is recommended by prescriptive grammars, but is not always realized.

The Welsh and French data make it clear that there may be an asymmetry between preverbal and postverbal negation in stages where discontinuous negation is moving toward postverbal negation. Tesnière's (1959) characterization of the asymmetry between *ne* and *pas* in modern French is well known:

> C'est l'ensemble du discordantiel et du forclusif qui constitue la négation française. [...] Le discordantiel ne forme pas à lui seul la negation. Il la prépare seulement. Et c'est ensuite le forclusif qui la réalise. [...] [L]e français [...] décroche d'abord sa pensée de la notion affirmative, puis il la raccroche à la notion negative, ce qui lui permet de nuancer le degré de la négation. C'est le discordantiel qui opère le décrochage, tandis que le forclusif exprime le raccrochage. (Tesnière 1959 : 224f)

> It is the combination of the preclusion and the contradictor that constitute French negation. The preclusion does not constitute negation by itself. It just prepares it. And it is then the contradictor that realizes it. French first detaches the thought from the affirmative notion, then attaches it to the negative notion, which allows the language to nuance the degree of negation. It is the preclusion that operates the detachment, whereas the contradictor expresses the attachment.[16]

[16]The English translation of the quote from Tesnière is mine.

The insight that French preverbal *ne* and incorporated *nid* in colloquial Welsh do not convey semantic negation plays a crucial role in the reworked analysis of discontinuous negation in French, based on the interaction of the negation marker with negative indefinites, which is proposed in Chapter 5 (Section 6), and the analysis of Welsh negation and negative indefinites in Chapter 5 (Section 7). Similar observations can be made for Afrikaans, where the sentence-final *nie* does not contribute a semantic negaton (Chapter 5, Section 10).

In his description of the diachronic patterns of Dutch negation, Zeijlstra (2004) refers to van der Horst and van der Wal's (1979) study of text frequencies of *en*-deletion in different constructions. Their results indicate that the use of preverbal *en* in the Dutch spoken in Holland gradually decreased between 1300 and 1600. By the end of the 17th century, the transition from a stage 2 to a stage 3 language was basically complete, and the preverbal marker of negation had practically disappeared from the language. As pointed out by an anonymous reviewer, preverbal *en* is occasionally found well into the 18th century. The reviewer provides the following example, taken from a text dating from 1785.

(20) Niemand was er die allengskens *niet en* begreep dat de qualiteyt daar van byzonder slegt moest zyn.
Nobody was there who by now SN SN understood that the quality that of particularly bad had be.
'There was nobody who by now didn't understand that its quality had to be particularly bad.'
(J.F. Tor, *Per koets naar Constantinopel*)

The historic data indicate a gradual process of change over a fairly long period of time, during which certain constructions were more likely to drop the preverbal marker than others. A stochastic OT analysis in terms of overlapping constraints, and interaction with construction specific constraints accounts for such an intermediate stage.

3.4.5 A Note on Pragmatics

According to the outline in this section, the Jespersen cycle has been completed in the northern varieties of Dutch, spoken in the Netherlands. In some dialects, the loss of the preverbal negator took place earlier than in others, and in Flemish dialects discontinuous negation is alive even today, as witnessed by the data in Haegeman and Zanuttini (1996), Breithbart and Haegeman (2008), and Van der Auwera and De Vogelaer (2008).

(21) Valère (*en*)-eet dienen boek *nie*. [West Flemish]
Valère (SN) has that book SN.
'Valère doesn't have that book.'

Given the persistent use of discontinuous negation over time, Breitbarth and Haegeman (2008) argue that Flemish *en* and *nie* have developed a different status in the system of negation. They claim that *nie* has become the neutral negator, whereas *en* does not semantically convey negation, but indicates affective polarity and has acquired an emphatic function. Emphasis in the Jespersen cycle is generally used to explain the rise of a postverbal negator, so the new pragmatic function of *en* suggests some sort of reanalysis of the original preverbal negator.

The observations made by Breitbarth and Haegeman suggest that there is more to the interpretation of discontinuous negation in Flemish than just the realization of truth-conditional negation (¬). Their observations are in line with claims made by Schwenter (2005) concerning the pragmatics of discontinuous negation in Brazilian Portuguese.

According to Schwenter (2005), all three forms of preverbal, discontinuous, and postverbal negation are available in Brazilian Portuguese (22).

(22) a. A Cláudia *não* veio à festa. [Brazilian Portuguese]
the Cláudia SN came to party
b. A Cláudia *não* veio à festa *não*.
the Cláudia SN came to party SN
c. A Cláudia veio à festa *não*.
the Cláudia came to party SN
'Cláudia didn't come to the party.'

There is no difference in the propositional meaning of the three negatives, so Schwenter pursues a pragmatic analysis. Preverbal negation is canonical negation in Brazilian Portuguese (and the only form found in European Portuguese). Schwenter rejects an analysis in terms of 'emphatic' negation for the discontinuous and postverbal forms, because of the difficulty of defining the relevant theoretical notions. Instead, he claims that discontinuous and postverbal negation is sensitive to information-structural properties of the discourse.

According to Schwenter, the proposition denied by discontinuous negation is contextually 'activated', but not necessarily believed by any of the interlocutors. Accordingly, it is often found in response to yes/no-questions, where the proposition is discourse old (either explicitly mentioned or inferable), as in (23A).

(23) Q: Você gostou da palestra da Paria?
'Did you like Maria's talk?'
A: Eu *não* fui *não*.
I SN went SN
'I didn't go.'
A': #Fui *não*.
went.1SG SN
A": Gostei *não*.
liked.1SG SN

The postverbal negation construction is even more restricted in that it requires the proposition to be directly and explicitly activated in the preceding discourse (23A" as opposed to 23A'). Schwenter (2006) extends these claims to discontinuous and postverbal negation in Italian and Catalan. See also Zanuttini (1996) for relevant data on Italian dialects with postverbal *mica* and their pragmatics, and Espinal (1993), Wheeler et al. (1999) for data and discussion of the pragmatics of discontinuous and postverbal negation in Catalan.

Given that this book focuses on the semantics of negation across languages, I leave the integration of the pragmatic insights from Breitbarth and Haegeman, and Schwenter and others into the OT analysis of the Jespersen cycle for a later occasion.

More discussion of the Catalan, French, and Flemish system of negation follows in Chapter 5 (Sections 5, 6 and 9 respectively). That chapter investigates co-occurrence restrictions on the negation marker and negative indefinites. It will turn out that discontinuous systems of negation impose special constraints on those combinations, because of possible asymmetries between the two markers.

3.4.6 Static and Dynamic Models of Language Change

Zeijlstra (2004: 56) models the transitions between the three main stages of preverbal, discontinuous, and postverbal negation as separate phases within the Jespersen cycle. This means that he describes six phases, whereas the OT model only defines three. According to the analysis advanced here, the intermediate phases involve an overlapping range of two constraints, which can be modeled in stochastic OT, whereas the three main phases involve a ranking that can be modeled in ordinal OT.

Empirically, the results of the two analyses are the same. Of course, the theoretical assumptions underlying the two proposals are quite different. In my opinion, the dynamics of the stochastic OT modeling provides a better insight into the optionality of the postverbal marker (in the transition from stage 1 to stage 2) and the preverbal marker (in the transition from stage 2 to stage 3) than the static phrase structure representation that Zeijlstra (2004: 175) proposes. In the OT model, the frequency effects correlate with the gradual change toward a new ordinal equilibrium. The contrast between stochastic OT and ordinal OT indicates that intermediate stages are unstable. This is reflected in the observation that intermediate stages of the Jespersen cycle are typologically rare (cf. Haspelmath 1997).

What is most relevant at this point is that it is possible to model the Jespersen cycle within the OT framework developed so far, and that this modeling provides new insights about this diachronic process. The three main stages establish a balance between two highly ranked constraints and one weak one. Transitions between main stages lead to intermediate stages, modeled in stochastic OT, which allows the transitions to be gradual, with no more than one constraint moving over another one at any step. The OT model sketched here then provides a viable alternative to the phrase structural analyses of the position of negation currently available in the literature.

3.5 Subordinate Clauses and Nonfinite Constructions

The analysis of the placement of negation in this chapter focused on main clauses. A more elaborate study of the position of negation should also take into account the different word orders found in main and subordinate clauses.

In Germanic languages, word order in main clauses is different from that in subordinate clauses. Verb second applies to main clauses, but subordinate clauses have an SOV pattern, with negation preceding the verb, as illustrated for Dutch and German in (24) and (25) respectively:

(24) a. …omdat Mare niet lachte. [Dutch]
…because Mare SN laughed
'…because Mare didn't laugh.'
b. *…omdat Mare lachte niet.
…because Mare laughed SN

(25) a. …weil Hans nicht kam. [German]
…because Hans SN came.
'…because Hans didn't come.'
b. *…weil Hans kam nicht.
…because Hans came SN

Discontinuous negation in subordinate clauses in older Dutch and Flemish places both negation markers before the finite verb (cf. 20) and discussion in Haegeman (1995, 1997).

Besides the distinction between main/subordinate clauses, the distinction between finite/nonfinite clauses can play a role in the placement of negation. For French, this is illustrated in (26) (cf. Pollock 1989 for discussion).

(26) a. Je n'ai pas invité Julie. [French]
I SN have SN invited Julie
'I didn't invite Julie.'
b. Il m' a prié de ne pas appeler la police.
He me.DAT has asked to SN SN call the police
'He asked me not to call the police.'

In finite clauses, *ne* precedes and *pas* follows the finite verb (26a), but in nonfinite constructions, both *ne* and *pas* precede the infinitival complement (26b).

Borsley and Jones investigate the role of finiteness in Welsh, on the basis of examples like (27).

(27) a. Nid yw hi 'n gweithio heno. [Formal Welsh]
SN be.PRES.3SG she PROG work tonight
'She is not working tonight.'
b. Dw i'n gisgwyl [i Mair beidio â mynd i Aberystwyth].
be.PRES.1SG I PROG expect to Mair SN with go to Aberystwyth
'I expect Mair not to go to Aberystwyth.'

Beidio in (27b) is analyzed as a nonfinite verb expressing negation. Negative verbs were excluded from the investigation in Chapter 1, but obviously, the analysis of negation in a language like Welsh is not complete until it implies a description of negative verbs.

The primary aim of this book is to describe the syntax–semantics interface of negation; a full analysis of the interaction of negation with word order in general, across the different clause types is outside the scope of this study. Even though the number of syntactic constraints in this book is higher than the number of semantic constraints, the syntactic analysis of negation developed so far for finite main clauses needs to be worked out in more detail in order to account for the patterns found in subordinate clauses and nonfinite constructions.

3.6 Conclusion

In this chapter, the markedness of negation has been grounded in an asymmetric frequency distribution between affirmative and negative speech acts. An evolutionary bidirectional OT model derives Horn's division of pragmatic labor, according to which marked forms pair up with marked meanings, and unmarked forms with unmarked meanings. Negation rather than affirmation ends up being formally marked, because negation is more infrequent. Given that all linguistic communities feel the need to express negation, negation emerges as a universal category of natural language.

The universal ranking FNEG >> *NEG might suggest that FNEG is not a violable constraint. The anecdotal evidence from Child Dutch, and the discussion of aphasic sign language users shows that it is useful to maintain FNEG as part of the OT system, because the order of the constraints FNEG and *NEG can be reversed in language acquisition and language breakdown.

The ranking FNEG >> *NEG determines that languages realizes negation in some form. The most frequent realization of negation is by means of a marker of sentential negation. This marker needs to be placed in the sentence. In the second half of this chapter, the position of sentential negation with respect to the verb was investigated. The interaction of the three constraints NEGFIRST, FOCUSLAST, and *NEG accounts for three main cases of typological variation: preverbal placement of negation, postverbal negation, and discontinuous negation.

The typological patterns were identified with the three main stages in the Jespersen cycle. The result is an OT model of typological variation that matches the pattern of diachronic change.

The analysis covered issues in the placement of negation in main clauses. The patterns observed for subordinate clauses and nonfinite constructions require a more elaborate syntactic analysis that is, however, outside the scope of this book, which focuses on the syntax–semantics interface.

Chapter 4
A Typology of Negative Indefinites

Introduction and overview Chapter 3 investigated ways in which languages convey negation by means of an expression that corresponds to the first-order logic connective ¬. In English this would be *not*. Negation can also be attracted to other expressions in the sentence, particularly indefinites in argument or adjunct position (Section 1). Negative attraction creates negative indefinites. The class of negative indefinites includes both negative quantifiers (English *nobody*, *nothing*, *nowhere*, *never*) and n-words.

Negative attraction is extended to multiple indefinites under negation in Section 2. Sentences involving a range of negative indefinites raise problems for the principle of compositionality of meaning, because some languages assign a double negation reading to such a sequence, and others a single negation reading. Chapter 1 argued that the compositionality problem cannot be solved in the lexicon, and exploited the polyadic quantifier analysis proposed by de Swart and Sag (2002) to offer a grammatical analysis. The argumentation is briefly summarized in Section 3, in preparation of the typology of double negation and negative concord languages.

The polyadic quantifier analysis developed by de Swart and Sag (2002) works well for French because it displays ambiguities between single and double negation readings in sentences that combine two negative expressions. However, in most other languages there is a strong bias toward either the double negation or the negative concord reading. Section 4 builds a bidirectional optimality theory (OT) built on top of the polyadic quantifier analysis in order to account for the systematic contrast between negative concord and double negation languages.

In negative concord languages, the functional motivation that favors marking of 'negative variables' in the syntax wins out. Double negation languages value first-order iteration in the semantics. Languages that display ambiguities have an overlapping range of constraints in a stochastic extension of the model developed in Chapter 6 (Section 3).

H. de Swart, *Expression and Interpretation of Negation: An OT Typology*, Studies in Natural Language and Linguistic Theory,
DOI 10.1007/978-90-481-3162-4_4,

As Section 4 emphasizes, the bidirectional setup is essential, for syntactic and semantic variation go hand in hand. Section 5 returns to the relation between negative concord and negative polarity, and offers a diachronic perspective while Section 6 concludes the chapter.

4.1 Negative Attraction

From a linguistic point of view, interesting questions arise in the expression and interpretation of clausal negation in sentences involving indefinites. This section investigates sentences containing a single negative indefinite, realizing the meaning phrased as $\neg\exists x$ in first-order logic. The expression of multiple indefinites under negation (realizing $\neg\exists x_1\exists x_2\ldots\exists x_n$ in first-order logic) is dealt with from Section 2 onward.

4.1.1 Neg-incorporation and Negative Attraction

Chapter 3 established that all natural languages have some way of marking sentential negation. This marker need not be an independent word, it can cliticize onto the verb (e.g., French *ne*) or even incorporate into the verb (cf. Chapter 3, Section 3). The incorporation of negation does not only occur in verbs, it is also seen in indefinite pronouns, adverbs, and conjunctions. In English, *never* is built out of the incorporation of *ne* into *ever*. Related forms are *nobody*, *nothing*, *nowhere*, *neither...nor* all of which contain an incorporated *ne*. Similarly, Dutch has the pairs *iemand-niemand* ('somebody'–'nobody'), *ergens–nergens* ('somewhere'–'nowhere'), and the negative conjunction *noch* ('neither'). Italian has *nessuno* ('nobody') and *niente* ('nothing'); Spanish has *nadie* ('nobody') and *nada* ('nothing'); Catalan has *ningú* ('nobody'); and French has *ni...ni* ('neither...nor').

Morphological incorporation of negation, also known as 'Neg-incorporation', is a widespread phenomenon, but it is not universal. Suppletive forms like the French *pas* 'not', *jamais* 'never', *rien* 'nothing'; Italian *mai* 'never'; Catalan *res* 'nothing'; and Greek KANENAN 'nobody' indicate that Neg-incorporation is not a prerequisite for the construction of indefinite forms that bear negative import.

This chapter focuses on the semantic class of negative indefinites, whether or not they involve negative incorporation. Because the emphasis is on variable binding operators, negative conjunctions are left aside, but see Horn (1989: 256–258), de Swart (2001b), Szabolcsi and Haddican (2004), and Doetjes (2005) for relevant discussion of the semantics of negative conjunctions.

Jespersen indicates that natural language has a strong tendency "to attract the negative notion to any word that can easily be made negative" (1917: 56). Following Mazzon (2004), I call this tendency Negative Attraction. Not all languages are equally susceptible to negative attraction, for negative attraction may conflict with other constraints governing the realization of negation in the sentence. Therefore, the tendency is best captured by means of a violable constraint that interacts with other constraints in the grammar of a language. I adopt the following formulation of the constraint NegAttract:[1]

- NegAttract
  Realize (clausal) negation on an indefinite in argument or adjunct position.

The forms *nobody*, *never*, *niemand*, *nessuno*, *rien*, *res*, *mai*, etc. all attract the negation in sentences involving an existentially quantified variable in the scope of negation.

Different expressions in a series (*nobody*, *nothing*, *nowhere*, *never*) impose restrictions on the ontological domain that the variable *x* bound by the quantifier may belong to (*x* has to be human or inanimate, or a place or a time, etc.). However, they share the same quantificational core (cf. Chapter 1, Section 4 for discussion).

NegAttract interacts with other constraints about the realization of negation in the sentence, in particular with the tendency to realize negation early in the sentence, formalized in Chapter 2 (Section 3) by means of the OT constraint NegFirst.

4.1.2 Interaction Between Negative Attraction and NegFirst

Jespersen (1917) observes that negative attraction may be in conflict with a tendency for the negation to be realized preverbally, as illustrated by the examples in (1) and (2):

(1) a. Nobody laughed.
    b. *Anybody didn't laugh.
(2) a. We didn't meet anybody [colloquial English]
    b. We met nobody [literary English]

The pattern shown in (1) is by no means restricted to English. Chapter 1 (Sections 3 and 4) discuss asymmetries between NPIs and n-words in several Romance languages and Greek. The grammaticality contrast in (1) has to do with the preverbal subject position, for the indefinite in postverbal object position (2) allows two options. Jespersen (1917: 56) qualifies (2a) as colloquial, and (2b)

[1]This formulation is narrower than the one found in Jespersen. Jespersen also talks about constructions in which negation is attracted to adjectives (*unhappy* versus *not happy*). The syntactic and semantic properties of affixal negation and negative indefinites are not the same, and my proposal focuses on argument structure, not on morphological operations. This motivates my narrower definition, but of course, one could adapt it to other needs. Compare Horn (1989: 273 sqq) for extensive discussion of affixal negation. See also Chapter 1 (Section 1) and Chapter 6 (Section 1) for remarks on the relation between affixal negation, sentential negation, and negative indefinites.

as more literary English. I follow this terminology, although nothing in my analysis depends on this being the right characterization.

Crucially, *nobody* in (2b) takes clausal scope, not constituent scope, so the sentence conveys the same propositional content as (2a). This does not necessarily exclude the idea that negative force is stronger if the negator comes later in the linear order (Mazzon 2004: 97) under the influence of FocusLast (see Chapter 3, Section 3). According to Quirk et al. (1985: 1033), sentences of type (2b) should indicate a stronger 'negativity' than (2a) because of this reason. Giannakidou (2006) develops similar ideas.

If the difference in meaning is viewed as substantial, the two candidates should be treated as optimal outcomes for different inputs that exist in parallel in the same language. This would imply a more discerning semantic representation than the first-order representation $\neg\exists$. Given the emphasis on truth-conditional negation in this book, I will not use the claim about the 'stronger' negativity of (2b) in my analysis. In this section, the two candidates (2a) and (2b) are treated as competitors for the same input meaning $\neg\exists$, as this is the propositional content they convey.

Chapter 3 established that modern English has a high ranking of NegFirst, albeit in a modified form. The English word *not* follows the auxiliary, but negation is required to precede the lexical verb. *Do*-support as in (2a) is used to satisfy this requirement. The definition of NegFirst (lv) is repeated from Chapter 3:

- NegFirst (lexical verb)
  Negation precedes the lexical verb.

NegAttract and NegFirst are two constraints that involve the realization of propositional negation in the syntax, but that are potentially conflicting.

A further relevant constraint is the markedness constraint *Neg. Its definition is also repeated from Chapter 3.

- *Neg
  Avoid negation in the output.

*Neg blocks unnecessary proliferation of negative forms in the OT syntax. As usual, the markedness constraint *Neg is ranked below the faithfulness constraint FNeg:

- FNeg
  Be faithful to negation, i.e., reflect the non-affirmative nature of the input in the output.

The universal ranking FNeg >> *Neg implies that negative sentences always incur at least one violation of *Neg (Chapter 3, Section 1). The comparison of the two candidates that comprise the candidate set for indefinite subjects in modern English in Tableau 1 shows that a negative indefinite in subject position satisfies NegFirst as well as NegAttract.

The negative indefinite *nobody* counts as a negative form with morphologically incorporated negation. As the violation patterns in Tableau 1 illustrate, *nobody* satisfies NegAttract, and it violates *Neg. A negative polarity item (NPI) such as *anybody* does not count as a negative form, because it is lexically classified as an

Tableau 1 Negative subjects (modern English)

Meaning ¬∃x Came(x)	Form	FNEG	*NEG	NEGFIRST (lv)	NEGATTR
	Somebody came	*		*	*
☞	Nobody came		*		
	Anybody didn't come		*		*
	Nobody didn't come		**		

Tableau 2 Preverbal negation with postverbal indefinites (colloquial English)

Meaning ¬∃x Meet(we,x)	Form	FNEG	*NEG	NEGFIRST (lv)	NEGATTR
	We met somebody	*		*	*
	We met nobody		*	*	
☞	We didn't meet anybody		*		*
	We didn't meet nobody		**		

Tableau 3 Negative attraction with postverbal indefinites (literary English)

Meaning ¬∃x Meet(we,x)	Form	FNEG	*NEG	NEGATTR	NEGFIRST (lv)
	We met somebody	*		*	*
☞	We met nobody		*		*
	We didn't meet anybody		*	*	
	We didn't meet nobody		**		

NPI carrying an existential meaning (cf. Chapter 1, Section 4). Accordingly, it does not satisfy NEGATTRACT, and it does not violate *NEG in Tableau 1.

English does not combine *nobody* with *not*, because this candidate incurs an extra violation of *NEG. This shows that *NEG functions as an economy constraint that blocks the unnecessary proliferation of negations. Many languages favor the combination of negative indefinites and the marker of sentential negation under the influence of special faithfulness constraints rising above *NEG (compare Chapter 5 for discussion). However, English is not such a language.

The example of negative indefinites in subject position leaves the ranking between the three constraints undecided, because the optimal outcome satisfies both NEGFIRST and NEGATTRACT. The conflict between NEGFIRST (lv) and NEGATTRACT is played out in examples involving postverbal indefinites such as (2a) and (2b).

The competition under the rankings NEGFIRST >> NEGATTRACT and NEGATTRACT >> NEGFIRST is spelled out in Tableaux 2 and 3. Under the former ranking (associated with colloquial English), the combination of negation plus an NPI constitutes the optimal form (Tableau 2). Under the second ranking (associated with literary English), a postverbal negative indefinite is the winner (Tableau 3).

As a result of the high ranking of NEGFIRST (lv) in Tableau 2, negation must be realized preverbally, and *anybody* comes out as the optimal expression of an indefinite under negation. In Tableau 3, *nobody* is the optimal form, because of the high ranking of NEGATTRACT. In both tableaux, the combination of *not* and *nobody* is ruled out as suboptimal, because this candidate incurs an extra violation of *NEG, which is unmotivated under this ranking. But see Chapter 5 (Section 3) for varieties in which utterances like 'we didn't meet nobody' constitutes the optimal outcome.

In Middle English spoken between 1500 and 1700, negation occurred in a position following the lexical verb as well as the auxiliary (cf. Mazzon 2004). The Shakespearian alternation between 'I love you not' and 'I do not love you' illustrates the low ranking of NEGFIRST in this variety of English (cf. Chapter 3, Section 4).

What Jespersen describes as literary English might very well be a remnant from this period. The rise of NEGFIRST (lv) in modern English contrasts with this earlier stage of the grammar, and motivates the grammaticalization of *do*-support. Accordingly, the two sentences in (2) are described as optimal outputs under two different grammars of English, reflecting two different registers or varieties of the language.[2]

The two varieties of English support the view that NEGATTRACT is a violable constraint, whose force varies from one grammar to the next. Haspelmath (1997: 206) discusses a similar conflict between NEGFIRST and NEGATTRACT for Baghdad Arabic.

(3) a. Saalim ma raħ i-šuf-ni hnak. [Baghdad Arabic]
Salim SN FUT he-see-me there
'Salim will not see me there.'
b. Ma-ħad kisər il šibbač.
SN-one broke the window
'No one broke the window.'
c. Saalim ma šaf ʔœy-waħid hnak.
Salim SN saw indef-one there
'Salim did not see anyone there.'

The marker of sentential negation occurs in preverbal position in (3a), so NEGFIRST is a highly ranked constraint in Baghdad Arabic. The incorporation of negation into the indefinite in subject position in (3b) satisfies both NEGFIRST and NEGATTRACT. (3c) shows that the grammar of Baghdad Arabic involves the ranking NEGFIRST >> NEGATTRACT, for the preverbal position of negation is maintained at the expense of negative attraction.

Similar data show that NEGATTRACT is ranked higher than NEGFIRST in Germanic languages such as Dutch and German.

(4) a. Niemand hat Maria eingeladen. [German]
nobody has Maria invited
'Nobody invited Maria.'

---

[2]As pointed out by an anonymous reviewer, *hardly anybody* and *almost nobody* seem to be in competition in roughly the same way as *not...anybody* and *nobody*. Thus the different rankings in Tableaux 2 and 3 interact with the selection of the intensifier.

b. Maria has niemand eingeladen.
Maria has nobody invited
'Maria invited no one.'
c. #Maria hat nicht jemand eingeladen. *¬∃/✓∃¬
Maria has SN someone invited.

(5) a. Niemand is gekomen. [Dutch]
Nobody has come
'Nobody came.'
b. #Iemand is niet gekomen. *¬∃/✓∃¬
Somebody has SN come
'Somebody didn't come.'

(6) a. #Wij hebben niet iemand uitgenodigd. *¬∃/✓∃¬
We have SN somebody invited
b. Wij hebben niemand uitgenodigd.
We have nobody invited
'We didn't invite anybody.'
c. *Wij hebben niet ook maar iemand uitgenodigd.
We have SN anyone invited.

According to Hoeksema (2000), negative attraction is obligatory in Dutch. The German and Dutch data support the grammar *NEG >> NEGATTRACT >> NEGFIRST for these languages. Under this ranking, the negative indefinites in (4a, b), (5a), and (6b) emerge as the optimal outputs, as illustrated in Tableau 4.

In English, the constraint ranking NEGFIRST >> NEGATTRACT is adopted in more literary varieties (Tableau 3). In modern, colloquial English, a version of NEGFIRST has risen above *NEG (Tableau 2). Dutch and German show that there is nothing inherently literary about this constraint setting: negative incorporation in preverbal as well as postverbal position is used in all varieties of the language (Tableau 4).

The suboptimal status of the candidate *niet ook maar iemand* in Tableau 4 shows that the replacement of a pronominal indefinite with an NPI does not have an effect

Tableau 4 Negative attraction in postverbal object position (Dutch)

Meaning ¬∃x Invite(we,x)	Form	FNEG	*NEG	NEGATTR	NEGFIRST
	Wij hebben iemand uitgenodigd We have somebody invited	*		*	*
☞	Wij hebben niemand uitgenodigd We have nobody invited		*		*
	Wij hebben niet iemand uitgenodigd we have SN somebody invited		*	*	
	Wij hebben niet ook maar iemand uitgenodigd We have SN anyone invited		*	*	
	Wij hebben niet niemand uitgenodigd we have SN nobody invited		**		

on the optimal outcome in Dutch. Sometimes, it is assumed that the adjacency of *niet* and *iemand* makes negative incorporation obligatory in Dutch (cf. Haeseryn et al. 1997). Although intervening material may help to elicit examples with full indefinites (cf. de Swart 2000), this is not always sufficient for pronominal indefinites, because of the high ranking of NEGATTRACT in Dutch.

The combination of a marker of sentential negation and a negative indefinite as in *niet niemand* is ruled out in Tableau 4 by the economy constraint *NEG. The suboptimal status of this candidate as a realization of the single negation meaning in Dutch is similar to the status of *not nobody* in English (Tableaux 2 and 3). It is due to the high ranking of *NEG in double negation languages such as Dutch and (standard) English, as will be argued extensively in Section 4, and Chapter 5 (Section 2).

The main interest in this section is the division of labor between the negation marker and negative indefinites in the realization of the input meaning $\neg\exists x$. There is crosslinguistic variation found in these patterns, depending on the relative strength of negative attraction and NEGFIRST.

Obviously, there is more to be said about the position of negative indefinites in the sentence in relation to other types of nominals (pronouns, definites, other quantifiers), and in constructions other than transitive sentences with a subject and a direct object. This syntactic investigation raises extensive problems, especially in Germanic languages like Dutch and German, which allow for extensive scrambling in the so-called Mittlefeld. But given the focus of this book on the syntax–semantics interface, I cannot delve into all the complexities here, and concentrate on the expression of the meaning $\neg\exists$ in basic intransitive and transitive sentences.[3]

---

[3]As an anonymous reviewer points out, adjacency and topicalization interacts with negative attraction in indefinite DPs, and leads to more options, as illustrated in (i)–(v):

(i) *Ik lust niet een koekje [Dutch]
I like SN a cookie

(ii) Ik lust geen koekje.
I like no cookie
'I don't like a cookie.'

(iii) Ik houd niet van een koekje.
I like SN of a cookie
'I don't like a cookie.'

(iv) ?Ik houd van geen koekje. (OK for some speakers, not others)
I like of no cookie
'I don't like a cookie.'

(v) Een koekje lust ik niet.
A cookie like I SN
'A cookie I don't like.'

An extension of the current analysis of Dutch negative attraction to full indefinites should be able to handle both the pronominal observations dealt with in the main text, and these supplementary data on full indefinites. An important reason to restrict the discussion in this book to pronominal and adverbial indefinites is that the cross-linguistic data on the interaction of negation with full indefinites are not widely available.

Chapter 3 presented NegFirst as a family of constraints, rather than a single constraint. The interaction of NegFirst and NegAttract implies that different versions of NegFirst have consequences for the use and position of negative indefinites. Some of these complexities are visible from constraints on negative attraction in English, as discussed in this section. Further evidence in favor of a flexible use of NegFirst comes from Swedish data discussed by Sells (2000, 2001).

4.1.3 Negative Attraction and NegFirst in Swedish

Swedish is a V2 language, just like Dutch and German. NegAttract has a high ranking in Swedish, just as in Dutch or German, as illustrated by the data in (7a, b) (from Sells 2000, 2001). The data in (7c-e) show that the situation is more complex.

(7) a. Ingen såg mig. [Swedish]
Noone saw me
'No one saw me.'
b. Jag såg ingen.
I saw no one
'I didn't see anyone.'
c. *Hon hade sagt ingenting.
she had said nothing
d. Hon hade inte sagt någonting.
she had not said anything
'She had not said anything.'
e. Hon hade ingenting sagt
she had nothing said
'She had not said anything.'

According to Sells (2000, 2001), the finite verb resides in I (the functional node Inflection) in Swedish, and nonfinite verb forms are in the VP. The data in (7) are then accounted for by the requirement that negative indefinites be outside the VP. In (7a), the negative indefinite is in subject position, so its position outside the VP is obvious. In (7b), the negative indefinite follows the finite verb, but is still outside the VP domain. (7c) is ungrammatical because the past participle is in the VP, which implies that *ingenting*, to the right of the participle, is in the VP too.

The examples illustrate a conflict between negative attraction and the requirement for negation to be expressed early (i.e., outside the VP). Chapter 3 describes NegFirst as a pattern corresponding with a family of constraints, rather than a single constraint. A flexible use of NegFirst suggests a special version of this constraint called NegFirst (VP) requiring negation to be expressed outside the VP.

- NegFirst (VP)
  Negation precedes the VP.

Tableau 5: Negative attraction and NEGFIRST (VP) in Swedish (production)

Meaning ¬∃x have-seen(she,x)	Form	FNEG	NEGFIRST (VP)	*NEG	NEGATTR	*SCRAMBLE
	Hon hade sagt någonting She had said anything	*				
	Hon hade sagt ingenting She had said nothing		*	*		
☞	Hon hade inte sagt någonting She had SN said anything			*	*	
☞	Hon hade ingenting sagt She had nothing said			*		*
	Hon hade inte sagt ingenting She had SN said nothing			**		

Under the assumption that NEGFIRST (VP) is operative in Swedish, the pair of sentences in (7d) and (7e) exemplifies two ways to resolve the conflict between NEGATTRACT and NEGFIRST (VP) in configurations that contain a past participle in VP. Either negation is expressed by a marker of sentential negation, and the argument is realized as an indefinite (7d), or the negative indefinite is scrambled over the participle so that it takes scope over the VP (7e).

Scrambling involves a marked word order. In an OT analysis, the position of *ingenting* to the left of the participle in (7e) involves the violation of a constraint penalizing scrambling as an instance of marked word order (cf. Hendriks et al. 2009, Chapter 3 for a discussion of scrambling in Germanic in these terms). Without delving into its exact formulation, I will call this constraint *SCRAMBLE.

If the ban against scrambling is as strong as the desire to attract negation to the indefinite, two optimal candidates emerge, as spelled out in Tableau 5.

The high ranking of NEGFIRST (VP) creates a conflict between NEGATTRACT and *SCRAMBLE in sentences that contain composed tense forms with a past participle (7c-e). On the one hand, negation should be attracted to the indefinite object, but on the other hand, negation should be realized outside of the VP. The two requirements cannot be reconciled when the object follows the past participle in the VP. In Tableau 5, the dotted line between NEGATTRACT and *SCRAMBLE indicates that the two constraints are equally strong. As a result, the tableau generates two optimal outputs for the same input.

The Swedish pattern shows that NEGFIRST interacts with other parts of the grammar governing clause structure. Given that a full theory of syntax is beyond the scope of this book, I leave it at this tentative analysis of the tension between NEGFIRST, negative attraction, and scrambling in Swedish. The example confirms the claim made in Chapter 3 (Section 3) that NEGFIRST is a pervasive pattern in natural language, which corresponds to a family of constraints, rather than a single one.

4.1.4 Toward Multiple Indefinites Under Negation

On the basis of the observations made in Chapter 3 and in Sections 1.1– 1.3 of this chapter, I conclude that there are two ways to realize the input meaning $\neg\exists x$, and express negation taking scope over a clause that involves an indefinite in argument position or an adjunct. Languages can combine a marker of sentential negation with an indefinite (a regular indefinite or an NPI), or use a negative indefinite (a negative quantifier or an n-word). In languages that use negative indefinites, the interaction of the constraint NEGATTRACT with other constraints governing word order and the placement of negation becomes relevant.

From a semantic point of view, the embedding of multiple indefinites under negation deserves special attention. The question addressed in Section 2 is whether languages realize the input $\neg\exists x\exists y\exists z$ by means of a sequence of negative indefinites, or whether the existential quantifiers binding argument positions lower in the chain are realized by means of regular indefinites or NPIs (special non-negative indefinites). The first option gives rise to the phenomenon of negative concord (cf. Chapter 1, Section 4), the second does not.

It is clear that negative concord can only arise in languages that have a high ranking of NEGATTRACT. Without a high ranking of NEGATTRACT, the language does not use negative indefinites, so negative concord does not arise. However, not all languages with a high ranking of NEGATTRACT in their grammar are negative concord languages. Dutch and German are examples (cf. Section 1.2). With multiple indefinites in the argument structure of the verb, a more complex situation arises, which requires the introduction of new constraints. The empirical classification is presented in Section 2. The semantics of negative concord in terms of resumptive negative quantification is set up in Section 3, while Section 4 formalizes the main findings in bidirectional OT.

4.2 Multiple Indefinites Under Negation: an Empirical Classification

Haspelmath (1997: 193–194) and Corblin and Tovena (2003) describe how natural languages express the meaning $\neg\exists x_1\exists x_2\ldots\exists x_n$. I follow their general classification, and distinguish three cases: indefinites, NPIs, and n-words.

4.2.1 Indefinites Under Negation

The simplest possible forms that express the meaning $\neg\exists x_1\exists x_2..\exists x_n$ involve a marker of sentential negation or a negative indefinite with $n/n - 1$ indefinites in its scope.[4]

[4] The Turkish example in (10) is from Haspelmath (1997: 193).

(8) Keiner hat etwas gesagt. [German]
No one has something said
'Nobody said anything.'

(9) Niemand heeft iets gezien. [Dutch]
Nobody has something seen.
'Nobody saw anything.'

(10) Bir şey duy-ma-dı-m. [Turkish]
Something hear-SN-PAST-1SG
'I didn't hear anything.'

So what seems to be the simplest possible formal combination from a (first-order) logical point of view is actually realized in several natural languages. However, not all languages allow this straightforward expression of indefinites under negation.

4.2.2 *Negative Polarity Items*

In the second case, NPIs are used to express existential quantification in the scope of negation. This situation often correlates with the observation that the simplest possible forms (as in Section 2.1) are blocked, because pronominal indefinites are positive polarity items (PPIs) that cannot be placed in negative contexts.

(11) a. #I did not buy something. [✓$\exists\neg$, *$\neg\exists$]
b. I did not buy anything.

(12) a. #Nobody saw something. [✓$\exists\neg\exists$, *$\neg\exists\exists$]
b. Nobody saw anything.
c. Nobody said anything to anyone.

Many NPIs occur in a wider range of contexts than negation and negative indefinites (Chapter 1, Section 3).

(13) a. If you saw anything, please tell the police.
b. Did anyone notice anything unusual?
c. Few people wrote down anything.

The examples in (13) illustrate that NPIs such as *anything* do not inherently carry a negative meaning. Rather they correspond with existential quantifiers with some additional meaning component (characterized as 'widening' of a set of alternatives by Kadmon and Landman 1993; as indicating the bottom of a scale by Fauconnier 1975, 1979, Krifka 1995, Israel 1996, de Swart 1998b; as sensitive to scalar implicatures by Chierchia 2006 or to a non-deictic interpretation of the variable Giannakidou 2008). NPIs occur in a wide range of languages (cf. Chapter 1, Section 3), so this is a typologically strong pattern.

4.2.3 N-words

In the third case, existential quantification in the scope of negation is expressed by means of 'n-words'. Just like in the second case, the simplest forms as in Section 2.1 are usually blocked, because indefinite pronouns are PPIs. N-words behave as negative quantifiers in isolation (14a, b), or in sentences in which they are the only expression of negation (15a, b), but express a single negative statement in combination with sentential negation (16a, b) or other n-words (17a-d), as has been observed in Chapter 1 (Section 4).[5]

(14) a. A ¿Qué viste? B: Nada. [Spanish]
A: What did you see? B: Nothing.
b. A: Quants en vas veure? B: Cap. [Catalan]
A: Now many did you see? B: None.

(15) a. Nessuno mangia. [Italian]
'Nobody ate.'
b. J'ai rien vu. [colloquial French]
I have nothing seen.
'I haven't seen anything.'

(16) a. No vi ninguno. [Spanish]
SN saw.1SG none
'I didn't see any.'
b. No en vaig veure cap. [Catalan]
SN PART saw.1sg none
'I didn't see any.'

(17) a. Nadie ha dicho nada. [Spanish]
Nobody has said nothing
'Nobody said anything.'
b. Nessuno ha detto niente. [Italian]
Nobody has said nothing
'Nobody said anything.'
c. Gnun a l'ha dit gnente. [Piedmontese]
Nobody has said nothing
'Nobody said anything.'
d. Personne n'a rien dit à personne [written French]
nobody SN has nothing said to nobody
'Nobody said anything to anyone.'

---

[5] The Catalan and Spanish data in (14) and (16) are from Vallduví (1994). The Spanish example in (17a), as well as the Italian and Piedmontese data in (15) and (17) are from Zanuttini (1991). The West Flemish data in (18) are from Haegeman and Zanuttini (1996). They also quote the Bavarian data in (18d) (from Bayer 1990). The Afrikaans example (18e) is from Kate van Gass (2007, personal communication). The Polish example in (19) is from Przepiórkowski and Kupść (1999) and the Greek example in (20) is from Giannakidou (1998). I gloss French *ne*, West Flemish *en*, and Afrikaans *nie* as markers of sentential negation (SN). Obviously, there are important syntactic and semantic differences between these markers, compare Chapters 3 and 5.

Negative concord is widespread in the family of Romance languages (16, 17), but it also occurs in various Germanic languages (18), in Slavic languages (19), Greek (20), and many other languages.

(18) a. dat Valère niemand kent. [West Flemish]
that Valère nobody knows
'that Valère doesn't know anybody.'
b. K'(en)-een niets nie gezien.
I (SN) have nothing SN seen
'I haven't seen anything.'
c. K'(en)-een an niemand niets gezeid.
I SN have to nobody nothing said
'I didn't say anything to anyone.'
d. und keinen andern nichd leihden wil. [Bavarian]
and no other SN tolerate wants
'and does not want to tolerate another'
e. Hij het nooit sy broer vergeven nie. [Afrikaans]
He has never his brother forgiven SN
'He has never forgiven his brother.'
(19) Nikt nigdy nikogo niczym nie uszczęśliwił. [Polish]
Nobody never nobody nothing SN made happy
'Nobody has ever made anybody happy with anything.'
(20) KANENAS dhen ipe TIPOTA. [Greek]
Nobody SN said.3SG nothing
'Nobody said anything.'

N-words differ systematically from NPIs (Ladusaw 1992; Vallduví 1994; Bernini and Ramat 1996; Haspelmath 1997; see Chapter 1, Sections 3 and 4 for extensive discussion). Importantly, they behave as negative quantifiers in isolation (14a, b), whereas NPIs behave as indefinites, and contribute an existential quantifier $\exists$ rather than a negative existential quantifier $\neg\exists$ (cf. 12).

Accordingly, NPIs like *anything* do not mean 'nothing' as the elliptical answer to a question (21), because they must be licensed by an operator with the right semantic properties (downward entailing or non-veridical, cf. Fauconnier 1975, 1979; Ladusaw 1979; Zwarts 1986; van der Wouden 1997; and Giannakidou 1997, 1998):

(21) Q: What did you see? A: Nothing.
*Anything.

The contrast between negative quantifiers and NPIs in isolation, illustrated for English in (21), is repeated for negative concord languages in (22a-e).[6]

[6] The Spanish example in (22a) is from Herburger (2001). The Catalan example in (22b) is from Vallduví (1994). The Greek example in (22c) is from Giannakidou (1998). The Polish example in (22d) is from Haspelmath (1997: 195).

(22) a.	Q: A quién viste? Who did you see?	A: A nadie. nobody. *A un alma. a soul	[Spanish]
b.	Q: Qui has vist? Who did you see? Q: Que queda sucre? Is there any sugar left?	A: Ningú. Nobody. A: *Gaire. Much.	[Catalan]
c.	Q: Pjon ihes? Who did you see?	A: KANENAN Nobody *kanenan Anybody	[Greek]
d.	Q: Kto przyszedł? who came?	A: Nikt. Nobody.	[Polish]

The data in (22) show that n-words like *nadie*, *ningú*, *KANENAN*, *nikt* mean 'nothing' as the elliptical answer to a question. NPIs like *un alma*, *gaire*, *kanenan*, etc. are not felicitous in elliptical contexts: the answer cannot convey a negative meaning in the absence of an overt marker of negation.

N-words participate in *negative concord*, the phenomenon whereby a sequence of seemingly negative expressions gets a single negation reading. Negative concord (NC) raises major questions for the principle of compositionality of meaning. Many existing proposals try to address this issue, for example, Zanuttini (1991), Ladusaw (1992), van der Wouden and Zwarts (1993), Corblin (1996), Déprez (1997a,b, 2000), Giannakidou (1998, 2000), Herburger (2001), de Swart and Sag (2002), Corblin et al. (2004), Zeijlstra (2004), and others. Chapter 1 (Sections 4 and 5) provided an overview of the main issues and some of the most important solutions that have been proposed in the literature. That discussion will not be repeated here. The OT analysis of double negation and negative concord developed in this chapter builds on the proposals made by de Swart and Sag (2002), so I will only refer to the polyadic quantifier analysis here.

4.3 Double Negation and Negative Concord as Instances of Polyadic Quantification

According to Jespersen (1917: 62) "when logicians insist that 'two negatives make an affirmative' their rule is not corroborated by actual usage in most languages. But it would be wrong to divide languages into some that follow this rule and others that do not, for on closer inspection we find that in spite of great differences between languages in this respect, there are certain underlying principles that hold good for all languages." The analysis of double negation and negative concord developed in this section follows the spirit of Jespersen's analysis in that I adopt a unified analysis of the two phenomena in which double negation and negative concord involve two ways of instantiating polyadic quantification.

4.3.1 Iteration and Resumption

The main semantic claims made by de Swart and Sag (2002) are that n-words are inherently negative, and that both double negation and negative concord involve polyadic quantification. Double negation involves iteration of monadic quantifiers (function application), and is first-order definable. (23) works out the iteration of two negative quantifiers *nobody* and *nothing* in polyadic quantifier theory:[7]

(23) a. Nobody has nothing to hide.
= Everyone has something to hide. [DN]
b. NO (HUM, $\{x|$ NO (INAN, $\{y|$ HIDE$(x,y)\}))\})$
c. $\neg\exists x \neg\exists y$ Hide(x,y)

(23b) spells out the function application of *nobody* to *has nothing to hide* in standard generalized quantifier theory. One negative quantifier is embedded under another one, which leads to the double negation reading spelled out in first-order logic as (23c).

Negative concord is interpreted in terms of resumptive negative quantification. De Swart and Sag (2002) adopt Keenan and Westerståhl's (1997: 879) definition of binary resumption:

- The binary resumption of a type <1,1> quantifier Q is the quantifier Q′ defined as follows:

$$Q'^{A,B}_{E} (R) = Q_{E2}^{A \times B} (R)$$

A type <1,1> quantifier Q is a standard determiner (*all*, *some*, *no*, etc.) establishing a relation between two sets, subsets of the universe of discourse E. In a binary resumption, the quantifier Q occurs twice, once with A and once with B as its restrictor. The resumptive quantifier Q′ ranges over pairs of individuals, with A × B as its restrictor, and the two-place relation R as its scope.

The application to an example like (17a), repeated here as (24a), shows the resumption of two n-words *nadie* and *nada*.

(24) a. Nadie ha dicho nada. [Spanish]
Nobody has said nothing
'Nobody said anything.'
b. $\text{NO}_{E2}^{\text{HUM} \times \text{INAN}}$ (SAY)
c. $\neg\exists x \exists y$ Say(x,y)

The Spanish sentence (24a) contains two n-words, *nadie* and *nada*. Both contribute the negative quantifier NO_E, which establishes a relation between two sets and functions as Q in the definition of resumption. *Nadie* ranges over human beings, and *nada* over inanimate things, so the restrictor A is HUM and the restrictor B is INAN. The two-place relation R is provided by the transitive verb *say*.

[7] (23) is an attested example found on the Internet.

The resumptive quantifier Q′ reads as $\text{NO}_{E2}^{\text{HUM} \times \text{INAN}}$ (SAY) in (24b). The polyadic quantifier ranges over pairs of humans and things, and requires no such pair to stand in the 'say' relation. If no pair of a human and a thing stands in the 'say' relation, there is not an individual *x* and a thing *y* such that *x* says *y*. The truth conditions of the polyadic quantifier in (24b) then boil down to the first-order formula $\neg\exists x \exists y\ Say(x,y)$ in (24c), and the negative concord reading gives rise to a world of silence.

Although the truth conditions of the negative concord sentence can be spelled out in first-order logic, the compositional semantics relies on a higher-order polyadic quantifier. This approach respects the inherently negative nature of both negative quantifiers (English *nobody*) and n-words (Spanish *nadie*), and locates the difference between double negation and negative concord in the grammar, rather than the lexicon.[8]

Following Keenan and Westerståhl, de Swart and Sag generalize the definition of resumptive quantification to a sequence of *k* monadic quantifiers Q′ binding just one variable each, and interpreted on the universe of discourse E, with a one-place predicate A as their restrictor, and taking a k-ary relation R as its scope.[9]

- Resumption of a k-ary quantifier.

$$Q'^{\,A1, A2, \dots Ak}_{E} (R) = Q_{Ek}^{A1 \times A2 \times \dots Ak} (R)$$

The resumptive quantifier is a polyadic quantifier binding *k* variables, interpreted in the universe of discourse E^k, taking the subset $A_1 \times A_2 \times \dots A_k$ of E^k as its restrictor, and the k-ary predicate R as its scope. The generalized definition is relevant for an example like (17d), repeated here as (25).[10]

(25) a. Personne n' a rien dit à personne. [written French]
Nobody SN has nothing said to nobody
'Nobody said anything to anyone.'
b. $\text{NO}_{E3}^{\text{HUM} \times \text{INAN} \times \text{HUM}}(\text{SAY})$
c. $\neg\exists x \exists y \exists z\ R(x,y,z)$

The sequence of n-words *personne*, *rien*, *personne* provides a series of monadic quantifiers NO_E, ranging over humans, things, and humans, respectively in the universe of discourse E. R is provided by the three-place predicate *say-to*. The resumptive quantifier reads as $\text{NO}_{E3}^{\text{HUM} \times \text{INAN} \times \text{HUM}}$ (SAY) in (25b), and ranges over triples of humans and things. No triple of a human, a thing, and a human stands in the 'say-to' relation according to (25b). The truth conditions of the resumptive quantifier are

---

[8] This book focuses on resumptive readings at the clausal level. For DP-internal resumptive negation in West Flemish, cf. Haegeman (2002a).

[9] Keenan and Westerståhl's definition is slightly more complex, because they want to generalize to the possibility of resumptive quantification with relational nouns. Given that the discussion in this book is restricted to pronominal and adverbial n-words binding a single variable, I maintain the easier definition for readability.

[10] For the moment, I ignore the role of the clitic *ne*, which will be analyzed in Chapter 5 as a marker co-occurring with negation, but not conveying semantic negation (Section 6).

equivalent to the first-order formula $\neg\exists x\exists y\exists z\, R(x,y,z)$ in (25c), which requires there not to be an individual *x*, a thing *y*, and an individual *z* such that *x* says *y* to *z*.

According to Keenan and Westerståhl, resumption only applies to a sequence of quantifiers that are somehow 'the same'. In the case of negative concord, resumption applies to a sequence of anti-additive quantifiers ('nobody', 'nothing', etc.). Quantifiers like *few, at most two* are monotone decreasing, but not anti-additive. They license NPIs, but do not participate in resumptive quantification, and do not lead to negative concord interpretations.

The syntax–semantics interface defines how DN and NC readings are accounted for in an HPSG framework. The syntactic details are not discussed here (cf. de Swart and Sag 2002, Chapter 1, Section 5). The HPSG grammar defines the space of possible meanings, but does not decide between DN and NC. In French, both readings are available for a sentence like (26).

(26) Personne n'aime personne. [French]
a. NO(HUM, $\{x|$NO(HUM, $\{y|x$ loves $y\})\})$ [iteration]
b. $\neg\exists x\neg\exists y$ Love(x,y) [DN]

(27) Personne n'aime personne. [French]
a. $\text{NO}_{E2}^{\text{HUM}\times\text{HUM}}(\text{LOVE})$
b. $\neg\exists x\exists y$ Love(x,y) [NC]

(26) and (27) are identical as far as the argument structure, the storing mechanism, and the relational interpretation of the word *love* is concerned. The difference resides in the interpretation of the polyadic quantifier upon retrieval from the N-store: iteration in (26), resumption in (27).

The HPSG grammar assumes no lexical difference between negative quantifiers and n-words, so in the rest of this book, I use the term 'Neg-expression' for both. The analysis relies on an extended notion of argument structure, so it works for n-words in both argument and adjunct position (thus, *nobody*, *nothing* as well as *never*, *nowhere*). Finally, it does not involve covert or empty negations (i.e., syntactically invisible but semantically potent negations, or syntactically visible but semantically inoperative negations). Thus the higher order construction of a polyadic quantifier respects the principle of compositionality of meaning in a surface oriented syntax (cf. Chapter 1, Section 4).

4.3.2 Cross-Linguistic Variation

Not all languages allow double negation and negative concord as freely as French does, although ambiguities are found in other languages as well (cf. Chapter 6, Section 3). The combination of two negative quantifiers in a language like English typically leads to a double negation reading (23), and resumption is only marginally available as an interpretive strategy. Languages like Spanish, Greek, Polish, and many other languages are on the other end of the spectrum. They are typical negative concord languages, which hardly ever realize the iteration version of the polyadic quantifier analysis, and normally display resumptive negative quantification (24). Unlike the French example (26/27), examples like (17a–c) in Spanish, Italian, and Piedmontese are not usually perceived as ambiguous, but exclusively display the negative concord reading.

Clearly, the HPSG analysis developed by de Swart and Sag (2002) does not predict crosslinguistic variation where it arises, as pointed out by Zeijlstra (2004: 207). Chapter 1 argued that the polyadic quantifier analysis needs enrichment with a typological dimension. My solution is to add an OT component to the existing HPSG analysis. De Swart and Sag's informal suggestion that languages may have a 'preference' for one interpretation over another can now be substantiated as a difference in constraint ranking. The OT analysis I propose in the remainder of this chapter builds on the grammar proposed by de Swart and Sag (2002), but locates the crosslinguistic variation in the bidirectional OT component.

4.4 Marking and Interpretation of Negation in Bidirectional OT

This section develops a bidirectional OT analysis of negation in two steps. Section 4.1 defines the OT syntax, and Section 4.2 combines this syntax with an interpretive mechanism, and spells out the bidirectional grammar. Section 4.3 offers some reflections on the space of typological variation in the bidirectional OT setup.

4.4.1 Generation of Double Negation and Negative Concord in OT

Jespersen (1917) observed that negation can be attracted to indefinites in argument position. Section 1 worked out an OT analysis of this phenomenon. Depending on the constraint ranking, the input meaning $\neg\exists$ can be realized by a marker of sentential negation plus an indefinite, or by a negative indefinite. The constraint driving the use of negative indefinites was called NEGATTRACT. Section 2 extended the empirical scope of the study to multiple indefinites under negation, and studied the ways languages realize the meaning $\neg\exists x_1, \exists x_2, \ldots \exists x_n$. Again, languages vary in their use of indefinites or negative indefinites.

Of course, only languages that have negative indefinites (Neg-expressions) can be discussed in a study of double negation and negative concord languages. This is clear from the semantic analysis in terms of polyadic quantification adopted in Section 3. However, NEGATTRACT is not enough to account for negative concord, for (standard) Dutch and German are languages with a high ranking of NEGATTRACT, but they are not negative concord languages. So there must be another constraint that drives the use of Neg-expressions in sentences involving multiple indefinites under negation.

According to Corblin and Tovena (2003: 326), natural languages frequently have linguistic means to indicate that an argument must be interpreted within the scope of negation. They refer to this as marking of 'negative variables'. Similarly, Haspelmath (1997: 231), building on Tanaka (1994), claims that the use of n-words is functionally motivated by the desire to mark the focus of negation, that is, the participants that are affected by the negation.

In terms of OT syntax, the use of n-words constitutes a case of marking an input feature in the output: the 'negative variable' is formally marked as such. In OT, this can be expressed by means of a Max constraint:

(28) MaxNeg
Mark 'negative variables' (i.e., mark indefinites in argument or adjunct position that are interpreted in the scope of an anti-additive operator such as negation, as formally negative).

MaxNeg favors the multiplication of negative indefinites within the scope of a marker of sentential negation or another negative indefinite. The functional motivation for marking the focus of negation explains why the use of n-words is widespread among natural languages.

However, the use of n-words is not universal: languages like Dutch, English, and Basque do not use n-words. This suggests that MaxNeg is not a hard constraint, and its position in the constraint ranking is not the same in every language. MaxNeg is in conflict with the markedness constraint *Neg introduced in Chapter 3, and repeated here as (29):

(29) *Neg
Avoid negation in the output.

MaxNeg and *Neg are conflicting constraints, because MaxNeg wants to reflect an input feature concerning negation in the output form, whereas *Neg wants to avoid negation in the output. The difference between languages with and without n-words can be accounted for in terms of the position of MaxNeg relative to *Neg. If *Neg is ranked above MaxNeg in the syntax, the optimal way to express the meaning $\neg\exists x_1 \exists x_2$ is by means of indefinite pronouns (examples 8–10) (Tableau 6). If MaxNeg is ranked higher than *Neg, n-words are used to express indefinites under negation (examples 16–20) (Tableau 7).

FNeg is always ranked at the top (see Chapter 3). The high ranking of FNeg makes it impossible to express indefinites under negation by means of non-negative indefinites

Tableau 6 Indefinite under negation in Dutch, Turkish, etc. (production)

Meaning $\neg\exists x_1 \exists x_2$	Form	FNeg	*Neg	MaxNeg
	indef + indef	*		**
☞	neg + indef		*	*
	neg + neg		**	

Tableau 7 N-word under negation in Spanish, Italian, etc. (production)

Meaning $\neg\exists x_1\exists x_2$	Form	FNEG	MAXNEG	*NEG
	indef + indef	*	**	
	neg + indef		*	*
☞	neg + neg			**

exclusively (in the absence of a marker of sentential negation). In Tableaux 6 and 7, the relevant candidates for comparison are those that somehow mark negation in the output. This invariably leads to a violation of *NEG. Two Neg-expressions are 'worse' than one, so the combination of two Neg-expressions incurs two violations of *NEG. This pattern of violations reflects the treatment of n-words as formally negative.

Languages that allow indefinites under negation (e.g., Dutch and Turkish), and languages that use n-words (e.g., Romance, Slavic, Greek, and Hungarian) differ in the relative ranking of the two constraints MAXNEG and *NEG. The ranking *NEG >> MAXNEG indicates that double negation languages value economy in the expression of negation over marking of negative variables (Tableau 6). The ranking MAXNEG >> *NEG reflects the formally negative marking of indefinites under negation found in negative concord languages (Tableau 7).

The question that immediately arises at this point concerns the interpretation of Neg-expressions in the two classes of languages. The combination of a Neg-expression with a sequence of indefinites permits recovery of the meaning $\neg\exists x_1\exists x_2 \ldots \exists x_n$ by application of the standard rules of first-order logic. However, for languages that mark negative variables by means of n-words, the issue of the interpretation of these structures is less trivial. Remember that the term 'Neg-expression' generalizes over negative quantifiers and n-words, and stands for an expression that has the lexical semantics $\neg\exists x$. Given that a sequence of two anti-additive quantifiers can be interpreted either in terms of iteration or in terms of resumption (cf. Section 3), the OT syntax needs to be mirrored in an OT semantics that determines the preference of one interpretation over another in the grammar of a language.

4.4.2 *Interpretation of Neg-expressions*

In isolation, it is not possible to determine whether a particular expression is a negative quantifier or an n-word, because they both contribute the meaning $\neg\exists$ as the fragment answer to a question (14a, b), (22), and in sentences where they constitute the sole expression with negative force (15a, b). Following de Swart and Sag (2002) and others, I take this to mean that n-words and negative quantifiers have the same lexical semantics (Chapter 1, Section 4). In Tableaux 6 and 7, the term 'Neg-expression' is the general term for expressions that are formally marked for negation but are interpreted either as negative quantifiers or as n-words.

If a sequence of Neg-expressions is interpreted in terms of resumption of anti-additive quantifiers, a single negation reading ensues (negative concord). This is the desired interpretation for the examples in (17a-d), repeated here in (30).

(30) a. Nadie ha dicho nada. [Spanish]
Nobody has said nothing
'Nobody said anything.'
b. Nessuno ha detto niente. [Italian]
Nobody has said nothing
'Nobody said anything.'
c. Gnun a l'ha dit gnente. [Piedmontese]
Nobody has said nothing
'Nobody said anything.'
d. Personne n'a rien dit à personne. [French]
Nobody SN has nothing said to nobody
'Nobody said anything to anyone.'

The interpretation of a sequence of two Neg-expressions by means of iteration results in a double negation reading. This is the desired interpretation of the English examples in (31), the Dutch example in (32), and the German example in (33). The examples in (31)–(33) were found on the Internet. Italics have been added to (32) and (33) to highlight the Neg-expressions; they do not indicate focus or stress.

(31) a. Nobody has nothing to hide.
b. Nobody is in prison for nothing.

(32) Het is een stad uit een opgewekt verhaal of film, waar iedereen, zelfs de schurk, beminnelijk is en warm; waar niemand veel geld heeft, maar waar iedereen rond komt; waar *niemand niets* te doen heeft, maar waar iedereen genoeg tijd over houdt; en waar een bedelaar te middernacht met een viool onder de arm naar huis keert, luid zingend, maar voor zichzelf, omdat hij er zin in heeft… [Dutch]
It is a city from a beat-up story or movie, where everyone, even the villain, is friendly and warm; where nobody has a whole lot of money, but where everyone has enough to make ends meet; where nobody has nothing to do, but where everyone has enough time left; and where a beggar goes home at midnight with a violin under his arm, singing loudly, but to himself, just because he feels like it..

(33) Wollen wir demnächst die Firmen verantwortlich machen, die Karteikarten, Kopierstifte und Tätowiertinte hergestellt haben? Wenn alle, nun auch im Ausland, schuldig sind, war *niemand* an *nichts* schuld.
Is the next step to hold responsible the companies that made index cards, copying pens, and tattoo inks? If all, now also those living abroad, are guilty, nobody is guilty of nothing.

The use of Neg-expressions in a generative OT system means that we run into the recoverability problem: from the expressions generated, multiple interpretations (DN and NC) can be derived, not just the intended one. This is not a problem for the French examples (26/27), but in most other languages sequences of Neg-expressions are not ambiguous. In the OT system developed in this chapter, recoverability is assured by the way the generation of negative sentences hangs together with their interpretation in a strong bidirectional OT analysis.

In the OT analysis, the familiar constraints FNEG and *NEG are 'double-edged' constraints that work in the generation as well as in the interpretation perspective. Semantically, FNEG requires a reflection of the negative form in a non-affirmative meaning. *NEG avoids a proliferation of negations in the semantics, and with a sequence of Neg-expressions this leads to a preference for resumption over iteration. *NEG needs to be balanced with a faithfulness constraint that prefers first-order interpretations and interprets every Neg-expression as contributing its own semantic negation. The faithfulness constraint needed to determine the interpretation of multiple Neg-expressions is formulated as INTNEG:

(34) INTNEG
Force Iteration (i.e., interpret every Neg-expression in the input form as contributing a semantic negation at the first-order level in the output).

INTNEG is motivated by the polyadic quantifier system set up in Section 3. Iteration is preferred, because it is first-order definable. However, it is in conflict with the markedness constraint *NEG: INTNEG favors multiplication of negation in the interpretation, whereas *NEG avoids negation in the output.

MAXNEG and INTNEG are mirror images of each other. Both maximize the reflection of input features in the output, MAXNEG in the syntactic form, INTNEG in the semantic interpretation. As semantic constraints, both FNEG and INTNEG are instances of the general constraint FAITHINT proposed by Zeevat (2000), and defined as a principle that forces the hearer to interpret all that the hearer has said. The next step is to show that the three constraints MAXNEG, *NEG, and INTNEG together account for double negation and negative concord languages.

FNEG outranks all the other constraints as usual. MAXNEG is a purely syntactic constraint that does not play a role in interpretation. So the constraints that need to be ordered in the OT semantics are the economy constraint *NEG and the faithfulness constraint INTNEG. If INTNEG is ranked higher than *NEG, a series of Neg-expressions is interpreted as multiple negation by forcing iteration (Tableau 8). If *NEG is ranked higher than INTNEG, a sequence of multiple Neg-expressions leads to a single negation meaning by resumption, which leads to the more economical single negation output (Tableau 9).

FNEG is ranked at the top, so a statement involving two Neg-expressions is never interpreted without a reflection of the non-affirmative meaning. As a result, the relevant candidates for comparison have at least one negation in the output, and always incur a violation of *NEG.

Tableau 8 Double negation in (standard) English, (standard) Dutch, etc. (interpretation)

Form neg + neg	Meaning	FNEG	INTNEG	*NEG
	$\exists x_1 \exists x_2$	*	**	
	$\neg\exists x_1 \exists x_2$		*	*
☞	$\neg\exists x_1 \neg\exists x_2$			**

Tableau 9 Negative concord in Spanish, Italian, etc. (interpretation)

Form neg + neg	Meaning	FNEG	*NEG	INTNEG
	$\exists x_1 \exists x_2$	*		**
☞	$\neg\exists x_1 \exists x_2$		*	*
	$\neg\exists x_1 \neg\exists x_2$		**	

Table 1 Bidirectional grammar

Negative concord languages	FNEG >> MAXNEG >> *NEG >> INTNEG
Double negation languages	FNEG >> INTNEG >> *NEG >> MAXNEG

The combination of two Neg-expressions leads to a double negation reading in languages like Dutch and English under the ranking INTNEG >> *NEG (Tableau 8). This ranking forces iteration of negative quantifiers at the expense of multiplication of negations in the semantics, which accounts for the double negation reading of the sequence of two Neg-expressions in examples (31)–(33). Under the ranking *NEG >> INTNEG (Tableau 9), single negation readings win over double negation readings in NC languages such as Spanish, Italian, Greek, and Polish. Semantic economy is valued higher than first-order interpretation, so resumption is preferred to iteration. This ranking accounts for the negative concord reading of the combination of two Neg-expressions in example (30).

The combination of generation and interpretation perspective derives the two rankings for negative concord and double negation languages in a strong bidirectional grammar (Table 1).

In this bidirectional grammar, the choice between resumption and iteration in the semantics is explicitly related to the functional desirability of marking negative variables in the syntax. In negative concord languages, MAXNEG is ranked above *NEG, so that the indefinites under negation are realized as Neg-expressions. The ranking of INTNEG under *NEG implies that the multiplication of negations in the form is absorbed in the semantics, and a single negation reading ensues by resumption. In double negation languages, INTNEG is ranked above *NEG, so every negative form is interpreted as contributing its own (first-order) negation to the semantics (iteration). MAXNEG is ranked below *NEG, so it is more important to avoid negative forms than to mark negative variables.

In both classes of languages, syntax and semantics work closely together. Thus, the bidirectional setup is a crucial feature of the analysis.

4.4.3 Reflections on the Bidirectional Grammar

The polyadic quantifier analysis developed by de Swart and Sag (2002) provides the space of possible meanings of a natural language negation system. The OT

Table 2 Factorial typology of three constraints

MAXNEG >> *NEG >> INTNEG	NC
MAXNEG >> INTNEG >> *NEG	Unstable
INTNEG >> MAXNEG >> *NEG	Unstable
*NEG >> MAXNEG >> INTNEG	Unstable
*NEG >> INTNEG >> MAXNEG	Unstable
INTNEG >> *NEG >> MAXNEG	DN

grammar makes it possible to encode the preference a language may have for resumption or iteration in the syntax–semantics interface.

*NEG is a double-edged constraint, which plays a role in syntax as well as semantics. The ranking of *NEG in relation to the syntactic faithfulness constraint MAXNEG determines whether negative variables are marked in the syntax or not. The ranking of the semantic faithfulness constraint INTNEG with respect to *NEG determines whether Neg-expressions each contribute a negation in the semantic representation or not.

The two directions of production and interpretation are combined into a strong bidirectional grammar of negation. This grammar provides a balance between form and interpretation of negation in the two classes of double negation and negative concord languages. The bidirectional grammar also raises new questions.

Even if the markedness of negation with respect to affirmation motivates the top ranking for FNEG along the lines of Chapter 3, more rankings need to be considered than the two orders given earlier. Aside from FNEG, the bidirectional grammar implies the three constraints *NEG, MAXNEG, and INTNEG, and three constraints permit six rankings, at least in principle. The factorial typology of the three constraints is spelled out in Table 2.

So far, the top and the bottom rankings have been established as grammars of a particular class of languages (double negation and negative concord). What about the other four rankings? I claim that the other four rankings do not represent stable negation systems, because generation and production are not well balanced, and natural languages only adopt grammars that show a good balance between syntax and semantics.

Let me substantiate this claim by looking in more detail at the relation between production and interpretation under the rankings characterized as 'unstable'. Observe that the ranking in Tableau 10 is unstable, because the meaning intended by the speaker is not recovered by the hearer.

The top row in Tableau 10 shows the input of a speaker who wants to express a single negation meaning. The ranking generates a sequence of two Neg-expressions as the optimal output for the single negation input. But the interpretation of a sequence of two Neg-expressions in the second half of the tableau leads to a double, rather than a single negation reading for the hearer. This means that the speaker's intended meaning is not recovered by the hearer. Under the assumption that a stable grammar should allow the hearer to recover the intended meaning as the optimal

Tableau 10 MAXNEG >> INTNEG >> *NEG (intended meaning not recovered)

Meaning $\neg\exists x_1\exists x_2$	Form	MAXNEG	INTNEG	*NEG
	neg + indef	*		*
☞	neg + neg			**
Form neg + neg	**Meaning**	**MAXNEG**	**INTNEG**	***NEG**
☞	$\neg\exists x_1\neg\exists x_2$			**
	$\neg\exists x_1\exists x_2$		*	*

Tableau 11 *NEG >> INTNEG >> MAXNEG (form not motivated)

Meaning $\neg\exists x_1\exists x_2$	Form	*NEG	INTNEG	MAXNEG
☞	neg + indef	*		*
	neg + neg	**		
Form neg + neg	**Meaning**	***NEG**	**INTNEG**	**MAXNEG**
	$\neg\exists x_1\neg\exists x_2$	**		
☞	$\neg\exists x_1\exists x_2$	*	*	

interpretation of the optimal form produced by the speaker, the bidirectional grammar in Tableau 10 cannot be the grammar of a natural language.

The ranking INTNEG >> MAXNEG >> *NEG is equally unstable. Given that there is no direct interaction between the semantic constraint INTNEG and the syntactic constraint MAXNEG, the argumentation is the same. I conclude that MAXNEG and INTNEG cannot both be higher than *NEG. The ranking in Tableau 11 illustrates that MAXNEG and INTNEG cannot both be ranked below *NEG either.

Here we have the reverse problem. An indefinite form is the optimal form for the speaker to choose in the expression of indefinites under negation. The hearer will not have a problem in interpreting this input. However, the use of two Neg-expressions also leads to a negative concord reading, even though the use of the n-word is not functionally motivated by the low ranking of MAXNEG. Under the assumption that the bidirectional grammar needs to balance production and interpretation, the ranking *NEG >> INTNEG >> MAXNEG in Tableau 11 cannot be the grammar of a natural language. The same problems arise with the ranking *NEG >> MAXNEG >> INTNEG, because MAXNEG and INTNEG do not interact directly.

The conclusion must be that only rankings where MAXNEG and INTNEG are distributed on either side of *NEG reflect viable options for a linguistic system that balances generation and interpretation of negative statements. This is of course exactly the bidirectional grammar proposed earlier in Section 4.2. In sum, the bidirectional grammar of negation in natural language takes one of two options:

- **Negative Concord**: MAXNEG >> *NEG >> INTNEG.
  If you mark 'negative variables' (MAXNEG >> *NEG in syntax), then make sure you do not force Iteration (*NEG >> INTNEG in semantics).
- **Double Negation**: INTNEG >> *NEG >> MAXNEG.
  If you force Iteration (INTNEG >> *NEG in semantics), then make sure you do not mark 'negative variables' (*NEG >> MAXNEG in syntax).

Even though a set of three constraints allow six rankings in principle, evidence of just two of these rankings is found in natural language. The other four rankings lead to unrecoverable meanings or unmotivated forms, and imply an unbalanced relation between syntax and semantics that does not match the grammar of a natural language.

Given that the two directions of optimization are tied together in the bidirectional grammar, my proposal is an instance of strong bidirectional OT, as proposed by Blutner (2000). As the optimal form-meaning pair, strong bidirectional OT picks out the pair which combines the best form with the best meaning in the two directions of optimization. The two rankings adopted for double negation and negative concord languages exemplify the process of strong bidirectional optimization.

4.5 Negative Concord and Negative Polarity

The relation between negative concord and negative polarity has been addressed in Chapter 1 (Section 4) and in this chapter (Section 3). Given that the focus of the book is on the expression and interpretation of negation, and NPIs are not Neg-expressions, a full analysis of negative polarity is outside of the scope of this study. However, some remarks on the relation between NPIs and n-words are called for.

It is well known that languages can have a double negation grammar or a negative concord grammar in different stages of their historical development (cf. Jespersen 1917; Horn 1989; Haspelmath 1997; de Swart and Sag 2002; Mazzon 2004; Falaus 2007a, b; and others). NPIs which strengthen negation and later develop into n-words often play a role in this process. In order to better comprehend this diachronic development, I propose a reformulation of the relation between negative concord and negative polarity, following insights offered by Szabolcsi (2004) in Section 5.1. Section 5.2 works out the contours of an OT analysis of the Jespersen cycle involving negative indefinites.

4.5.1 Negative Polarity in OT

Several analyses of negative concord take this phenomenon to be a version of negative polarity (e.g., Laka 1990, Ladusaw 1992). Chapter 1 (Sections 3 and 4) offers a critical discussion on these proposals. Szabolcsi (2004) turns the picture around, and models NPIs as similar to n-words in certain ways. She proposes a new account of the relation between English *some-no-any*, based on insights from Moltmann (1995), Postal (2000, 2004, 2005), and de Swart and Sag (2002). Importantly, she interprets sentences like (35a, b) in terms of resumption:

(35) a. Nobody said anything.
b. No man talked with any man but Bill.
c. *I didn't always say anything.

According to Szabolcsi, *nobody* and *anything* form a resumptive quantifier $\text{No}_{x,y}$ in (35a). The binary resumptive quantifier licenses the exception phrase in (35b) (cf. also Moltmann 1995). The intervention of a scope-bearing operator such as *always* blocks the construction of the resumptive quantifier and leads to ungrammaticality (35c).

De Swart and Sag's analysis of negative concord and Szabolcsi's treatment of negative polarity *any* show clear similarities: both rely on the semantics of resumption in a polyadic quantifier framework. According to Szabolcsi, this is a natural move: "Treating negative polarity and negative concord with the same semantic device seems quite natural. After all, they are variations on the same meaning." (Szabolcsi 2004: Section 8.3).

The main difference between negative concord and negative polarity is that NPIs need to be licensed, whereas n-words are 'self-licensing'. Under Szabolcsi's treatment, this means that with the polarity item *any*, the formation of a resumptive quantifier is mandatory, and this calls for a licensor to form a resumptive quantifier with. With n-words, resumption arises if a sequence of n-words is found in the N-store, but n-words can also stand alone (as in fragment answers).

Szabolcsi's analysis is attractive for various reasons. First, it allows a natural account of the free mixing of NPIs and n-words in sentences like (36) (from Muller 1991) and (37) (from Ponelis 1985, quoted by van der Wouden 1994):

(36) a. Personne n'a rien dit à personne. [French]
Nobody SN has nothing said to no one
b. Personne n'a rien dit à qui que ce soit.
Nobody SN has nothing said to anyone
c. Personne n'a dit quoi que ce soit à personne.
Nobody SN has said anything to no one
d. Personne n'a dit quoi que ce soit à qui que ce soit.
Nobody SN has said anything to anyone
'Nobody said anything to anyone.'

(37) a. Sy is met niks ooit tevreden nie. [Afrikaans]
She is with nothing ever happy SN
'She is never happy with anything.'
b. Niemand is ooit tevrede nie.
Nobody is ever content SN
'Nobody is ever satisfied.'

If NPIs like *qui que ce soit, ooit* and n-words like *personne, niemand* both contribute to a resumptive quantifier, the semantics of all sentences in (36) and (37) involves the triadic quantifier $NO_{x,y,z}$. The truth-conditional equivalence of the four sentences in (36) is thereby explained.

4.5.2 From Negative Polarity to Negative Concord: a Diachronic Analysis

Szabolcsi's analysis allows an easy transition between a double negation language using NPIs to a negative concord language using n-words, which could facilitate diachronic change. When French *rien*, *personne*, etc. emerged in Old/Middle French as expressions that strengthen the preverbal negative particle *ne*, they naturally appeared in a context that satisfies their licensing conditions, namely in the c-command domain of *ne*. The lexical bleaching which shifts their interpretation from regular NPIs to the type of NPI that requires resumption, affects the semantics, not the syntax.

Once the resumptive semantics is in place, the next change the NPI needs to undergo is to loosen up its condition of licensing, so that it can stand alone, and occur in subject position. In an intermediate stage, the item may already have adopted quite a few features of an n-word, but still allow for negative polarity uses. Negative polarity uses of n-words are rare in French (Muller 1991) and in Greek (Giannakidou 1997, 1998), but easier to obtain in Italian (Zanuttini 1991) and Spanish (Herburger 2001), even if they are perceived as archaic in these languages (cf. Chapter 1, Section 4).

Szabolcsi's insights can be modeled in the OT framework by means of a variation on the constraint MaxNeg. The intuition behind this constraint (Chapter 4, Section 5) is that Neg-expressions mark the focus of negation, by using a special form of the indefinite that occurs in an argument position within the scope of negation. Up to now, I have assumed that marking of negative variables involves a negative form, so I adopted the following formulation:

- MaxNeg
  Mark 'negative variables' (i.e., mark indefinites in argument or adjunct position that are interpreted in the scope of an anti-additive operator such as negation, as formally negative).

Accordingly, every form that satisfies the requirements for MaxNeg and FNeg incurs a violation of *Neg at the same time. But suppose now that there is a weaker form of MaxNeg, called MaxPol, which requires negative variables to be marked with a 'special' polarity sensitive form, but not necessarily a formally negative form:

- MaxPol
  Mark 'negative variables' (i.e., mark indefinites in argument or adjunct position that are interpreted in the scope of an anti-additive operator such as negation with a special polarity sensitive form).

The concept of MAXPOL is broader than that of MAXNEG. The special polarity sensitive form that satisfies MAXPOL could be an n-word or a polarity item. In the first case, it would incur a violation of *NEG, in the second case it would not. Languages with a high ranking of MAXPOL block the use of regular indefinites under negation: these indefinites behave as PPIs.[11]

Standard English can then be characterized as a language that adopts the ranking {MAXPOL, INTNEG} >> *NEG. This grammar leads to the patterns in Tableau 12 as the realization of the input $\neg\exists x$ Buy(Mary,x).

The ranking MAXPOL >> *NEG in the syntax of English favors the use of a 'special' polarity sensitive form of the indefinite under negation, and blocks the use of the regular indefinite *something*. The ranking INTNEG >> *NEG in the semantics makes sure this form is not overtly negative, and does not incur a violation of *NEG. This favors the use of *anything* rather than *nothing*. In a double negation language such as English, the use of an overtly negative form in combination with the marker of sentential negation leads to a double negation interpretation, and the original single negation meaning would not be recoverable (cf. Section 4.3). The bidirectional analysis explains the grammaticalized use of NPIs in standard modern English.

The Jespersen cycle of diachronic change relates the weakening of negation to the use of items that mark the focus of negation. NPIs are one attested source of expressions that reinforce negation (Haspelmath 1997). In the OT system, this process can be viewed as a change in ranking from *NEG >> MAXPOL to MAXPOL >> *NEG in the OT syntax. The next diachronic stage models the change from a negative polarity system to a negative concord system.

This step corresponds with a change in the licensing status of the item, that is, a change from a 'special' not necessarily negative indefinite to a Neg-expression, which incurs a violation of *NEG in the OT syntax. This change in the lexical status of the 'special' indefinite correlates with a change in ranking in terms of OT semantics from INTNEG >> *NEG to *NEG >> INTNEG, in order to preserve the single negation meaning of a sequence of Neg-expressions.

Tableau 12 Negative polarity items in English (production)

Meaning $\neg\exists$x Buy(m,x)	Form	MAXPOL	INTNEG	*NEG
	Mary didn't buy something	*		*
☞	Mary didn't buy anything			*
	Mary didn't buy nothing			**

[11] A language which uses regular indefinites under negation ranks MAXPOL below a constraint blocking multiplication of forms with the same (indefinite) meaning. I will assume such an economy constraint exists but will not propose a formulation here. Note that MAXPOL is formulated for strongly negative contexts, created by anti-additive operators. Of course, NPIs occur in a much wider range of monotone decreasing or non-veridical contexts (cf. Chapter 1, Section 3), so other constraints triggering the use of these expressions have to be formulated. However, the diachronic development from NPIs to n-words takes place in contexts involving anti-additivity or anti-morphism, so the restriction of MAXPOL is appropriate for the issue at hand.

An intermediate stage of overlapping interpretive constraints (*NEG 0 INTNEG) can facilitate the transition. The overlap between INTNEG and *NEG affects 'special' polarity sensitive forms that are on their way to be perceived as n-words (i.e., items like French *personne*), not items that remain polarity items even in the next stage of the development (i.e., items like French *quoi que ce soit*). Which items develop into n-words is a lexical matter, and cannot be predicted on the basis of their internal constitution. Negative concord languages that have lost the possibility of using NPIs in anti-additive contexts have restricted their syntactic constraint from MAXPOL to MAXNEG.

Negative concord languages can switch to a double negation system in two steps. Overlap between the semantic constraints *NEG and INTNEG leads to the type of ambiguities observed for French in Section 3. Chapter 6 (Section 3) shows that the pattern occurs in Hungarian, Welsh, and Romanian as well, and provides a formal analysis. If there is too much overlap, the system loses its stability, because {MAXNEG, INTNEG} >> *NEG is not a stable grammar, as noticed in Section 4. The only way this instability can be remedied is to reduce the overlap, and return to a negative concord system or to raise INTNEG and demote MAXNEG at the same time. This implies the transition from a negative concord to a double negation system.

The Jespersen cycle of double negation/negative concord can be summarized as a system with six stages (Table 3).

The cycle starts with a double negation system (Stage 1), but given the fact that the cycle repeats itself, it could have started anywhere. Dutch and German use regular indefinites in anti-additive contexts (cf. examples in Section 2), so they constitute Stage 1 languages. The NPIs they use are not grammaticalized as a necessary support of a weak negation.

The emergence of MAXPOL above *NEG induces a strengthening of negation by means of polarity items (Stage 2). The obligatory presence of *any*-pronouns in anti-additive contexts characterizes English as a Stage 2 language, in line with Szabolcsi's (2004) analysis of examples (35a, b), cf. also Tableau 12.

When the interpretive constraint INTNEG is lowered, certain NPIs start oscillating between a denotation in terms of existential quantification, or as Neg-expressions (Stage 3). In Stage 4, n-words and polarity items freely mix, but the lexical status of individual items as either an n-word or an NPI is fixed. Greek is a

Table 3 The Jespersen cycle of negative polarity/negative concord

Stage 1	INTNEG >> *NEG >> {MaxNEG, MAXPOL}	Regular indefinites under negation
Stage 2	{INTNEG, MAXPOL} >> *NEG >> MAXNEG	Weakening of negation, strengthening with polarity items
Stage 3	MAXPOL >> INTNEG 0 *NEG >> MAXNEG	Polarity items developing into n-words
Stage 4	{MAXPOL, MAXNEG} >> *NEG >> INTNEG	Free mixing of polarity items and n-words
Stage 5	MAXNEG >> *NEG >> {INTNEG, MAXPOL}	'Strict' negative concord, excluding polarity items
Stage 6	MAXNEG >> *NEG 0 INTNEG >> MAXPOL	Ambiguities with sequences of Neg-expressions

good example of a Stage 4 language: it uses n-words as well as NPIs (38a) in anti-additive contexts, but it excludes existential readings of n-words (a Stage 3 characteristic) in all monotone decreasing contexts (38c, d), except with non-veridical *before* (38b) (examples from Giannakidou 1997).

(38) a. Dhen idha tipota/ TIPOTA [Greek]
SN saw.1SG anything/nothing
'I didn't see anything.'
b. O papus pethane prin na dhi
the grandfather died.3SG before SUBJ see.3SG
kanena/KANENA apo ta egonia tu.
any/ none from the grandchildren his
'My grandfather died before seeing any of his grandchildren.'
c. Pijes pote/*POTE sto Parisi?
Went.2SG ever/never in-the Paris
'Have you ever been to Paris?'
d. I Ilektra ine poli kurasmeni ja na milisi se kanenan/*KANENAN.
the Electra be.3SG very tired for SUBJ talk.3SG to anyone/ no one
'Electra is too tired to talk to anybody.'

Ambiguities between single/double negation readings of sequences of Neg-expressions (a Stage 6 characteristic) are not discussed by Giannakidou (1997, 2000, 2006), who suggests they are not available in Greek (cf. Chapter 6, Section 3).

The fact that languages like Italian, Spanish, and Catalan still allow NPI uses of n-words, as has been established by Zanuttini (1991), Vallduví (1994), and Herburger (2001) suggests that these negative concord languages are basically in Stage 4, but may have an overlap with Stage 3. Italian examples of non-negative interpretations of n-words have been provided in Chapter 1 (Section 4). The following examples of NPI uses of Spanish n-words are from Herburger (2001).

(39) a. Dudo que vayan a encontrar nada. [Spanish]
doubt.1SG that will.3SG.SUBJ find nothing
'I doubt that they will find anything.'
b. Es damasiado tarde para ir a ninguna parte.
Is too late for go to nowhere
'It is too late to go anywhere.'

It is generally acknowledged that the non-negative uses of Romance n-words are archaic, which supports the view that they are basically Stage 4 languages.

Stage 5 relies exclusively on n-words, because MAXPOL has been restricted to MAXNEG. Slavic languages are in between Stage 4 and Stage 5 languages. Progovac's (1994) study of Serbo-Croation shows that this language contains two sets of expressions that occur in negative statements, the *ni*-pronouns in (40a), and the *i*-pronouns in (40b).

(40) a. Marija ne voli ni(t)koga. [Serbo-Croatian]
Mary SN loves nobody
'Mary doesn't love anyone.'
b. Milan ne tvrdi [da Marija poznaje i(k)koga].
Milan SN claims that Mary knows anyone
'Milan does not claim that Mary knows anyone.'

The *ni*-pronoun occurs in the same clause as the negation marker *ne* (40a), and cannot be licensed long distance (41a). The *i*-pronouns are incompatible with clause mate negation (41b), and appear with superordinate negation only (40b).

(41) a. *Ne verujem da Marija voli ni(k)koga. [Serbo-Croatian]
SN claim that Mary loves nobody
b. *Milan ne poznaje i(t)koga.
Milan SN knows anyone.

Following Progovac, I assume that *ni*-pronouns are n-words, and *i*-pronouns are NPIs. The data in (40) and (41) suggest the ranking MAXNEG >> MAXPOL >> *NEG >> INTNEG for Serbo-Croatian. Because of the ranking MAXNEG >> MAXPOL, n-words are preferred over NPIs whenever possible. Accordingly, the NPI *i(t)koga* in (41b) is blocked, because the n-word *ni(t)koga* in (40a) is a better candidate.

However, n-words cannot be licensed long distance as in (41a), because of the clauseboundedness of negative concord. This restriction has its roots in the interpretation of negative concord as resumptive quantification, an instance of polyadic quantification (cf. Chapter 6, Section 1 for discussion). Because n-words are blocked in this environment (at least under a single negation reading of the sentence) by a hard constraint that is ranked above MAXNEG, MAXPOL becomes relevant in complex clause constructions, and the use of the NPI *i(k)koga* in (40b) comes out as the optimal form.

This blocking account of NPIs by n-words extends to Hungarian. Surányi (2006b) points out that Hungarian has two series of pronouns showing up in negative environments, the *s*-pronouns (*senkit* 'nobody', *semmit* 'nothing', *sehova* 'nowhere') and the *vala*-pronouns (*valaki* 'anybody', *valami* 'anything'). The *s*-pronouns allow negative interpretations in isolation, and participate in negative concord, so they are properly characterized as Neg-expressions. The *vala*-pronouns occur in a wide range of non-veridical contexts, and are classified as NPIs. According to Surányi, *vala*-NPIs cannot be clausemate to negation. Surányi doesn't provide examples, but Tóth (1999) illustrates the relevant contrasts in (42) and (43).

(42) *s*-expressions [Hungarian]
a. Pál nem látott senkit.
Paul SN saw nobody.ACC
'Paul didn't see anybody.'
b. *Mária nem mondta hogy Pál látott senkit.
Maria SN said that Paul saw nobody.ACC

(43) *vala*-NPIs [Hungarian]

a. *Pál nem mondott valamit is.
   Paul SN said anything.ACC

b. Pál nem mondta, hogy Mária valakit is látott.
   Paul SN said that Mary anybody.ACC saw
   'Paul did not say that Mary saw anybody.'

S-expressions (42a), but not *vala*-NPIs (43a) are used with clausemate negation. When negation is in the matrix clause, a *vala*-NPIs must be used in the subordinate clause (43b), and an *s*-expression is infelicitous (42b). Thus the blocking effects found in Hungarian are similar to those illustrated in (40) and (41) for Serbo-Croatian. These observations support the same grammar MAXNEG >> MAXPOL >> *NEG >> INTNEG for Hungarian as was posited for Serbo-Croatian.

In Stage 6, the overlap between interpretive constraints leads to ambiguities with sequences of Neg-expressions. Eventually a collapse into the double negation system of Stage 1 becomes an option. It is not strictly necessary to go through Stage 5 before reaching Stage 6. Given the ambiguities found in French and Welsh, it is quite feasible to have overlapping interpretive constraints in a system that freely mixes n-words and NPIs (Stage 4). Note that NPI uses of n-words in French are more exceptional and even more strongly perceived as archaic, which provides evidence that French is further advanced in the cycle than Italian or Spanish.

Szabolcsi's proposals rely on a semantics of *any* that involves no more than $\neg$ and $\exists$ as the semantic input. The 'bleached' semantics of *any* is crucial in identifying whether negative polarity and negative concord involves the same mechanism of resumption. As Szabolcsi points out, many polarity items have a referential semantics or trigger implications so that they are not as 'bleached' of meaning as the word *any*. Szabolcsi's system is not set up to account for NPIs that clearly induce widening or scalar implicatures. The constraint MAXPOL does not come into play either, for the expressions relate to a different input. In specific cases, it may be hard to decide whether a polarity item has a lexical semantics of its own or whether it is completely 'bleached' of meaning.

Vlachou (2007) claims that the French polarity item *qui que ce soit* is not synonymous to *personne* as it implies a notion of widening that is absent from the n-word. But Muller (1991) assumes that the sentences in (36) have the same truth conditions. Under Muller's position, *qui que ce soit* is sufficiently 'bleached' in the sense that Szabolcsi (2004) uses the term, French can be classified as a Stage 4 language, in which NPIs and n-words freely mix in negative statements.

Obviously, the OT analysis of the Jespersen cycle of negative polarity/ negative concord presented in this section is fairly schematic and many details remain to be worked out. I leave this for further research, and restrict the discussion to negative concord and double negation in Chapters 5 and 6.

4.6 Conclusion

The investigation of the marking and interpretation of negation in this chapter illustrates how a bidirectional version of optimality theory offers new perspectives on the range of variation found in natural language for the expression and meaning of negative indefinites. Negation can be realized on an indefinite argument or an adjunct because of the pressure of negative attraction, as observed by Jespersen (1917). In the expression and interpretation of multiple negations, languages display variation, and allow either double negation or negative concord systems. The compositionality problems raised by the semantics of negative concord are solved by the introduction of a higher order polyadic quantifier (iteration vs. resumption). The typological variation is analyzed in a bidirectional OT analysis built on top of the polyadic quantifier analysis.

A high ranking of MAXNEG in the syntax leads to a preference for the marking of negative variables by means of Neg-expressions. A high ranking of INTNEG in the semantics leads to a preference for first-order negation, and thereby to a double negation reading. The combination of MAXNEG, INTNEG, and *NEG in a bidirectional OT grammar leads to the ranking MAXNEG >> *NEG >> INTNEG for negative concord, and INTNEG >> *NEG >> MAXNEG for double negation. Constraint rankings other than the two postulated for these classes of languages are unstable, because they do not strike a balance between generation and interpretation. Accordingly, they cannot provide the grammar of a natural language.

In accordance with the polyadic analysis developed by de Swart and Sag (2002), the OT analysis developed in this chapter relies on the interaction of variable binding Neg-expressions, rather than the marker of sentential negation to account for single and double negation. Chapter 5 will focus on the interaction of Neg-expressions with the marker of sentential negation. The account of strict and non-strict negative concord languages developed there relies mainly on the bidirectional OT grammar advanced in this chapter.

The ambiguities in French examples such as (26/27) originally motivated the polyadic analysis developed by de Swart and Sag (2002), but seem to be lost in the ordinal OT analysis developed in this chapter. A stochastic extension of the analysis is needed to describe resumptive readings in double negation languages, and double negation readings in negative concord languages. This issue will be addressed in Chapter 6 (Section 3).

Chapter 5
Sentential Negation and Negative Indefinites

Introduction and overview This chapter integrates the results on sentential negation (from Chapter 3) with the analysis of negative concord and double negation (from Chapter 4). Section 1 develops the classification of co-occurrence restrictions between sentential negation and negative indefinites in negative concord and double negation languages. The grammar of negative spread supports the claims made by de Swart and Sag's (2002) that in the presence of n-words, the marker of sentential negation in negative concord languages is semantically redundant (Section 2).

Even though the marker of sentential negation is not needed to convey negation in negative concord contexts, many languages combine it with n-words. In line with much current literature, I take sentential negation to serve as a scope marker in such cases. Two syntactic constraints governing the scope of negation account for the contrast between strict and nonstrict negative concord languages. Section 3 shows how the preverbal/postverbal asymmetry exploits NegFirst. Section 4 introduces a new constraint in order to capture strict negative concord languages in which a marker of sentential negation always accompanies an n-word.

Sections 5 through 10 treat in more detail the complex situations found in Catalan, French, Welsh, Hungarian, West Flemish, and Afrikaans. Section 11 concludes the chapter.

5.1 Classification of Co-occurrence Restrictions

Haspelmath (1997) identifies three types of co-occurrence restrictions between negative indefinites and the marker of sentential negation. His classification is presented in this section, but similar distinctions have been established by den Besten (1986), Hoeksema (1997), van der Wouden (1994, 1997), Giannakidou (1997, 1998), and Zeijlstra (2004). I integrate the terminologies as much as possible.

H. de Swart, *Expression and Interpretation of Negation: An OT Typology*, Studies in Natural Language and Linguistic Theory,
DOI 10.1007/978-90-481-3162-4_5,

Type I languages establish strict negative concord and require the presence of a marker of sentential negation in all sentences involving n-words. Type II languages do not combine the marker of sentential negation with n-words in the expression of single negation. They are either double negation languages or languages displaying negative spread. Type II languages establish nonstrict negative concord, and display preverbal/postverbal asymmetries.

5.1.1 Type I: Obligatory Presence of Negation Marker (Strict NC)

Haspelmath (1997: 201) distinguishes three types of co-occurrence restrictions between Neg-expressions and markers of sentential negation. The first type involves negative indefinites (Neg-expressions, NEG) that always co-occur with verbal negation (SN), independently of their position in the sentence. Examples are provided by the Polish *ni*-series (*nikt* 'nobody', *nic* 'nothing', etc.) in (1) (cf. Przepiórkowski and Kupść 1999, Richter and Sailer 2004). Similar examples are found in other Slavic languages, in Greek, Romanian (2), Hungarian (3), etc. The examples in (1) are from Haspelmath (1997: 201), the examples in (2) from Corblin and Tovena (2003), and the examples in (3) from de Groot (1993):

(1) a. Nikt *(nie) przyszedł. [Polish]
nobody *(SN) came.
'Nobody came.'
b. *(Nie) widziałam nikogo.
*(SN) saw nobody.
'I saw nobody.'

(2) a. Nimeni *(nu) a venit. [Romanian]
Nobody *(SN) has come.
'Nobody came'
b. *(Nu) a venit nimeni.
*(SN) has come nobody.
'Nobody came'

(3) a. Senki *(nem) olvas. [Hungarian]
Nobody *(SN) read-3SG
'Nobody read.'
b. *(Nem) jön senki.
*(SN) come-3SG nobody
'Nobody is coming.'
c. Sehol *(nem) lát-t-am senki-t.
Nowhere *(SN) see-past-1SG nobody-ACC
'I did not see anybody anywhere.'

The type SN + V + NEG is the most frequent type in Haspelmath's (1997) language sample. He refers to Tanaka (1994) for evidence that this type is functionally motivated, because both the scope and the focus of negation are marked. The close connection between the verb and sentence negation is expected if Aristotle's and Jesperson's view of negation as predicate denial is adopted, as argued extensively in Horn (1989). The attraction of negation to indefinite arguments marks the focus.

den Besten (1986), Hoeksema (1997), van der Wouden (1997), Giannakidou (1998), and Zeijlstra (2004) refer to type I languages as 'negative doubling', 'proper' or, 'strict' negative concord.

5.1.2 Type II: No Co-occurrence of Negative Indefinites and Negation Marker

In type II languages, negative indefinites never co-occur with verbal negation in the expression of a single negation reading, e.g. the English *no*-series (4), the Dutch *niets* ('nothing'), *niemand* ('nobody'), etc. (5). Co-occurrence of the marker of sentential negation with a negative indefinite invariably leads to a double negation reading (6).

(4) a. Nobody came.
 b. I saw nobody.

(5) a. Niemand kwam. [Dutch]
 Nobody came.
 b. Ik zag niemand.
 I saw nobody.

(6) a. Common people are not nothing.
 b. In tegenstelling tot het lege gebaar van Sun is dit niet niets [Dutch]
 'In contrast to the empty gesture by Sun, this is not nothing.'

According to Haspelmath (1997: 202), type II (V + NEG) is rare in cross-linguistic distribution. In his language sample, only European languages represent this type. Note however, that Mandarin Chinese also counts as a type II language, according to the following corpus data from Xiao and McEnery (2008).

(7) a. Tian'anmen Guangchang zhouwei de jianyi
 Tian'anmen Square around DE simply:equipped
 shuiguan wu shi bu zai gongshui
 tap no moment SN ASP supply:water
 'The simply equipped taps around the Tian'anmen Square were never not supplying water (i.e. were supplying water all the time).'
 b. meiyou yi ge Zhongguoren bu zhidao na zhi kangkai de ge
 not:have one CL Chinese SN know that CL fervent DE song
 'No Chinese does not know that fervent song.'

Double negation is emphatic in both (6) and (7), as expected under the view that double negation is marked relative to single negation (cf. Chapter 1, 11).

Even though type II is not restricted to European languages, it is relatively infrequent outside Europe. Haspelmath explains the relative rarity of type V + NEG as the result of a discrepancy between the semantics (which requires clausal scope of negation), and the surface expression of negation (which is on a participant, rather than on the verb in this type.)

Type II is the normal situation for double negation languages like English, Dutch, etc. However, the pattern also arises in systems that display negative spread. Haspelmath does not discuss this category, but the patterns in a language like spoken French are similar to those in English and Dutch when it comes to the combination of the marker of sentential negation *pas* and n-words like *personne, rien*. *Pas* is the standard expression of sentential negation in sentences like (8). However, *pas* is not required with n-words, be they preverbal (9a) or postverbal (10a) in the expression of a single negation reading. The insertion of *pas* in (9b), (10b) (from Tovena, Déprez, and Jayez 2004) does not lead to ungrammaticalities, but the sentences get a double negation reading. So the sentences in (9b) and (10b) pattern like those in (6) and (7).

(8) a. Il vient pas. [spoken French]
He comes SN.
'He doesn't come.'
b. J'ai pas vu Pierre.
I have SN seen Pierre.
'I haven't seen Pierre.'

(9) a. Il est venu pour rien.
He is come for nothing
'He came for nothing.'
b. Il est pas venu pour rien.
He is SN come for nothing
≠ 'He didn't come for anything.' [NC]
= 'He didn't come for nothing.' [DN]

(10) a. J'ai rien dit.
I have nothing said
'I didn't say anything.'
b. J'ai pas rien dit.
I have SN nothing said
≠ 'I didn't say anything.' [NC]
= 'I didn't say nothing.' [DN]

It is often assumed that if Neg-expressions do not co-occur with verbal negation in a language, they also do not co-occur with each other in the expression of a single negation reading. Haspelmath (1997: 219) emphasizes that this is not true. In spoken French, a combination of two negative indefinites is possible, as in (11). The sentences are ambiguous (cf. Chapter 4, 3), but one of the interpretations is the single negation reading characteristic of negative concord.

(11) a. Personne a rien dit.
Nobody has nothing said
= 'Nobody said anything.' [NC]
= 'Nobody said nothing.' [DN]

b. Personne est le fils de personne.
Nobody is the son of nobody
= 'Nobody is the son of anybody.' [NC]
= 'Nobody is the son of nobody.' [DN]

In contrast to (11), sentences (9b) and (10b) never exhibit a single negation reading. In Section 6, I will provide an analysis of French, including the behavior of the clitic *ne* in written French.

Spoken varieties of modern French indicate that true type II negative concord languages exist. The Occitan data in (12) (from Bernini and Ramat 1996) and the West Flemish data in (13) (from Haegeman and Zanuttini 1996) confirm this.

(12) a. Ai ges d'amic. [Occitan]
have.1SG nothing of friend
'I have no friend.'
b. Degun vegué ren.
Nobody saw nothing
'Nobody saw anything.'

(13) a. da Valère niemand kent. [West Flemish]
that Valère nobody knows
'that Valère doesn't know anybody.'
b. K'een an niemand niets gezeid.
I-have to nobody nothing said
'I didn't say anything to anyone.'

No marker of sentential negation is required to support the postverbal negative indefinite in sentences such as (12a) and (13a) (although such a support is optionally available in West Flemish, cf. Section 9). Still, the sequence of Neg-expressions in (12b) and (13b) expresses a single negation reading, so the sentences display negative concord.

The existence of negative concord interpretations without the support of a marker of sentential negation provides strong support for the analysis defended in de Swart and Sag (2002), adopted in Chapter 4, in which negative concord is driven by negative quantifiers, rather than by sentential negation. However, according to Haspelmath (1997: 218), type II languages displaying negative concord are rare, so it is highly relevant to give an account of type I and type III languages in this chapter.

5.1.3 Type III: Preverbal/Postverbal Asymmetry (Nonstrict NC)

Type III languages are languages in which negative indefinites (NEG) sometimes co-occur with verbal negation (SN) and sometimes do not, e.g. the Italian, Spanish, and Portuguese *n*-series.

(14) a. Ninguém veio. [E. Portuguese]
Nobody came

b. Não veio ninguém.
SN came nobody.
'Nobody came.'

The general observation is that a preverbal marker of sentential negation is obligatory when the n-word is postverbal (as in 14b), whereas no preverbal marker is present when the n-word is preverbal (as in 14a). Type III, characterized as (SN) + V + NEG by Haspelmath (1997) is strong in Romance, but rare elsewhere. According to Zanuttini (1991: 151–153) and Ladusaw (1992), the functional motivation for this type is that postverbal n-words in Romance are unable to take sentential scope. A preverbal expression of negation (n-word or SN) is thus motivated by the desire to express negation at the clausal (propositional) level.

den Besten (1986), Hoeksema (1997), van der Wouden (1994, 1997), Giannakidou (1997, 1998), and Zeijlstra (2004) refer to type III languages as 'improper' or 'non-strict' negative concord. In the remainder of this chapter, I will provide an account of all three types of negative concord languages in an extension of the bidirectional OT grammar developed in Chapter 4.

5.2 Toward an Analysis

Section 1 showed that type II languages may imply either double negation or negative concord (negative spread). Nonstrict and strict varieties of negative concord are classified as type I and type III languages. In this section, I show that a straightforward extension of the bidirectional grammar developed in Chapter 4 explains why double negation languages like Dutch and English cannot be type I or III, but must be type II languages.

5.2.1 Lack of Co-occurrence

Chapter 3 showed that the marker of sentential negation incurs a violation of the constraint *NEG in the OT syntax, which is motivated by the need to satisfy the stronger constraint FNEG. Chapter 4 argued that a negative indefinite also incurs a violation of the constraint *NEG. If both expressions incur a violation of *NEG in the syntax, the combination of a marker of sentential negation and a negative indefinite (in some or in all cases) in the expression of a single negation reading involves two violations of *NEG in the syntax. In order to avoid conveying a double negation, the multiplication of negative forms must be balanced in the OT semantics with the ranking *NEG >> INTNEG. This ranking is the hallmark of negative concord languages (cf. Chapter 4). Accordingly, Neg-expressions in type I and type III languages are n-words, and all double negation languages are type II languages in Haspelmath's classification.

Neg-expressions in type II languages are either negative quantifiers in double negation languages such as English or n-words in negative 'spread' languages like (spoken) French or Occitan, in which a single negation may be conveyed by a sequence of negative indefinites, but not by a negative indefinite in combination with the marker of sentential negation (cf. the examples in 8–13) in Section 1.2.

A straightforward extension of the OT analysis developed in Chapter 4 accounts for the English, Dutch, and Mandarin Chinese data in (4)–(7). NEGATTRACT is operative in the OT syntax, and this constraint favors the realization of negation on an argument. In the absence of other faithfulness constraints driving the insertion of a negation marker, *NEG blocks the combination of a marker of sentential negation and a negative indefinite in the expression of a single negation reading. Tableau 1 illustrates.

Chapter 3 adopted the ranking FNEG >> *NEG as the universal order of these constraints in all natural languages. The high ranking of FNEG blocks 'indef V' as the optimal candidate for the expression of a negation meaning.

With the candidate 'indef SN V', the marker of sentential negation (SN) satisfies FNEG, and incurs a violation of *NEG, just like a Neg-expression. The candidate 'neg V' is a better candidate, because it satisfies NEGATTRACT as well as FNEG. The combination of the marker of sentential negation with a Neg-expression as in the candidate 'neg SN V' satisfies all the relevant faithfulness constraints, but incurs an extra violation of *NEG in comparison to the candidate 'neg V'. Economy then dictates that that 'neg V' is a better candidate.

This grammar thus accounts for the observation that the form 'Nobody came' (English) or 'Niemand kwam' (Dutch) is the optimal expression of the meaning $\neg\exists x Come(x)$ in examples (4) and (5) (cf. Chapter 4, Section 1).

As far as the interpretation is concerned, the high ranking of INTNEG leads to a double negation reading as the optimal interpretation of sentences like (6) and (7), in which a marker of sentential negation combines with a negative indefinite. Tableau 2 illustrates.

Tableau 1 Negative indefinite without a marker of sentential negation (production) (English, Dutch)

Meaning $\neg\exists xV(x)$	Form	FNEG	*NEG	NEGATTR	MAXNEG
	indef V	*		*	*
	indef SN V		*	*	*
☞	neg V		*		
	neg SN V		**		

Tableau 2 Double negation in English, Dutch, etc. (interpretation)

Form SN V neg	Meaning	FNEG	INTNEG	*NEG
	$\exists x\ V(x)$	*	**	
	$\neg\exists x\ V(x)$		*	*
☞	$\neg\neg\exists x\ V(x)$			**

For the combination of a marker of sentential negation and a Neg-expression, the ranking FNEG >> *NEG implies that the semantic contribution of the negation marker in the input form needs to be acknowledged. The competition between a single negation and a double negation reading is decided by the relative order of INTNEG and *NEG.

For double negation languages like English, Dutch, and Mandarin Chinese, Chapter 4 adopted the ranking INTNEG >> *NEG. Both the marker of sentential negation and the Neg-expression contribute a negation in the semantics in order to fully satisfy INTNEG. The double negation reading incurs two violations of *NEG, but the low ranking of this constraint in the semantics makes DN the optimal interpretation in the grammar of Dutch and English.

The grammar exemplified in Tableau 2 then obtains the double negation reading of the combination of a marker of sentential negation and a Neg-expression in the same way as we obtained the double negation reading from the combination of two Neg-expressions in Chapter 4. The bidirectional grammar FNEG >> INTNEG >> *NEG >> MAXNEG developed for double negation languages like Dutch and English thus extends in a straightforward way to cases in which a marker of sentential negation co-occurs with a Neg-expression.

This accounts for one half of the set of type II languages. Economy reasons will also turn out to explain the French data in (9) and (10), and their Occitan and Flemish counterparts in (12) and (13). However, given that these are negative concord languages, and not double negation languages, this will be easier to show once I have dealt with Haspelmath's type I and type III languages.

If the competition between NEGATTRACT, MAXNEG, and *NEG was all there was, the analysis would predict that negative concord languages never combine the marker of sentential negation with a Neg-expression. Negative concord languages would then all be of type II, and display negative spread. But of course, combinations of sentential negation and n-words are the norm in type I and type III languages. Type I languages are even the most frequent variety of negative concord. So other constraints are active, and come into play in these languages. Before I work out the OT analysis, it is useful to explain why cross-linguistic variation in the use of the marker of sentential negation in concord languages is not unexpected.

5.2.2 Sentential Negation in Resumptive Negative Quantification

As pointed out by de Swart and Sag (2002: 401) and discussed in Chapter 1 (Section 5), the fact that the role of sentential negation in negative concord is subject to considerable cross-linguistic variation constitutes a significant problem for approaches to negative concord in which sentential negation plays an important role as the licensor of the n-word (Laka 1990, Ladusaw 1992, Przepiórkowski and Kupść 1999, Giannakidou 1998, Zeijlstra 2004 and others).

Licensing conditions on weak, medium, and strong negative polarity are by and large the same across languages. If negative concord involves a mechanism by means of which the n-word is licensed by a marker of sentential negation or a negative head, Haspelmath's co-occurrence restrictions seem to imply that each language has its own set of licensing conditions on n-words. After all, in some negative concord languages the n-words are incompatible with the marker of sentential negation (French, Occitan), in other languages n-words always require the marker of sentential negation (Slavic, Greek, Hungarian, Afrikaans) and still other languages display a preverbal/postverbal asymmetry (Portuguese, Spanish, Italian). According to Ladusaw (1992), a proliferation of licensing conditions is not very attractive.

Zeijlstra (2004) fully endorses the consequences of the licensing approach, and claims that negative markers in different types of negative concord languages have different negation features. Variation thus resides in the lexicon (cf. also Giannakidou 2000, 2006, Bošković 2008). In Chapter 1, I argued that lexical approaches to negative concord face serious empirical and conceptual problems. In order to overcome these problems, I posited that the distinction between double negation and negative concord languages resides in the grammar.

Under the polyadic quantifier analysis developed by de Swart and Sag (2002), negative concord is viewed as semantic agreement (cf. Chapter 4, Section 3). The single negation reading of a sequence of Neg-expressions is the result of a resumptive interpretation of a sequence of monadic negative quantifiers. The double negation reading arises from the iterative interpretation of a sequence of monadic negative quantifiers.

Formally, a resumptive negative quantifier interprets a sequence of anti-additive quantifiers $Q^1 \ldots Q^k$ of type <1,1> as one complex negative quantifier, Res_Q, of type $<1^k,k>$. The resumptive quantifier maps a series of *k* one-place predicates and one *k*-ary predicate onto a proposition. As such, it binds the sum of all the variable of the composing quantifiers. The resumptive interpretation of the Italian sentence (15a) is repeated here from Chapter 1.

(15) a. Nessuno ha detto niente. [Italian]
Nobody has said nothing.
'Nobody has said anything.'

b. $\text{NO}_{E2}^{\text{hum}\times\text{thing}}$ (SAY) [NC]

c. $\neg\exists x\exists y$ Say(x,y)

In order to account for mixed cases in which a sequence of Neg-expressions combines with a marker of sentential negation, de Swart and Sag (2002) extend the construction of the resumptive polyadic quantifier to nonvariable binding operators such as sentential negation, but also *without* (cf. Chapter 6, Section 1).

Sentential negation is a propositional operator, not a variable binding operator. In terms of the polyadic quantifier theory, this means that it is an expression of a different type. Full NPs (or DPs) denote functions from the power set of the universe of discourse provided by a one-place predicate to truth values; they are defined as (Lindström) type <1> quantifiers. Determiners map a one-place predicate onto a DP, so they denote functions from the power set of the universe of discourse to type <1> quantifiers; they are defined as type <1,1> quantifiers.

A sentential operator like negation denotes a function from propositional entities into truth-values. Accordingly, de Swart and Sag treat a nonvariable binding operator such as negation as a quantifier with adicity zero, or a quantifier of type <0>. The extended definition of resumptive quantifiers of different types is repeated from Chapter 1 in (16).

(16) Resumption of a sequence of *k* type <1,1> quantifiers *Q* and *l* type <0> quantifiers *Q'* leads to the construction of a resumptive quantifier *Q"* of type <1^k, k>, such that:

$$Q"_E{}^{A1...Ak}(R) = Q_{E^k}{}^{A1 \times A2 \times ...Ak}(R)$$

Where $A_1 \ldots A_k$ are subsets of the universe of discourse E, and $A_1 \times A_2 \times \ldots A_k$ and R are subsets of E^k.

Resumption is only defined for quantifiers that are somehow 'the same'. The resumptive negative quantifier is only defined for anti-additive quantifiers such as *nobody*, *nothing*, etc. As an antimorphic operator, *not* subsumes anti-additivity, so the negation marker can participate in the resumptive negative quantifier. As a type <0> quantifier, however, it does not affect the adicity of the polyadic quantifier.

Resumptive quantification of quantifiers of different types is worked out in (17) for Italian postverbal n-words.

(17) a. Mario non ha parlato di niente con nessuno. [Italian]
Mario SN has talked about nothing to nobody
'Mario didn't talk to anyone about anything.'
b. $NO_{E2}{}^{INAN \times HUM}(TALK_M)$
c. $\neg\exists x\exists y$ Talk(m,x,y)

Italian is a nonstrict negative concord language which requires the obligatory presence of the marker of sentential negation with postverbal n-words as in (17a), but not with preverbal ones (cf. 15a). In terms of the polyadic quantifier approach, the n-words *niente* and *nessuno* provide two type <1> quantifiers. The negation marker *non* provides a type <0> quantifier. The verb applied to the subject (written as $TALK_M$) denotes a two-place relation. The resumptive quantifier is spelled out in (18b) as $NO_{E2}{}^{INAN \times HUM}(TALK_M)$, and ranges over pairs of things and humans. No pair should be such that the inanimate member of the pair is talked about to the animate member of the pair by Mario. The first-order truth conditions are spelled out in (18c) as $\neg\exists x\exists y$ Talk(m,x,y), and require there to not be nothing that Mario talked about to anybody.

The negation marker *non* is absorbed in the mixed resumptive quantifier (17b), and leaves no reflection in the truth conditions (17c). Given that sentential negation is not a variable binding quantifier, it does not add any variables to the sum of variables. This supports de Swart and Sag's (2002: 401) conclusion that the marker of sentential negation is semantically redundant in negative concord contexts.

Of course, both strict and nonstrict negative concord languages have a marker of sentential negation. They employ it to express negation in propositions that do not involve 'negative variables' (i.e. indefinites in argument or adjunct position within the scope of negation) (Chapter 3, Section 1). Sentences like (18) illustrate the role of the marker of sentential negation in conveying propositional negation in nonstrict negative languages (Italian, 18a), and strict negative concord languages (Polish and Hungarian, 18b and c). The examples are from Zanuttini (1991), Przepiórkowski and Kupść (1999), and de Groot (1993).

(18) a. Gianni non mangia. [Italian]
Gianni SN eat
'Gianni doesn't eat.'

b. Janek nie pomaga ojcu. [Polish]
John SN helps father
'John doesn't help his father.'

c. János nem dohányz-ik. [Hungarian]
János SN smoke.3SG
'János doesn't smoke.'

In examples like (18), the negation marker is responsible for the syntactic marking of negation, as well as for the semantic interpretation of the sentence as expressing a negative proposition. If we rule out the option of lexical ambiguities, there is no doubt that the negation marker has the semantics of a negation operator in double negation and negative concord languages alike.

The syntactic and semantic status of the marker *non* in (17a) is strictly the same as that of *non* in (18a). In fact, *non* in (17a) has to have the semantics of a truth-functional operator, otherwise it could not participate in the resumption of a sequence of anti-additive quantifiers (17b). It is just that in the course of the resumption process, the negation contributed by the sentential negation marker is absorbed in the polyadic negative quantifier, so there is no separate contribution of *non* in the first-order truth conditions spelled out in (17c). In the absence of an n-word, no resumptive negative quantifier is built in (18a), so the semantic contribution of *non* is directly reflected in the truth conditions. Thus the claim that negation is semantically redundant is restricted to sentences involving one or more n-words.

In sum, under to the polyadic quantifier analysis, the marker of sentential negation is always semantically potent (it denotes the first-order propositional operator $\neg$). The process of resumption implies that sentential negation is absorbed in the polyadic quantifier, and is semantically redundant in the context of an n-word, because n-words are inherently negative. This analysis predicts that languages are free to exclude the marker of negation from concord constructions (negative spread as observed in spoken French, Occitan, and Flemish cf. 9–12) or include the negation marker in the concord system, and exploit it for syntactic purposes (scope marking in type I and III languages) (examples 1–3 and 14, respectively).

Haspelmath's (1997) classification reflects the typological patterns found in the languages of the world. In this chapter, the polyadic quantifier analysis proposed by de Swart and Sag (2002) is coupled with an OT grammar that accounts for the typological dimension, as an extension of the analysis developed in Chapter 4. Section 3 will show that nonstrict (type III) negative concord languages exploit the constraint NEGFIRST, introduced in Chapter 3 to account for the cross-linguistic preference for preverbal negation. Strict (type I) negative concord languages require the introduction of a new constraint, which will be called MAXSN (Section 4).

5.3 Preverbal/Postverbal Asymmetry

As outlined in Section 1.3, type III languages are characterized by a preverbal/postverbal asymmetry. Since Zanuttini (1991) and Ladusaw (1992), it is well known that n-words in these languages can occur without negation in preverbal position, but need the support of a marker of sentential negation to mark clausal scope when they occur in postverbal position:

(19) a. Mario *(non) ha parlato di niente con nessuno. [Italian]
Mario *(SN) has talked about nothing to nobody
'Mario didn't talk to anyone about anything.'
b. Nessuno (*?non) ha parlato con nessuno.
Nobody (*?SN) has talked with nobody.
'Nobody talked to anyone.'
(20) a. *(No) ha visto a nadie. [Spanish]
*(SN) has seen nobody
'He hasn't seen anybody.'
b. Nadie (*?non) ha dicho nada.
Nobody (*?SN) has said nothing
'Nobody said anything.'
(21) a. Não veio ningém [European Portuguese]
SN came nobody.
'Nobody came.'
b. Ninguém (*não) veio.
Nobody (*SN) came
'Nobody came.'

As these examples indicate, negation must be preverbal, but it does not matter whether it is expressed by a marker of sentential negation (19a, 20a, 21a), or by an n-word (19b, 20b, 21b). When the preverbal negation is expressed by a Neg-expression, a marker of sentential negation is excluded. Insertion of a preverbal marker of sentential negation in combination with a preverbal n-word generally leads to ungrammaticality, and marginally to double negation readings (Zanuttini 1996, Corblin and Tovena 2003, cf. also Chapter 6).

According to Horn (1989: 449), Haspelmath (1997: 211), and Corblin and Tovena (2003: 332), the desire to express negation early in the sentence, called NegFirst, provides the key to the Romance pattern. NEGFIRST has been defined as a violable constraint in Chapter 3 (Section 3).

- NEGFIRST
  Negation precedes the finite verb.

The preverbal expression of negation guarantees clausal scope, so the sentence conveys a negative proposition.

Variants of NEGFIRST are discussed in the literature, e.g. Jespersen (1917, 1933), Dahl (1979), Horn (1989), Haspelmath (1997), Corblin and Tovena (2003), and Corblin et al. (2004). As argued in Chapter 3, the constraint NEGFIRST is functionally motivated by the desire 'to put the negative word or element as early as possible, so as to leave no doubt in the mind of the hearer as to the purport of what is said' (Jespersen 1933: 297 as quoted by Horn 1989: 293, who dubs this principle NEGFIRST).

Languages that place the marker of negation in a preverbal position have a high ranking of NEGFIRST in the OT syntax. NEGFIRST can be satisfied either by a marker of sentential negation or by a negative indefinite. In Chapter 4 (Section 1), I exploited this insight to explain the alternation between (22a) and (b) in English, and the alternation between negative indefinites and negative polarity items in Swedish.

(22) a. We didn't meet anyone.
 b. We met no one.

The claim that Neg-expressions satisfy NEGFIRST proves crucial for nonstrict negative concord languages. Following Zanuttini (1991), Ladusaw (1992), and Corblin and Tovena (2003), I claim that NEGFIRST is operative in Haspelmath's type III languages, where it accounts for the asymmetry between preverbal and postverbal n-words.

The type III languages in Haspelmath's sample are all SVO languages. In the OT analysis, the distinction between preverbal and postverbal n-words correlates with the contrast between clausal or VP scope of negation. In the input meaning, I write clausal negation as ¬∃xV or ¬V∃x. This notation reflects that negation takes scope over the proposition as a whole, but the focus of negation is on an indefinite in preverbal or postverbal position. The high ranking of MAXNEG in the syntax of negative concord languages promotes the multiplication of negative indefinites. The usual constraint ranking MAXNEG >> *NEG for concord languages is complemented with a highly ranked constraint NEGFIRST for languages like Portuguese, Spanish, and Italian.

In sentences with a preverbal n-word, the n-word satisfies NEGFIRST, so the marker of sentential negation is not needed. This is illustrated in Tableau 3.

The first candidate illustrates the high ranking of FNEG as usual. The candidates 'indef SN V' and 'neg V' show that the marker of sentential negation (SN) and the n-word (neg) both satisfy the constraint NEGFIRST. Because of the high ranking of MAXNEG in negative concord languages, the candidate 'indef SN V' is suboptimal, and 'neg V' is preferred. The combination of a Neg-expression with the marker of sentential negation in the last candidate of the tableau is suboptimal, for the negation marker is not necessary to satisfy any of the faithfulness constraints. Therefore, the extra violation of the markedness constraint *NEG is not motivated in any way, and economy makes 'neg V' a better candidate than 'SN negV'.

Tableau 3 illustrates that a grammar with a high ranking of NEGFIRST accounts for the data in (19b), (20b), and (21b). Note further that NEGFIRST and MAXNEG are not in direct competition, so their mutual order is irrelevant (indicated by the dotted line), as long as they are both ranked above *NEG.

Tableau 3 Preverbal n-word without marker of negation in type III languages (Italian, Spanish, etc.) (production)

Meaning ¬∃xV	Form	FNEG	MAXNEG	NEGFIRST	*NEG	INTNEG
	indef V	*	*	*		
	indef SN V		*		*	
☞	neg V				*	
	neg SN V				**	

Tableau 4 Preverbal marker of negation with postverbal n-word in type III languages (Spanish, Italian, etc.) (production)

Meaning ¬V∃x	Form	FNEG	MAXNEG	NEGFIRST	*NEG	INTNEG
	V indef	*	*	*		
	SN V indef		*		*	
	V neg			*	*	
☞	SN V neg				**	

Tableau 5 Preverbal negation with postverbal n-word in type III languages (Spanish, Italian, etc.) (interpretation)

Form SN V neg	Meaning	FNEG	MAXNEG	NEGFIRST	*NEG	INTNEG
	V∃x	*				**
☞	¬V∃x				*	*
	¬V¬∃x				**	

When the indefinite under negation is postverbal, the outcome of the optimization process is different. As (19a), (20a), and (21a) indicate, a preverbal marker of sentential negation is obligatory when the n-word is in postverbal position. The high ranking of NEGFIRST in the syntax is responsible for this outcome, as shown in Tableau 4. The ranking of INTNEG below *NEG in the semantics signals the preference for resumptive polyadic quantification, so this grammar leads to the desired single negation interpretation (Tableau 5).

FNEG and *NEG are double-edged constraints that are operative in the syntax as well as in the semantics. Accordingly, they play a crucial role in the production Tableaux 3 and 4, as well as the interpretation Tableau 5. INTNEG does not play a role in the production Tableaux 3 and 4, because it is a purely interpretive constraint. MAXNEG and NEGFIRST do not play a role in the interpretation Tableau 5, because they are purely syntactic constraints. All the constraints are visible in the three tableaux, so that they show the full bidirectional grammar at work. Tableau 4 indicates that 'V neg' is not an optimal expression of the meaning ¬V∃x, because the postverbal n-word does not satisfy NEGFIRST. The n-word cannot be replaced by a regular indefinite, for the candidate 'SN V indef' does not satisfy MAXNEG. In order to satisfy both NEGFIRST and MAXNEG, the combination of a preverbal marker of sentential negation and a postverbal n-word is called for. Accordingly, 'SN V neg' emerges as the optimal realization of the input meaning.

The comparison of Tableau 4 to Tableau 3 reveals that the preverbal n-word satisfies two constraints (MAXNEG and NEGFIRST), whereas the postverbal n-word satisfies only one constraint (MAXNEG). The marker of sentential negation must be inserted in examples (20a), (21a), and (22a) in order to satisfy the faithfulness constraint NEGFIRST (Tableau 4), whereas it is blocked in (20b), (21b), and (22b) for economy reasons (Tableau 3).

Tableau 5 confirms that the OT semantics does not care how many negations there are in the form. The ranking FNEG >> *NEG >> INTNEG adopted for negative concord languages implies that only one of them is realized in the interpretation. According to the polyadic analysis, the propositional negation is absorbed in the resumptive quantifier without affecting the adicity of the quantifier (cf. Section 2.2). Thus the preverbal/postverbal asymmetry is a syntactic phenomenon.

In line with the claims made by Zanuttini (1991), Ladusaw (1992), and de Swart and Sag (2002), I take preverbal negation to mark the clausal scope of negation. Even though the negative marker is semantically redundant, its co-occurrence is functionally motivated by the desire to express negation 'early' in the sentence. The preverbal position is exploited to guarantee clausal scope of the VP-internal n-word.

In the OT perspective, the presence of the marker of negation has to be functionally motivated, because the n-word is inherently negative, and takes care of the negative force of the sentence. If the negation marker is not necessary to satisfy a particular syntactic faithfulness constraint, it is not used, and negative spread ensues. Tableau 3 shows that the markedness constraint *NEG is at work even in negative concord languages that are generally in favor of multiplication of negative forms in the syntax. OT constraints are minimally violated, and the "best" candidate is the winner (Chapter 2, Section 1). In the absence of a functional motivation for the use of a marker of sentential negation in sentences with a preverbal n-word, it is left out by an appeal to economy.

NEGFIRST is operative in several Romance languages, including Spanish, Italian, Sardinian Portugese (Posner 1984), and old Romanian (Falaus 2007a, b, 2008), but it is not a family trait of Romance (cf. Martins 2001). On the one hand, NEGFIRST is not sufficient to account for formal, written French, modern Romanian, Old Spanish, or Old Portuguese, which are all strict concord languages. On the other hand, NEGFIRST is also operative outside of the family of Romance languages, for instance in New Testament Greek, older varieties of several Slavic languages (which are class I languages in their modern varieties, cf. Haspelmath 1997: 210–212), and certain varieties of (white) vernacular English (Labov 1972).

Haspelmath (1997: 211–213) cites Old Russian examples (23) (from Křížková 1968: 24) and examples from New York city white vernacular English (24) (from Labov 1972: 785–786) in which NEGFIRST is operative.

(23) a. I ne idjaše s nimi nikto že [Old Russian]
and SN went with them nobody PART
'And nobody went with them.'
b. Ničego že sja bojat' běsi, tokmo kresta.
nothing PART self fear demons only cross.
'The demons are afraid of nothing, except the cross.'

(24) a. We don't ever see none of them guys. [NYCWVE]
b. *Nobody don't know.

Note that the pattern in (24) is different from the one found in African-American English (cf. 35). Adger, Wolfram, and Christian (2007) provide a range of data on negation in different varieties of American English, and show that some of them display strict negative concord and others nonstrict negative concord.

From the description in Hoyt (2006), it can be inferred that Palestinian Arabic is also a nonstrict negative concord language (25).

(25) a. ma-ʕind-hæ wɛla nɪtfɪt χağal. [Palestinian Arabic]
SN-at-her not.even bit shame
'She doesn't have the least bit of shame.'
b. wɛla ḥada fi:-hʊm šæ:f-ni.
not.even one.MS in-them saw.3MS-me
'Not even one of them saw me.'
c. wɛla yo:m ma-ʕağabni l-ɛkɪl. [DN]
not.even day SN-pleased.3MS-me the-food
'There wasn't one day the food didn't please me.'

Recall from Chapter 1 that *wɛla*-DPs behave like n-words. The preverbal/postverbal contrast in (25a, b) marks Palestinian Arabic as a nonstrict negative concord language. The combination of the preverbal *wɛla*-DP with the marker of sentential negation in (25c) leads to a double negation reading.

As Zanuttini (1991) and Corblin and Tovena (2003) point out, insertion of a preverbal marker of sentential negation in sentences containing a preverbal n-word, as in (19b), (20b), and (21b), generally leads to ungrammaticality, but marginally a double negation reading seems to be available. These findings are reminiscent of the double negation readings in French combinations of *pas* and an n-word (examples 9b and 10b in Section 2). The grammar developed so far cannot account for these sentences, because it relies on a strong version of bidirectional OT. In Chapter 6 (Section 5), I offer an account of the double negation reading in negative concord languages, which exploits weak bidirectional OT.

In sum, NEGFIRST is operative in a variety of not necessarily related languages. Micro-variation and diachronic patterns suggest that languages alternate between type I and type III grammars. Given that many strict negative concord languages have a preverbal marker of sentential negation, there are strong correlations between NEGFIRST and the constraint MAXSN that will be introduced in Section 4, which render such alternations possible.

5.4 Obligatory Marker of Sentential Negation

Type I languages are strict negative concord languages, which require the presence of a marker of sentential negation in all negative sentences. Polish, Romanian, and Hungarian are strict negative concord languages, as illustrated in (1)–(3) in Section 1.3. Examples from Greek (26) (Giannakidou 1997, 1998), Japanese (27) (Watanabe 2004), Russian (28) (Zeijlstra 2004), and Serbo-Croatian (29) (Progovac 1993, 1994) are added here.[1]

---

[1]Watanabe (2004) glosses *nani-mo* as 'what-mo', because it consists of a *wh*-item followed by the focus particle *mo*. In line with the way n-words are written in this book, I gloss it as 'nothing'.

(26) a. *(Dhen) ipa TIPOTA. [Greek]
*(SN) said.1SG nothing.
'I didn't say anything.'
b. KANENAS *(dhen) idhe ti Roxani.
Nobody *(SN) saw.3SG the Roxanne.
'Nobody saw Roxanne.'

(27) a. John-wa nani-mo tabe- *(nak)-atta. [Japanese]
John-TOP nothing eat- *(SN)- PAST
'John didn't eat anything.'
b. Dare-mo monku-o iwa- *(nak)-atta.
nobody complaint-ACC say- *(SN)- PAST
'Nobody complained.'

(28) a. Nichego *(ne) rabotaet [Russian]
Nothing *(SN) works.
'Nothing works.'
b. *(Ne) rabotaet nichego
*(SN) works nothing
'Nothing works.'

(29) a. Milan *(ne) voli niko-ga [Serbo-Croatian]
Milan *(SN) loves no one-ACC
'Milan doesn't love anyone.'
b. Milan nikada *(ne) vozi
Milan never *(SN) drive.3SG
'Milan never drove.'

Just like type III languages, type I languages require a marker of sentential negation with a postverbal n-word (26a–29a). Unlike type III languages, type I languages also require such a marker when the sentence contains a preverbal n-word (26b–29b).

A high ranking of NEGFIRST would be sufficient to account for the configuration in (26a–29a), but it would not account for sentences like (26b–29b). The new constraint that accounts for the obligatory presence of the marker of sentential negation in all sentences containing an n-word is formulated as MAXSN.

- MAXSN
  negative clause (i.e. a clause that conveys a negative proposition) must bear a marker of sentential negation.

MAXSN is a faithfulness constraint that is functionally motivated by the desire to realize clausal negation by means of a sentential operator. MAXSN is inspired by Richter and Sailer's (2006) 'Neg Criterion'.[2] Richter and Sailer (2006) work in the constraint-based framework of HPSG. They posit the Neg Criterion as a language-specific

[2]Richter and Sailer (2006) call their principle the Neg Criterion, due to its similarity in effect to the Neg Criterion of Haegeman and Zanuttini (1996). Given the differences in theoretical framework between Haegeman and Zanuttini's generative syntax and Richter and Sailer's HPSG formulation, the details of the formulation are not the same.

constraint operative in Polish (and presumably other strict negative concord languages). In their formulation, the Neg Criterion requires the marker of sentential negation to be part of the sentence when the logical form of the sentence conveys propositional negation. I generalize this idea here to a universal constraint, which is, however, soft, so its force depends on the position of MAXSN in the OT ranking of constraints.

The negation marker in type I languages signals that negation ranges over the proposition as a whole, and the verb is in the scope of negation. The close connection between the verb and sentence negation is expected if Aristotle's and Jesperson's view of negation as predicate denial is adopted (Horn 1989). The behavior of the marker of sentential negation in raising contexts supports the view that it operates as a scope marker (cf. Section 6 for French, and Chapter 6, Section 2.3 for other strict negative concord languages).

The marker of sentential negation co-occurs with the n-word, which marks the focus of negation by negative attraction to the indefinite. The high ranking of MAXSN in negative concord languages captures Haspelmath's (1997) intuition that class I languages are functionally motivated, because they mark both the scope and the focus of negation, and mark them by means of separate expressions (SN and n-words respectively). MAXSN is operative in languages if the grammar ranks it above *NEG. In type I and II languages, this constraint is ranked below *NEG. The addition of MAXSN below *NEG to Tableaux 1, 3, and 4 would not change the optimization process.

Many strict negative concord languages have a preverbal marker of sentential negation. This suggests that NEGFIRST is also ranked high in these languages. The insertion of a negation marker in sentences with postverbal and preverbal n-words in a grammar with the ranking {NEGFIRST, MAXSN} >> *NEG is spelled out in Tableaux 6 and 7.

Tableau 6 Strict negative concord with preverbal negation and a postverbal n-word (Greek, Polish, Russian, Romanian, ...) (production)

Meaning $\neg V\exists x$	Form	FNEG	MAXNEG	MAXSN	NEGFIRST	*NEG
	V indef	*	*	*	*	
	SN V indef		*			*
	V neg			*	*	*
☞	SN V neg					**

Tableau 7 Strict negative concord with preverbal n-word and preverbal negation marker (Greek, Polish, Russian, Romanian, ...) (production)

Meaning $\neg\exists xV$	Form	FNEG	MAXNEG	MAXSN	NEGFIRST	*NEG
	indef V	*	*	*	*	
	indef SN V		*			*
	neg V			*		*
☞	neg SN V					**

Candidates with purely existential forms are ruled out by the universally high ranking of FNEG. Given that both MAXSN and MAXNEG are ranked above *NEG, the sentence needs to contain an n-word as well as a marker of sentential negation in order to express a single negation reading, whether the input contains a negative variable in preverbal or postverbal position. MAXSN and MAXNEG are not in direct competition, so their mutual ranking in Tableaux 6 and 7 is irrelevant, as indicated by the dotted line. It suffices that MAXSN and MAXNEG are both ranked higher than *NEG.

Unlike NEGFIRST, MAXSN is not sensitive to the position of the marker of sentential negation: it can be preverbal or postverbal, as long as it is there. In Type I languages with a preverbal marker of sentential negation, NEGFIRST and MAXSN are both ranked above *NEG, but they each play their own role. Both a negation marker and a preverbal n-word satisfy NEGFIRST (cf. Tableaux 3 and 4), so both can be used to mark clausal scope of negation in nonstrict NC languages. In strict NC languages, the marker of sentential negation is exclusively used to mark clausal scope as propositional negation, for n-words do not satisfy MAXSN (cf. Tableaux 6 and 7). In type I languages, the n-word is licensed by MAXNEG, just like in any concord language, but MAXSN can only be satisfied by the presence of a negation marker.

With a preverbal marker of sentential negation, alternations between type III and type I languages are frequent. In a synchronic perspective, alternations between strict and nonstrict negative concord are found in the family of Romance languages (compare Italian and Spanish to Romanian, as well as the varieties of American English discussed in Adger, Wolfram, and Christian 2007). In a diachronic perspective, transitions occur from type III to type I languages, or vice versa. Old Spanish was a type strict NC language, but its modern variant is nonstrict. Old Romanian was a nonstrict NC language (30a), but the modern variant exemplifies strict NC, as shown by Falaus (2008) (2a, repeated as 30b).

(30) a. Nimea are a sedea de-a dereapta. [Old Romanian]
nobody have.3SG to sit of right
'Nobody will sit on the right (side).'
b. Nimeni *(nu) a venit. [Modern Romanian]
Nobody *(SN) has come.
'Nobody came'

The change from a strict to a nonstrict negative concord language is currently taking place in Catalan, as argued in Section 5. The patterns indicate that this type of diachronic change is not unidirectional (A. Jäger 2008).

In languages that have a preverbal marker of sentential negation, the constraint MAXSN can easily be weakened to NEGFIRST in such languages (leading to a change from strict to nonstrict negative concord), or NEGFIRST can be strengthened to MAXSN (leading to a change from nonstrict to strict negative concord). Re-ranking of constraints is the appropriate mechanism to use in both typological and diachronic variation (cf. Chapter 2), so these patterns fit the OT analysis of negation.

In Tableaux 6 and 7, I did not consider any candidate that did not realize the marker of negation preverbally, because strict negative concord was illustrated in (26–29) with languages that have a preverbal marker of sentential negation.

However, MaxSN cannot generally be reduced to a version of NegFirst. The constraints are independent in NC languages that satisfy the constraint MaxSN with a postverbal marker of SN such as Afrikaans and certain Flemish dialects.

Afrikaans exemplifies negative doubling, with an obligatory postverbal negation marker *nie* in every sentence that contains an n-word (31a, b). Similar observations have been made for certain Flemish dialects (31c) (from Van der Auwera and de Vogelaer 2008).

(31) a. Jan het gehoop dat *niks* met hom sou gebeur *nie*. [Afrikaans]
Jan has hoped that nothing with him would happen SN
'Jan hoped that nothing would happen to him.'
b. Sy hou *nooit* op met werk *nie*.
She holds never up with work SN
'She never stops working.'
c. Ik heb Jan *niet* gezien *niet*. [Flemish, Aarschot dialect]
I have Jan SN seen SN
'I have not seen Jan.'

The sentence-final marker *nie* in Afrikaans is licensed by a constraint that favors a position late in the sentence (FocusLast, cf. Chapter 3). This means that there is no interaction between NegFirst and MaxSN in this strict negative concord language.

Typologically speaking, there are fewer type I languages that behave like Afrikaans than that behave like Slavic, Greek, etc. Probably, the typological distribution finds its source in the pervasive influence of NegFirst in natural language (cf. Chapter 3, Section 3). However, this does not make the situation in strict negative concord languages with a postverbal negation marker any less stable. Compare Biberauer (2006, 2007) and references therein for more on the status of sentence-final *nie*, and Section 10 for some remarks on FocusLast negative spread and negative doubling in Afrikaans.

Note that the high ranking of the constraint MaxSN favors candidates that incur additional violations of *Neg. Given that the ranking *Neg >> IntNeg favors resumptive negative quantification, the proliferation of the negations does not have any consequences for the semantics. Tableau 8 illustrates this for the interpretation of a sentence with a preverbal n-word in a type I language.

As Tableau 8 illustrates, it does not matter how many negative forms the sentence contains in a strict negative concord language. Under the ranking FNeg >> *Neg >> IntNeg in the OT semantics, they all get absorbed into one resumptive negative quantifier. The marker of sentential negation is required to satisfy MaxSN

Tableau 8 Preverbal negation with preverbal n-word in type I languages (Greek, Slavic, Hungarian, etc.) (interpretation)

Form neg SN V	Meaning	FNeg	MaxNeg	MaxSN	*Neg	IntNeg
	$\exists x V$	*				**
☞	$\neg\exists x V$				*	*
	$\neg\exists x \neg V$				**	

in the OT syntax, and thereby mark propositional negation, but it is semantically redundant in a context which contains one or more Neg-expressions.

I conclude that both type I and type III languages support the view put forward by de Swart and Sag (2002) that the marker of sentential negation, though inherently negative, does not affect the semantics of negative concord (Section 2.2). As a result, languages can use the negation marker for different syntactic purposes. Whether or not a concord language has a (pre)verbal marker of sentential negation depends on functional considerations of clausal scope marking, which give rise to constraints like NEGFIRST or MAXSN.

5.5 Mixed Cases

Catalan and Brazilian Portuguese exemplify a mixture of class I and class III properties (Ladusaw 1992, Vallduví 1994, Zanuttini 1996, Zeijlstra 2004):

(32) a. En Pere *(*no*) ha fet *res*. [Catalan]
The Peter *(SN) has done *nothing*.
b. *Ningú* (*no*) ha vist en Joan.
Nobody (SN) has seen John.

(33) a. *(*Não*) vi *ninguem*. [Br. Portuguese]
*(SN) saw nobody.
'I didn't see anybody.'
b. *Ninguem* (*não*) disse *nada*.
Nobody (SN) said nothing.
'Nobody said anything.'

As (32) and (33) indicate, a preverbal marker of sentential negation is required with a postverbal n-word. With a preverbal n-word, the negation marker is optional. This indicates that NEGFIRST is ranked above *NEG, but MAXSN somehow interacts with *NEG. The ranking MAXNEG >> NEGFIRST >> {MAXSN, *NEG} accounts for these data. Suppose that MAXSN and *NEG are ranked equally high (in ordinal OT) or have a strongly overlapping range (in stochastic OT, cf. Chapter 3, Section 4), but NEGFIRST is ranked above both.

The high ranking of NEGFIRST generates a preverbal marker of sentential negation with postverbal n-words, just like in a type III language (32a, 33a, Tableau 9). With preverbal n-words (32b, 33b), NEGFIRST is satisfied by the n-word, so the competition

Tableau 9 Generation of Catalan/Brazilian Portuguese with postverbal n-word

Meaning ¬V∃x	Form	MAXNEG	NEGFIRST	MAXSN	*NEG	INTNEG
	V neg		*	*	*	
☞	SN V neg				**	

Tableau 10 Generation of Catalan/Brazilian Portuguese with preverbal n-word

Meaning $\neg\exists x_1 V \exists x_2$	Form	MAXNEG	NEGFIRST	MAXSN	*NEG	INTNEG
☞	neg V neg			*	**	
☞	neg SN V neg				***	

between MAXSN and *NEG is decisive. If MAXSN and *NEG are ranked equally high, the grammar generates two optimal outputs (Tableau 10).

The difference between preverbal and postverbal n-words is accounted for by the high ranking of NEGFIRST. However, Catalan is not a full type III language, because MAXSN is not ranked (strictly) below *NEG. It shares features with type I languages in allowing rankings in which MAXSN wins over *NEG. Thus a marker of sentential negation optionally shows up in outputs for the expression of preverbal n-words.

As pointed out by Vallduví (1994), the optionality of a preverbal marker of sentential negation in combination with a preverbal n-word does not have a semantic effect. In the OT semantics, *NEG is ranked above INTNEG, so all negative meanings of the individual Neg-expressions and the negation marker are collapsed into the single negation of the resumptive quantifier (cf. Tableaux 5 and 8).

Vallduví does not discuss the factors that differentiate the two varieties of Catalan. Maria Teresa Espinal (2005, personal communication) suggests that the older generation typically uses the type I variety, and the younger generation uses the type III variety. She relates this to the influence of Spanish. If her observations are correct, there are actually two grammars of Catalan, one that ranks MAXSN >> *NEG (capturing the language of the older generation), and one that ranks *NEG >> MAXSN (for the younger generation). Both are negative concord languages with the ranking *NEG >> INTNEG in the semantics, so the sentences (32a, b) with and without a marker of sentential negation are interpreted as conveying single negation under both grammars. Under this analysis, Catalan is moving away from a type I language toward a Spanish-like type III language. As argued in Section 4, negative concord languages with a preverbal negation marker easily permit these diachronic patterns.

Optional negation markers with preverbal n-words are also attested in Basque Spanish, the Spanish spoken by Basque bilinguals (34a, from Franco and Landa 2006), and Corrientes (Argentine) Spanish (Cuervo and Mazzaro 2008).

(34) a. Aqui nadie no sabe sobre eso [Basque Spanish]
here nobody SN knows about that
'Here nobody knows about that.'
b. Nunca no nos pasó nada.
Never SN us happened nothing [Corrientes Spanish]
'Nothing ever happened to us.'

Haspelmath (1997: 211, 213) observes that the Catalan pattern is also found in Old Church Slavonic, and in several (mostly African-American) dialects of English. Haspelmath quotes the following examples from Labov (1972: 785–786), which indicates that the negation marker is optionally present with n-words in subject position in African American English (AAE).

(35) a. *Nobody* do*n't* know where it's at. [AAE]
b. *Nobody* fights fair.

Under the assumption that there is free variation in AAE between the constructions in (35a) and (35b), this language displays a mixed case as well.

Berber also displays features of a type I as well as a type III language. Ouali (2003, 2005) observes that postverbal n-words require the presence of the preverbal negation *ur*, whereas with preverbal n-words, *ur* is optional in some varieties. Ouali discusses different Berber dialects, with slightly different restrictions. Example (36) illustrates the combination of the negation marker and n-words in Tamazight Berber.

(36) a. *ur* as-wshi.x *walu* [Tamazight Berber]
SN him-give.PER.3SG nothing
'I did not give him anything.'
b. *urdgin* (*ur*) dix gher Frans
never (SN) go.PERF.1SG to France
'I have never been to France.'

It is hard to account for such mixed systems in a linguistic theory that locates the distinction between strict and nonstrict negative concord in the lexicon. Such an approach can postulate a lexical ambiguity of the marker of sentential negation between a negative and a non-negative meaning (Zeijlstra 2004, Bošković 2008) or posit lexical ambiguities for n-words (Giannakidou 2006), but that does not really predict when the marker of negation is showing up and why. In the OT analysis, no unmotivated lexical ambiguities of n-words need to be postulated, and the marker of sentential negation is always interpreted as contributing a semantic negation. When languages have a negation system in between a type I and a type III language, the OT system can model this as a balanced system that reflects the interaction of NEGFIRST, MAXSN, MAXNEG, and *NEG.

5.6 Two Varieties of French

The (modern, continental) French language occupies an intermediate position in the typology. As far as the negation system is concerned, two varieties should be distinguished. In the formal (written) language, all negative sentences contain the preverbal marker *ne*, even those that contain the postverbal marker of sentential negation *pas*, so we are dealing with a type I (strict negative concord) language displaying discontinuous negation. In the spoken language, *ne* is generally dropped, and sequences of n-words display negative spread (type II negative concord). Section 6.1 provides an overview of the differences between written and spoken French, as far as their negation system is concerned. Section 6.2 provides an analysis of the co-occurrence restrictions between negation and negative indefinites in both varieties. This leads to a revision of the analysis of discontinuous negation proposed in Chapter 3 in Section 6.3.

5.6.1 Negation in Written and Spoken French

In formal (written) French, all negative sentences contain the preverbal marker *ne*, even those that contain the postverbal marker of sentential negation *pas* (37a). Multiple n-words allow an interpretation in terms of negative concord (37d). Sentences in which negation is conveyed by an n-word do not contain *pas*. The combination of *pas* with an n-word leads to a double negation reading (37e). Because of the obligatory presence of *ne*, this variety of French is a clear instance of a type I language.

(37) a. Il n'est *(pas) venu. [written French]
He SN is *(SN) come.
'He didn't come.'
b. Il n'a rien vu.
He SN has nothing seen.
'He hasn't seen anything.'
c. Personne n'est venu.
Nobody SN is come.
'Nobody came.'
d. Personne n'a rien dit.
Nobody SN has nothing said.
'Nobody said anything.'
e. Il n'est pas venu pour rien.
He SN is SN come for nothing.
'He has not come for nothing.' [DN]

In spoken French, the clitic *ne* is frequently dropped, even in the higher registers (Ashby 1981, 2001). The postverbal marker *pas* or a single n-word is sufficient to make the sentence express a negative statement (38a-c). A sequence of n-words can express negative concord (38d) (negative spread). The combination of *pas* with an n-word conveys double negation (38e), just as in the written variety (37e). Because of the incompatibility of *pas* with an n-word under the single negation reading, spoken French qualifies as a type II negative concord language (cf. Section 1.2).

(38) a. Il vient pas. [spoken French]
He comes SN.
'He doesn't come.'
b. J'ai rien dit
I have nothing said.
'I didn't say anything.'
c. Personne a dit ça.
Nobody has said that
'Nobody said that.'
d. J'ai rien dit à personne.
I have nothing said to nobody.
'I haven't said anything to anyone.'

e. Il est pas venu pour rien.
He is SN come for nothing.

The main empirical difference between written (37) and spoken French (38) involves the use of the preverbal clitic *ne*. The standard view, defended by Bréal (1897, 1900), Clarke (1904), Tesnière (1959), Gaatone (1971), Horn (1989), Godard (2004) and others is that *pas, personne, rien*, etc. are the bearers of sentential negation in the modern language. *Ne* is perceived as semantically non-negative, and labeled as a co-negative marker (Corblin and Tovena 2003, cf. also the quote from Tesnière 1957 in Chapter 3, Section 4). The view of *ne* as a scope marker is visible in raising contexts such as (39) (from Kayne 1984):

(39) a. Paul accepte de ne recevoir personne.
Paul agrees to SN receive no one
'Paul agrees to not see anybody.'
b. Paul n'accepte de recevoir personne.
Paul SN agrees to receive no one.
'Pal does not agree to see anybody.'

Raising verbs allow a 'low' and a 'high' construction of negation with infinitival complements and sometimes with subjunctive complements as well. As Kayne (1984) and Godard (2004) point out, the interpretation of the sentence depends on whether the negative indefinite is in the same verbal domain as the *ne*-verb of not. In (39a), Paul agrees that it is not the case that there is someone whom he will receive, whereas in (39b) it is not the case that there is someone that Paul agrees to see. The difference in the position of *ne* implies that the negation is interpreted downstairs in (39a), and upstairs in (39b). So even though *ne* does not carry semantic negation, it determines the scopal domain of negation. Other negative concord languages display a comparable behaviour in raising contexts, compare Chapter 6, Section 2.3.

5.6.2 Co-occurrence Restrictions Between Negation and Negative Indefinites

The development of French negation over time is often taken as the prime illustration of the Jespersen cycle (cf. Chapter 3, Section 4). The diachronic change from 'Jeo ne dis' (old French) via 'Je ne dis pas' (written French) into 'Je dis pas' (modern spoken French in France) is a perfect fit to the three main stages of the Jespersen cycle. The original preverbal marker of negation (*ne*) is first reinforced by a postverbal marker of negation (*pas*), leading to the discontinuous negation *ne..pas*. When *ne* looses its negative force, only the postverbal marker *pas* remains.

Chapter 3 (Section 3) proposed two constraints that govern the position of the marker of sentential negation: NEGFIRST and FOCUSLAST. As far as the expression of propositional negation is concerned, NEGFIRST favors a preverbal marker of negation (Romance, Slavic, etc.) and FOCUSLAST a postverbal one (Germanic, Afrikaans, spoken

Tableau 11 Discontinuous negation in written French (first version, cf. Chapter 3)

Meaning ¬Come(x)	Form	NegFirst	FocusLast	*Neg
	Il ne vient		*	*
	Il vient pas	*		*
☞	Il ne vient pas			**

Tableau 12 Postverbal marker of negation in spoken French (first version, cf. Chapter 3)

Meaning ¬Come(x)	Form	FocusLast	*Neg	NegFirst
	Il ne vient	*	*	
☞	Il vient pas		*	*
	Il ne vient pas		**	

French). If NegFirst and FocusLast are both ranked above *Neg, a discontinuous marker of negation emerges as the optimal outcome (Old English, informal Welsh, written French, etc.). Tableaux 11 and 12 spell out the placement of negation in the two varieties of French according to the analysis proposed in Chapter 3.

Under this view, the two grammars of spoken French and written French reflect the second and third stages of the Jespersen cycle (discontinuous and postverbal negation). I want to argue that the expression of propositional negation is more complex than that reflected in Tableaux 11 and 12. The evidence will come from the co-occurrence restrictions between the marker of sentential negation and negative indefinites discussed in this section. A revised analysis of discontinuous negation in written French is offered in Section 6.3.

Note that the interaction of negation with preverbal n-words in (37b-d) indicates that *ne* is licensed to satisfy MaxSN, as well as NegFirst. Written French is a type I language, and not a type III language, because it does not display a preverbal/postverbal asymmetry. Of course, most strict negative concord languages involve a preverbal marker of sentential negation, as observed in Section 4. But the ranking of MaxSN above *Neg actually allows NegFirst to be below *Neg (rather than above it as in Tableau 11), and still have an effect on the placement of negation (cf. Tableaux 13–16).

Note also that there is a crucial difference between between *ne* and *pas*. *Ne* is compatible with n-words, and its presence is required in the written French sentences (37b-e). This is an indication that MaxSN is at work in this variety of the language. But *pas* does not combine with n-words in the expression of a single negation reading, in either written French (37b-d) or spoken French (38b-d). The fact that *pas* is not used with preverbal n-words (37c, 38c) indicates that *pas* does not satisfy MaxSN, and that FocusLast is ranked too low to play a role in sentences involving n-words in French. This suggests that FocusLast is in fact ranked below *Neg, rather than above it, as in Tableaux 11 and 12. These observations suggest the ranking {MaxSN, MaxNeg} >> *Neg >> NegFirst >> FocusLast for written French.

Tableau 13 Generation of sentences with preverbal n-word in written French

Meaning ¬∃xV	Form	MaxSN	MaxNeg	*Neg	NegFirst	FocusLast
	neg V	*		*		*
☞	neg *ne* V			**		*
	neg V *pas*	*		**		
	neg *ne* V *pas*			***		

Tableau 14 Generation of sentences with preverbal n-word in spoken French

Meaning ¬∃xV	Form	MaxNeg	*Neg	MaxSN	FocusLast	NegFirst
☞	neg V		*	*	*	
	neg *ne* V		**		*	
	neg V *pas*		**	*		
	neg *ne* V *pas*		***			

Tableau 15 Generation of sentences with postverbal n-words in written French

Meaning ¬V∃x	Form	MaxSN	MaxNeg	*Neg	NegFirst	FocusLast
	V neg	*		*	*	
☞	*ne* V neg			**		
	V *pas* neg	*		**	*	
	ne V *pas* neg			***		

Tableau 16 Generation of sentences with postverbal n-words in spoken French

Meaning ¬V∃x	Form	MaxNeg	*Neg	MaxSN	FocusLast	NegFirst
	ne V neg		**			
	ne V *pas* neg		***			
☞	V neg		*	*		*
	V *pas* neg		**	*		*

Tableaux 13 and 14 illustrate how (37c) and (38c) emerge as the optimal form for the expression of a single negation reading with a preverbal negative indefinite. Tableaux 15 and 16 do the same for postverbal n-words (as in 37b, 38b).

In order to simplify the presentation, Tableaux 13–16 only consider candidates that satisfy FNeg and MaxNeg. Preverbal *ne* satisfies MaxSN as well as NegFirst, but *pas* does not come into play to satisfy MaxSN.

The insertion of *pas* in a sentence that already contains an n-word obligatorily leads to a double negation reading in both varieties of the language (37e, 38e). Chapter 6 (Section 10) will provide an account of the double negation reading of sentences like (37e) and (38e) in a weak version of bidirectional OT. Here, I concentrate on the single negation readings in (37a–d) and (38a–d).

French *pas* is not the only postverbal marker of negation which displays this behavior. Zanuttini (1994: 433–434, 441–442) and Espinal (2000) provide examples from other Romance varieties that do not require the co-occurrence of a negative marker with postverbal n-words.

(40) a. Maria a mangia nen. [Piedmontese]
Maria CL_{sub} eats SN
'Maria doesn't eat.'
b. I l'hai vist gnun.
I have seen nobody
'I haven't seen anybody.'
c. El l'ha mina scrivuu. [Milanese]
he has SN written
'He hasn't written.'
d. L'ha mangiaa niént.
he has eaten nothing.
'He hasn't eaten anything.'

(41) a. Jo he vist ningú. [Roussillon Catalan]
I have seen nobody
'I saw nobody.'
b. He menjat res
Have eaten nothing
'I ate nothing.'

In principle, there is no reason why a postverbal marker of negation could not satisfy MAXSN. Afrikaans *nie* has been argued to be an instance of a postverbal negation marker that satisfies MAXSN (example 30 in Section 4, compare also the examples from certain Flemish dialects in Section 9). In standard modern French, *pas* is not used to satisfy MAXSN, and neither are the postverbal negation markers in the Romance languages exemplified in (40) and (41). However, examples (42)–(46) illustrate that postverbal negation markers in some Romance varieties satisfy MAXSN.

Zanuttini (1991: 185 sqq) attempts to relate the observations about French to the status of *pas* as a specifier, rather than a head, but her arguments have been refuted by Déprez (1997a, b). As an alternative, Déprez (1997a, b, 2000) proposes an analysis in terms of the nature of the n-words involved, and Vinet (1998) assigns distinct lexical features to the negator *pas* in different varieties of French. Neither Zanuttini's analysis nor Déprez's proposal is compatible with the OT analysis developed so far, but we can work out a version of Vinet (1998).

What the contrast between *ne* and *pas* in written French tells us is that there may be lexical markers that can be used as a scope marker (and satisfy MAXSN) (*ne*), and others that don't (*pas*). I take this to be an issue that is lexically decided in the distribution of labor between *ne* and *pas*. This distribution may very well be due to the historical origin of *ne* as a negation marker, and of *pas* as a reinforcement of negation. But as far as the synchronic grammar of modern French is concerned, all we have is a lexical difference between the two markers, which interacts with the grammar, as reflected in the violation pattern in Tableaux 13–16.

Independent evidence in favor of a shallow, lexical difference between the 'old' preverbal and the 'new' postverbal marker comes from Romance languages in which a postverbal marker of negation does satisfy MaxSN. In older stages of French, it was in fact possible to combine *pas* with n-words. An example from Molière is cited in Bernini and Ramat (1996: 174), and repeated here as (42a). This situation survives in certain dialects of continental French (42b, from Muller 1991), and in Québec French (42c, from Muller 1991). An Internet search (restricted to the .fr domain) revealed examples of the type illustrated in (42d), suggesting possibly systematic violations of the rule against the combination of *pas* with n-words. I leave the question open whether such examples indicate innovation or a return to older stages of the language that survived in certain dialects. I am interested in them as an instance of micro-variation within modern French.

(42) a. Ne faites pas semblant de rien. [17th century French]
SN make.IMP SN semblance of nothing.
'Do not fake anything.'
b. S'il y a quelque chose, il fera pas d'cadeau à personne. [Frenchdialect]
If there is something, he give.FUT SN of present to nobody.
'If there is a problem, he will not grant anyone a favor.'
c. Il y a pas personne en ville. [Québécois]
There is SN nobody in town.
'There is no one in town.'
d. Enfin, j avais pas de pb avec personne. [internet French]
Anyway, I had SN of problems with nobody
'Anyway, I didn't have problems with anybody.'

The examples cited involve a postverbal n-word. Since both the n-word and the negation marker are postverbal, the co-occurrence with the marker of sentential negation cannot be explained as an effect of NegFirst. So the presence of *pas* in these sentences must be licensed by MaxSN. The absence of *ne* in (42b–d) supports this view, but the presence of both *ne* and *pas* in (42a) requires further analysis (cf. Section 9).

Déprez (2000: 259) and Martineau and Déprez (2004) strengthen the evidence, and show that the postverbal negation marker (*pa* or *pas*) is obligatory with both preverbal and postverbal n-words in Louisiana French Creole and Martinique Creole.

(43) a. Mo te pa wa pe(r)son. [Louisiana French Creole]
I have SN seen nobody.
'I did not see anybody.'
b. A(r)jen gruj pas.
Nothing moves SN.
'Nothing moves.'
c. Personn *(pa) wè anyen. [Martinique Creole]
No one *(SN) saw nothing
'Nobody saw anything.'

Zanuttini (1991: 181–182) provides the examples in (44) and (45), which illustrate the situation in Lengadocian varieties of Occitan and in Valdotain.

(44) a. Degun es pas vengut. [Occitan]
Nobody is SN come.
'Nobody has come.'
b. Vendrà pas jamai.
Come.FUT.3SG SN never.
'He will never come.'

(45) a. Dze si ren. [Valdotain]
I know nothing
'I don't know anything.'
b. L'est pas ren.
It is SN nothing.
'It is nothing.'

Schwenter (2005) points out that postverbal *não* can optionally double n-words in Brazilian Portuguese, as illustrated in (46).

(46) Q: Nossa! Reúne todo mundo assim em festa de natal? [Br. Portuguese]
'Wow! Everyone gets together like that for Christmas parties?'
A: Lá ninguém dá festa não. Lá é muito difícil ter festa.
there no one gives parties SN. There is very difficult to party
Ninguém lá gosta de festa. Festa só na casea dos outros.
No one there likes of parties. Parties are in the houses of the others
'There, no one gives parties. It's very difficult to have parties there.'
'Nobody there enjoys parties. Parties only happen in someone else's house.'

In the French varieties exemplified in (42), in Valdotain (45), and in Brazilian Portuguese (46), the postverbal marker of negation is optional, but in Louisiana French Creole (43) and Occitan (44) it is obligatory. The data support the claim that these languages have a high ranked constraint MAXSN, similar to the grammar sketched for written French in Tableau 12, and that it is the postverbal negative marker that satisfies MAXSN. The optionality in Québecois, Valdotain, and Brazilian Portuguese can be explained as the result of an overlapping range of the constraints *NEG and MAXSN. If *NEG and MAXSN are ranked equally high, or overlap in range, the output oscillates between the presence and the absence of the negation marker, without any changes in meaning (compare the treatment of Catalan and Brazilian Portuguese preverbal negation markers in Section 5).

The claim that *ne* satisfies MAXSN in Tableaux 13–16, but *pas* does not provide an immediate explanation for the absence of *pas* from sentences containing an n-word in both written French (37b–d) and spoken French (38b–d). In spoken French (Tableaux 14 and 16), only MAXNEG is ranked above *NEG. This means that multiplication of n-words is functionally motivated, but of course, MAXNEG does not target the marker of sentential negation. Under the OT analysis, there is no semantic motivation for a negation marker in sentences containing an n-word, for the n-word conveys semantic negation and satisfies FNEG. Tableau 14 illustrates that the candidate 'neg V pas' incurs an extra violation of *NEG compared to 'neg V'. The candidate 'neg V' is better, because it avoids this extra violation, which is unmotivated under the ranking at hand. Thus, *pas* is left out of sentences involving one or more Neg-expressions for reasons of economy.

According to the grammar of written French (Tableaux 13 and 15), the presence of *ne* is motivated by the functional need to satisfy MAXSN. However, the lexicon determines that *pas* does not satisfy this constraint in modern written French, although it may in other varieties of the language. The candidate 'neg *ne* V *pas*' in Tableau 13 incurs an extra violation compared to 'neg *ne* V'. For economy reasons then, *pas* is blocked in sentences containing Neg-expressions in both written and spoken French.

5.6.3 Discontinuous Negation Revisited

The fact that *pas* does not occur in sentences containing n-words underscores the relevance of the markedness constraint *NEG as an economy constraint. From a semantic point of view, it does not matter how many negations the sentence contains, for they are all absorbed in the negative polyadic quantifier under the ranking *NEG >> INTNEG. Even in negative concord languages, though, the plurification of negative forms is not random, but needs to be motivated by functional considerations. In languages that have a high ranking of *NEGFIRST or MAXSN, the marker of sentential negation only shows up in sentences containing n-words to mark clausal scope. And in languages with discontinuous negation, only the marker that satisfies MAXSN combines with n-words: in modern written French, this is *ne*, but not *pas*.

As argued in Section 6.2, I follow the literature in analysing *ne* as a co-negative marker that functions as a scope marker in the syntax, but is semantically non-negative, rather than the bearer of semantic negation in modern French. I implement this view as the failure of *ne* to satisfy FNEG in OT syntax and semantics.

In Tableaux 13–16, the constraint FNEG was left out to simplify the presentation. N-words are inherently negative, so all Neg-expressions satisfy FNEG, and no candidate in Tableaux 13–16 violates this constraint. However, the addition of FNEG to the tableaux governing the expression of propositional negation shows how discontinuous negation can be brought in line with the constraint ranking in Tableaux 13–16. In the grammar {FNEG, MAXSN} >> MAXNEG >> *NEG, the clitic *ne* is licensed by the high position of MAXSN in the syntax. But the status of *ne* as a co-negative marker not bearing semantic negation makes it impossible to have *ne* stand alone as the marker of propositional negation. *Pas* is the bearer of semantic negation, and thus the marker that satisfies FNEG in the discontinuous negation *ne..pas*.

Discontinuous negation then arises from the collaboration between a syntactic and a semantic negative marker. Instead of Tableaux 11 and 12, I adopt Tableaux 17 and 18 as the final account for the expression of propositional negation in the two varieties of French.

Ne is a negative form, so it incurs a violation of *NEG. *Ne* does not convey a negative meaning in modern French anymore, and its status as a co-negative translates as a violation of FNEG. In accordance with this view, the candidate form 'il ne vient' violates FNEG, just as 'il vient' does in both tableaux. *Pas* is used as a last resort strategy: it is inserted to express semantic negation in those contexts in which no Neg-expression is available to convey a negative proposition.

Ne satisfies MAXSN, which is relevant in Tableau 17, with the ranking MAXSN >> *NEG. In spoken French, the demotion of MAXSN implies that there is no more functional motivation for *ne*. The ranking *NEG >> MAXSN implies that *ne* is left out in this variety of French for economy reasons (Tableau 18).

In written French, a postverbal rather than a preverbal position of *pas* is optimal to satisfy FOCUSLAST, in a situation where NEGFIRST has already been satisfied by the preverbal clitic *ne* (Tableau 17). FOCUSLAST plays a role in the strengthening of negation (cf. Chapter 3, Section 2). In spoken French, the demotion of MAXSN is accompanied by a demotion of NEGFIRST (Tableau 18), maintaining the postverbal placement of *pas*.

Four important insights emerge from the discussion of French negation. First, the presentation of the Jespersen cycle in Chapter 3 captured the competition in the placement of the marker of sentential negation in sentences expressing propositional negation, but did not consider the more complex interaction with n-words. In fact, the analysis developed here suggests that a crucial condition for the development of a discontinuous negation along the lines of French is for the language to be a type III language, and display strict negative concord. It would be worth exploring this issue in more detail, but currently I do not have all the cross-linguistic data needed to substantiate this claim, so the connection is left for future work.

Tableau 17 Discontinuous negation in written French (production) (final version)

Meaning ¬Come(x)	Form	FNEG	MAXSN	*NEG	NEGFIRST	FOCUSLAST
	Il vient	*	*		*	*
	Il ne vient	*		*		*
	Il vient pas		*	*	*	
☞	Il ne vient pas			**		
	Il ne pas vient			**		*
	Il vient ne pas			**	*	

Tableau 18 Postverbal negation in spoken French (production) (final version)

Meaning ¬Come(x)	Form	FNEG	*NEG	MAXSN	FOCUSLAST	NEGFIRST
	Il vient	*		*	*	*
	Il ne vient	*	*		*	
☞	Il vient pas		*	*		*
	Il ne vient pas		**			

The main difference between the original and the revised analysis is that Chapter 3 adopted a symmetric treatment of the preverbal and postverbal negation, and analyzed the three stages of the Jespersen cycle in terms of the interaction between NEGFIRST, FOCUSLAST, and *NEG only. The addition of MAXSN and the asymmetric treatment of French discontinuous negation under which *ne* does not satisfy FNEG offer a new view of the Jespersen cycle. Although the revised analysis has rebalanced the constraints, it maintains the insight that discontinuous negation satisfies NEGFIRST as well as FOCUSLAST, at the expense of an additional violation of *NEG. Tableau 17 illustrates that the interaction of FNEG and MAXSN makes it possible for the competition between NEGFIRST and FOCUSLAST to be played out even when both are ranked below *NEG. Although the details of the analysis are different from Chapter 3, the main line of the argumentation has thus been preserved.

Second, French displays another instance of economy in a language that favors multiplication of negation. Just as in Spanish and Italian, the presence of the marker of sentential negation *pas* is governed by economy principles where the presence of Neg-expressions and the preverbal marker *ne* is not. Outside the family of Romance languages, Berber exemplifies a similar situation.

Ouali (2003, 2005) observes that certain Berber dialects have a discontinuous negation. For Tamazight Berber this is illustrated in (47a). The preverbal marker *ur* is obligatory, and the postverbal marker *sha* is optional, so with respect to the expression of propositional negation, the situation is not exactly the same as in French. The negation system of Tamazight Berber is in a transitional stage between a stage I and a stage II language in the Jespersen cycle (cf. Chapter 3, Section 4.3). As far as the combination of the negation marker with negative indefinites is concerned, Tamazight Berber does not use the postverbal marker in combination with an n-word (47b). The insertion of *sha* renders the sentence ungrammatical (47c).

(47) a. *ur* ssex (*sha*). [Tamazight Berber]
SN drink-PERF.2SG (SN)
'I don't drink.'
b. *agidge ur*-as-t iwshan
nobody SN-him-it give.IRE.3SG
'No one gave it to him.'
c. **agidge ur*-as-t iwshan *sha*
nobody SN-him-it give.IRE.3SG SN

The ungrammaticality of (47c) is an effect of economy. If the preverbal marker of sentential negation *ur* in (47a) is weakened, emphatic negation licences the postverbal marker *sha* (Chapter 3, Sections 3 and 4). The presence of *sha* is not necessary in (47b), because the n-word *agidge* is inherently negative. As (47b) is a more economical expression than (47c), it is preferred. *Sha* is then blocked in (47c) for the same reason that *pas* is blocked in the expression of single negation in (37b, c).

Third, the data indicate that the behavior of preverbal markers of negation in concord languages is much more consistent than that of postverbal markers of negation. This asymmetry has already been observed by Zanuttini (1991) and Zeijlstra (2004), but it gets a new interpretation in the OT system developed here. The presence of preverbal markers of sentential negation in negative concord constructions is governed by the two functionally motivated constraints NEGFIRST and MAXSN. The presence of postverbal markers of sentential negation in concord languages can be motivated by MAXSN (as argued for Afrikaans *nie*, and the postverbal negative markers in French Creole and Occitan), but it can also be motivated by the fact that an existing preverbal marker is too weak to express semantic negation, and the postverbal marker comes in to satisfy FNEG (as argued for French *pas*). Even in the postverbal domain, the influence of NEGFIRST is felt, as observed in Chapter 3 (Section 3.4), and worked out for Welsh in Section 7.

Fourth, the analysis confirms that that there is no reason why a type II language could not display negative concord. The reason that (spoken) French behaves like a class II language (negative spread), rather than a class I or a class III language is that there are no additional syntactic constraints that trigger the insertion of a marker of sentential negation in a sentence that contains one or more Neg-expressions. Note that a DN language like Dutch would end up with the same candidate V + neg as the optimal output for the input in Tableau 16. This illustrates the fundamental claim that the status of Neg-expressions is not determined by the lexicon but by the grammar, and that the marker of sentential negation plays an ancillary role in negative concord.

The last remaining issue concerns the double negation readings that arise in the combination of *pas* and n-words. Sentences like (37e) and (38e) are not ungrammatical in standard modern French, even though they are unidirectionally suboptimal candidates in the generation Tableaux 12–16. In Chapter 6, I will explain the double negation readings these sentences give rise to as an effect of weak bidirectional optimality or 'superoptimality' in Blutner's (2000, 2004) terms.

5.7 Negation and Negative Indefinites in Welsh

Chapter 3 (Section 4.3) discussed the complex situation of the expression of sentential negation in formal and colloquial Welsh. Here, the focus is on the co-occurrence restrictions of the negation marker and negative indefinites.

Welsh is a VSO language. Borsley and Jones (2005) show that in formal Welsh, the postverbal particle *ddim* is optionally used to reinforce the preverbal marker of negation *nid* in sentences expressing propositional negation (48a, b). In informal Welsh, the presence of *ddim* is mandatory in such contexts (48c, d).

(48) a. Nid oedd Sioned yn gweithio. [formal Welsh]
SN be.IMPF.3SG Sioned PROG work.
'Sioned was not working.'

b. Ni soniodd Sioned ddim am y digwyddiad. [formal Welsh]
SN mention.PAST.3SG Sioned SN about the event
'Sioned did not talk about the event.'

c. Doedd Gwyn *(ddim) yn cysgu. [informal Welsh]
SN.be.IMPF.3SG Gwyn *(SN) PROG sleep.
'Gwyn was not sleeping.'

d. Fydd Sioned ddim yn hapus. [informal Welsh]
be.FUT.3SG Sioned SN PRED happy.
'Sioned will not be happy.'

In formal Welsh (48a), the preverbal particle *nid* negates the entire clause. The negative particle *ddim* optionally reinforces propositional negation (48b). In the informal Welsh example (48c), negation has been incorporated into the verb (*oedd-doedd*), but the negative form of the verb is not strong enough to convey semantic negation. The particle *ddim* in postverbal position bears the negative force of the sentence. Not all verb forms in informal Welsh have special negative forms. With nondistinctive forms, *ddim* is the only visible marker of sentential negation (48d).

N-words co-occur with the preverbal marker of sentential negation in formal Welsh (49a), and with verbs that incorporate negation in informal Welsh (49b). In informal Welsh they can also contribute negative force on their own (49c). A sequence of n-words gives rise to negative concord (49d).

(49) a. Ni soniodd neb am y digwyddiad. [formal Welsh]
SN mention.PAST.3SG nobody about the event.
'No one mentioned the event.'

b. Does neb yn yr ardd. [informal Welsh]
SN.be.PRES.3SG nobody in the garden.
'No one is in the garden.'

c. Welish i neb. [informal Welsh]
see.PAST.3SG I nobody.
'I saw no one.'

d. Fydd neb byth yma.
be.FUT.3SG no-one never here.
'No one will ever be here.'

As far as formal Welsh is concerned, the fact that we are dealing with a negative concord language indicates that the bidirectional grammar has the ranking FNEG >> MAXNEG >> *NEG >> INTNEG. A high ranking of either MAXSN or NEGFIRST accounts for the obligatory presence of the preverbal negative particle in all negative sentences (48a, b as well as 49a). Given that the negative particle *ni(d)* appears in a preverbal position, and both subject and object are postverbal, it is impossible to decide which of the two constraints applies, but at least one of them must have a ranking higher than *NEG.

With a high ranking of MAXSN (or NEGFIRST) in formal Welsh, an example like (49a) is the optimal outcome of the ranking in Tableau 19.

The ranking of MAXNEG above *NEG guarantees the use of the n-word, and the high ranking of MAXSN guarantees the use of the preverbal particle *ni(d)*. Accordingly, formal Welsh behaves like a type I (strict negative concord) language.

Tableau 19 Generation of preverbal *ni(d)* with a postverbal n-word (formal Welsh)

Meaning ¬∃xMention (x, the_event)	Form	FNEG	MAXSN	MAXNEG	*NEG
	Soniodd rywun am y digwyddiad. 'Mentioned someone the event'	*		*	
	Ni soniodd rywun am y digwyddiad. 'SN mentioned someone the event'			*	*
	Soniodd neb am y digwyddiad 'Mentioned no one the event'		*		*
☞	Ni soniodd neb am y digwyddiad 'SN mentioned no one the event'				**
	Ni soniodd neb ddim am y digwyddiad 'SN mentioned no one SN the event'				***

The preverbal particle in formal Welsh survives in part in informal Welsh in verb forms that have incorporated *ni(d)*, as in (48c, 49b). However, the incorporation does not extend to all verbs (48d, 49c, d), and the incorporated form is unable to express semantic negation by itself (48c). This situation is reminiscent of the difference between the realization of *ne* in written and spoken French discussed in Section 6, and suggests that in informal Welsh, the constraint MAXSN is being lowered. An overlapping range between MAXSN and *NEG accounts for the optional dropping of the preverbal negation marker. In those cases in which the verb does not have a special negative form anymore, the n-word is the sole expression carrying the negative force of the sentence (49c).

Informal Welsh displays a complex interaction between n-words and the marker of sentential negation *ddim*. Welsh *ddim* is not allowed after a negative subject, so a sentence like (50a) is ungrammatical. Adverbial n-words are fine in this position, as illustrated by (49d), repeated as (50b), so there is a strict separation between the negation marker and an adverbial n-word:

(50) a. *Fydd neb ddim yma.
be.FUT.3SG no one SN here
'No one will be here.'
b. Fydd neb byth yma.
be.FUT.3SG no-one never here.
'No one will ever be here.'

Ddim is not incompatible with n-words in general, so it is different from French *pas* (cf. Section 6). *Ddim* co-occurs with n-words in (51a) and (b):[3]

[3]The expression *dim byd* literally means 'no world', but it is used as a fixed expression for 'nothing'.

(51) a. Na'th Emrys ddim gweld dim byd.
do.PAST.3SG Emrys SN see NEG world
'Emrys didn't see anything.'
b. Fuo' fo ddim yn gweithio erioed.
be.PAST.3SG he SN PROG work never.
'He has never worked.'

In fact, the examples (51a) and (b) would be ungrammatical without *ddim*, as (52a) shows, unless *erioed* is moved to a position immediately following the subject (52b):

(52) a. *Fuo' fo'n gweithio erioed.
be.PAST.3SG he PROG work never.
b. Fuo' fo erioed yn gweithio.
be.PAST.3SG he never PROG work.
'He has never worked.'

The fact that *ddim* is required with a sentence-final adverbial n-word (51b), but is missing with a negative subject (50b) or when the n-word is in an adverbial or object position close to the verb (as in 49c or 52b) shows that negation must be expressed 'high enough' in the structure to take clausal scope. The need for negation to take clausal scope suggests that a version of NEGFIRST is at work here. However, it cannot be the 'standard' version of NEGFIRST that requires negation to precede the verb, because informal Welsh has lost the preverbal particle *ni(d)*, and *ddim* is postverbal.

In Chapter 3 (Section 3.4), I already suggested that the exact formulation of NEGFIRST might vary from one language to the next. For Afrikaans, I proposed a version of NEGFIRST that structures the postverbal domain. I extend that proposal to Welsh, and formulate variations that reflect different degrees of closeness to the verb or hierarchical order as relevant for propositional scope. The Welsh phrase structure is VS(A)O(A), with an adverbial position immediately following the subject, and another one at the end of the sentence. The first version of NEGFIRST mirrors the negative dependency constraint of Borsley and Jones (2005: 45):[4]

- NEGFIRST (Welsh version 1) (also called NEGFIRST1)
  Negation must precede everything following VS(A)O.

NEGFIRST1 immediately explains why *ddim* has to occur in postsubject position in the expression of propositional negation: the sentence-final adverbial position would be too low in the structure. Adverbial n-words like *byth*, *erioed* can appear in either adverbial position. In postsubject position, they satisfy NEGFIRST1, so (52b) is fine.

[4] The constraints are intended to account for finite main clauses. The situation of negation in non-finite and subordinate clauses in Welsh is different, and will not be treated here, because the ramifications for the syntax are beyond the scope of this book (cf. Chapter 3, Section 5 for some discussion, but no analysis). Borsley and Jones (2005) offer extensive data and an HPSG analysis.

In sentence-final position, they don't satisfy NEGFIRST1, so (52a) is ungrammatical. An adverbial n-word in sentence-final position requires the support of *ddim* in postsubject position in order to satisfy NEGFIRST1 (51b). Tableau 20 illustrates.

N-words in subject or object position satisfy NEGFIRST1, so (49b-d) do not need the support of *ddim*. The presence of *ddim* in postsubject position is redundant according to NEGFIRST1, and therefore ungrammatical (50a). Tableau 21 illustrates.

The status of *ddim* as a marker of sentential negation makes it dispensable, and therefore sensitive to economy arguments, hence the contrast between Tableaux 20 and 21. An n-word like *byth* is not dispensable in this way, because it is a variable binding quantifier. The extra violation of *NEG *byth* incurs in (50b) is legitimate because the n-word satisfies the higher ranked constraint MAXNEG. The extra violation of *NEG *ddim* incurs in (50a) is not legitimate, because the presence of the marker of sentential negation is not needed to satisfy any faithfulness constraint. The contrast between (50a) and (50b) thus confirms the general asymmetry between markers of sentential negation and adverbial n-words, illustrated for French in Section 6 and for type III (nonstrict negative concord) languages like Spanish in Section 3.

NEGFIRST requires negation to be 'close enough' to the verb, and 'high enough' in the structure to take propositional scope. In SVO and V2 languages, 'high enough' typically correlates with 'outside of the VP' (Zanuttini 1991, Ladusaw 1992, Sells 2001). In a VSO language like Welsh, there is no evidence for a VP, according to Borsley (1996). However, Welsh still displays subject-object asymmetries: whereas (50a) is downright ungrammatical, (51a) alternates with (49c). That is, an n-word in object position does not require the presence of *ddim* (49c), but does not block it either (51a).

The Welsh version 1 of NEGFIRST allows (49c), but would block (51a) for economy reasons. The well-formedness of (51a) suggests that a stricter version of

Tableau 20 Generation of postverbal *ddim* + sentence-final adverbial n-word

Meaning ¬∃t Work(he,t)	Form	NEGFIRST1	MAXNEG	*NEG
	Fuo' fo'n gweithio erioed. 'Was he working never'	*		*
☞	Fuo' fo ddim yn gweithio erioed. 'Was he SN working never'			**

Tableau 21 Generation of n-word in subject position without *ddim*

Meaning ¬∃x Here(x)	Form	NEGFIRST1	MAXNEG	*NEG
☞	Fydd neb yma 'Will be no one here'			*
	Fydd neb ddim yma 'Will be no one SN here'			**

NEGFIRST is operative in certain varieties of Welsh. I formulate it here as NEGFIRST (Welsh version 2), also called NEGFIRST2.

- NEGFIRST (Welsh version 2) (also called NEGFIRST2)
  Negation must precede everything following VS(A).

NEGFIRST2 allows (51a), but blocks (49c), as illustrated in Tableau 22.

The main difference between versions 1 and 2 of the Welsh constraint NEGFIRST is that NEGFIRST1 treats all the arguments of the verb on a par (excluding sentence-final adverbs as outside of the argument structure), whereas NEGFIRST2 is more focused on closeness to the verb, treating objects as well as sentence-final adverbs as 'too far away' from the verb to ensure propositional scope of negation.

From the presentation of the Welsh data by Borsley and Jones, it is not entirely clear to me whether (49c) and (51a) are freely available for all speakers, or whether this is an instance of dialectal or individual variation. Free variation would suggest an oscillation of the system between a stricter and a more relaxed interpretation of versions 1 and 2 of NEGFIRST. This situation could be modeled by ranking the two constraints at the same level (in ordinal OT), or allow them to overlap (in stochastic OT). The strength of the two constraints could be slightly different for different speakers, possibly depending on the regional variety of Welsh they are speaking. A third option would be to assume the variation between (49c) and (51a) to reflect a diachronic change in progress. This could also be modeled in a stochastic version of the OT analysis. The data do not allow me to decide one way or another, but the insight that the distribution of *ddim* in Welsh involves a version of NEGFIRST seems validated in either case.

Dialectal variation is reported for the combination of two negative adverbs in postsubject position. According to Borsley and Jones (2005: 98), South Wales speakers accept examples like (53) and (54), but North Wales speakers do not:

(53) Dw i ddim byth yn mynd i wisgo 'r sgidie 'ma eto.
be.PRES.1SG I SN never PROG go to wear the shoes here again.
'I'm never going to wear these shoes again.'

(54) Dw i ddim erioed wedi gweld hyn o 'r blaen.
Be.PRES.1SG I SN never PERF see this of the front
'I've never seen this before.'

The unacceptability of (53) and (54) for North Wales speakers is accounted for by NEGFIRST2. With the adverbial n-word (*byth* or *erioed*) in postsubject position,

Tableau 22 Generation of *ddim* with an n-word in object position

Meaning ¬∃x See(e,x)	Form	NEGFIRST2	MAXNEG	*NEG
	Na'th Emrys gweld dim byd. 'Emry sees nothing'	*		*
☞	Na'th Emrys ddim gweld dim byd. 'Emry sees SN nothing'			**

NegFirst2 has been satisfied. *Ddim* is redundant in such cases, and is blocked for economy reasons. The South Wales speakers who allow (53) and (54) adopt an even stricter version of NegFirst, which does not allow the adverbial n-words *byth* and *erioed* to satisfy NegFirst. I formulate it here as NegFirst (Welsh version 3), also called NegFirst3.

- NegFirst (Welsh version 3) (also called NegFirst3)
  Negation must precede everything following VS.

Examples (53) and (54) satisfy NegFirst3 by means of the marker of sentential negation *ddim*, which appears in a position immediately following the subject, but preceding all other adverbial material. Adverbial n-words such as *byth* and *erioed* can follow *ddim* as instances of MaxNeg, but they do not satisfy NegFirst3. Under version 2 of NegFirst, the insertion of *ddim* is blocked for economy reasons, but under version 3 the presence of *ddim* is functionally motivated.

The variation spelled out in (53) and (54) is then another instance of variation in the exact formulation of what counts as 'high enough' in the structure or as 'close enough' to the verb to be able to express sentential scope of negation. The data provided by Borsley and Jones support dialectal variation in the ranking of NegFirst versions 2 and 3.

Independent evidence in favor of the analysis defended here comes from sentences that violate NegFirst, but that are not ungrammatical. Borsley and Jones (2005: 133) report double negation readings for sentences combining two negative adverbs in postsubject position, the second of which is *ddim*:

(55) a. Dw i ddim ddim yn poeni.
be.PRES.1SG I SN SN PROG worry.
'I don't not worry.'
b. Dw i byth ddim yn poeni.
be.PRES.1SG I never SN PROG worry.
c. Dydy hi erioed ddim wedi helpu.
SN.be3SG she never SN PERF help
'She has never not helped.'

Note that in (53) and (54), *ddim* is followed by the adverbial n-word, whereas in (55b, c), an adverbial n-word is followed by *ddim*. Under NegFirst1 or NegFirst2, the first negative adverb following the subject satisfies NegFirst. In all cases, the second negative adverb following the subject is the marker of sentential negation *ddim*. If *ddim* is not required to satisfy NegFirst, its presence means that the candidate incurs an extra violation of *Neg. Syntactically, it is then a less optimal form than its counterpart without the marker of sentential negation. Therefore, sentences like (55) are predicted to be ungrammatical under the OT analysis developed in this section, but obviously they are not. In Chapter 6, I will show that the combination of an n-word with a marker of sentential negation whose presence is not motivated by constraints like NegFirst of MaxSN leads to a double negation reading as the result of a weak bidirectional optimization process.

The NEGFIRST versions 1 through 3 offered in this section spell out the range of variation in propositional scope for the negation marker in a VSO language that has an adverbial position immediately following the subject. If these constraints are viewed as instances of NEGFIRST albeit in the postverbal domain, the Welsh data provide additional evidence for the pervasiveness of NEGFIRST in natural language.

5.8 Negation and Negative Indefinites in Hungarian

Hungarian has a designated position immediately preceding the verb, which hosts incorporated nominals (Farkas and de Swart 2003), but also negation and focus expressions. Example (56) is from Puskás (2006).

(56) Pál nem evett. [Hungarian]
Paul.NOM SN eat.PAST.3SG
'Paul did not eat.'

The marker *nem* co-occurs with Neg-expressions in preverbal (57a) as well as postverbal position (57b) (de Groot 1993, Puskás 2006).

(57) a. Pál senkivel nem beszélt.
Paul nobody.INSTR SN speak.PAST.3SG
'Paul didn't speak with anybody.'
b. Pál nem beszélt senkivel.
Paul SN speak.PAST.3SG nobody.INSTR
'Paul didn't speak with anybody.'

The obligatory presence of *nem* in all sentences containing a Neg-expression indicates that Hungarian is a strict negative concord (type I) language. In the OT system, the grammar of Hungarian involves a highly ranked constraint MAXSN.

Sequences of Neg-expressions can occur in preverbal and/or postverbal position (58a, b from Puskás 2006 and Surányi 2006b). In elided contexts, they have negative force in the absence of a negation marker (58c, from Puskás 2006).

(58) a. Senki sehova nem jött el.
nobody.NOM nowhere.to SN come.PAST.3SG PREF
'Nobody came along anywhere.'
b. Senkivel nem beszél semmiröl
nobody.INSTR SN speak.PAST.3SG nothing.DELAT
= He didn't speak about anything with anybody. [NC]
= He didn't speak with anybody about nothing. [DN]
c. A: Kit láttál? B: Senkit.
A: who.ACC see.PAST.2SG B: nobody.ACC
A: 'Who did you see? B: Nobody.'

Pronouns like *senki* are labeled *s*-pronouns by Surányi (2006a, b). *S*-pronouns qualify as Neg-expressions rather than NPIs, because they are negative in isolation (for instance as the elliptical answer to a question, 58c), and because they allow a double negation reading as well as a negative concord interpretation in contexts like (58b).

Besides the series of *s*-pronouns illustrated in (57) and (58), Hungarian has a series of *sem*-expressions, which also participate in negative concord, but show a slightly different behavior.[5] Surányi (2006b) points out three important differences between *s*-pronouns and *sem*-expressions. First, *sem*-expressions display a preverbal/postverbal asymmetry, as illustrated in (59).

(59) a. Nem érkezett senki sem.
SN arrive.PAST.3SG nobody.NOM SEM
'There hasn't arrived anybody.'
b. Senki sem jött el sehova.
nobody.NOM SEM come.PAST.3SG PREF nowhere-to
'Nobody came along anywhere.'
c. *Senki sem nem jött el.
nobody SEM SN come.PAST.3SG PREF
d. Senki sem ma nem jött el. [DN]
nobody SEM today SN come.PAST.3SG PREF
'Nobody is such that it's today that he did not come along.'

With the postverbal *sem*-expression in (59a), the presence of the preverbal marker *nem* is obligatory, whereas the preverbal *sem*-expression in (59b) does not require the support of *nem*. In fact, it would be ungrammatical to add *nem* in most cases, as illustrated by (59c). The only way to render the combination of a preverbal *sem*-expression and the preverbal negation marker *nem* felicitous is by according the sentence a double negation reading (59d).

The data in (59a–d) show that negation needs to be expressed preverbally in Hungarian. In the OT system, this implies a high ranking for NEGFIRST. The patterns in (59) are similar to those observed for Italian and Spanish (cf. Section 3). However, Italian and Spanish are nonstrict negative concord (type III) languages, and on the basis of the data in (57) and (58), I just established the status of Hungarian as a strict negative concord (type I) language. This is the first puzzle raised by *sem*-expressions.

The second puzzle involves the behavior of multiple *sem*-expressions. Preverbal *sem*-expressions co-occur with postverbal *s*-expressions (59b) as well as *sem*-expressions (60a) under a negative concord reading. However, it is impossible to have more than one *sem*-expression in preverbal position (60b). It is possible

[5] Surányi (2006a,b) glosses the *sem* part of the *sem*-expressions as SEM. I follow his glosses, even though I adopt a different semantics of *s*-pronouns and *sem*-expressions, as will become clear shortly.

to combine *s*-expressions with *sem*-expressions in the preverbal domain, as long as the *sem*-expression is the last one in the sequence (60c).

(60) a. Senki sem jött el sehova sem.
nobody.NOM SEM come.PAST.3SG PREF nowhere-to SEM
'Nobody came along anywhere.'
b. *Senki sem sehova sem jött el.
nobody.NOM SEM nowhere-to SEM come.PAST.3SG PERF
c. Senki soha sehova sem jött el.
nobody.NOM never nowhere-to SEM come.PAST.3SG PERF
'Nobody ever came along anywhere.'

Surányi's account of the differences in distribution between *s*-pronouns and *sem*-expressions relies on a characterization of *sem*-expressions as carrying logical negation, and of *s*-pronouns as non-negative. This route is closed to me, for *s*-pronouns are clearly Neg-expressions, according to the criteria of elided answers (58c), and possible double negation readings (59b). Accordingly, *sem*-expressions as well as *s*-pronouns carry semantic negation in my analysis.

However, Surányi's basic insight is that there is a lexical distinction between *s*-pronouns and *sem*-expressions. This view can be reinterpreted in the OT framework in the following way. Hungarian is a negative concord language, so it has the bidirectional grammar MAXNEG >> *NEG >> INTNEG. Furthermore, I propose the ranking {MAXSN, NEGFIRST, MAXNEG} >> *NEG in the OT syntax. Accordingly, all negative sentences need to involve a preverbal marker of sentential negation (*nem* in 56–58). This grammar treats Hungarian as a strict negative concord language, so the syntax and semantics of *s*-pronouns as Neg-expressions is taken care of.

As far as *sem*-expressions are concerned, I follow Surányi's proposal that they incorporate negation. The particle *sem* historically derives from the merger of *is* 'also' and *nem* 'not' (Surányi 2006b: 281). Thus, *sem*-expressions are close to minimizers, which, in Hungarian, are also preferably constructed with *sem*, as shown in (61).

(61) Egy cseppet sem érdekli Pétert.
a drop.ACC SEM interest.3SG Peter.ACC
'It doesn't interest Peter a bit.'

Given the historical origin of *sem* as *is* 'also' plus *nem* 'not', Surányi (2006b: 306) claims that *sem*-expressions grammaticalize the scalar nature of n-words by means of an appended scalar additive particle. It is by no means unusual to find that the diachronic source of a Neg-expression involves an additive scalar particle, as emphasized by Haspelmath (1997: 222). However, the same particle *sem* is also incorporated in the series of *s*-pronouns. The important assumption I make is that the diachronic source of the two series of Neg-expressions is the same, but synchronically the *s*-series has lost its sensitivity to focus, whereas *sem*-expressions have maintained it (so far).

If the *sem* particle is still compositionally present in the *sem* expressions, this implies that *sem*-expressions satisfy MaxNeg as well as MaxSN. Under this assumption and the ranking {MaxSN, NegFirst, MaxNeg} >> *Neg, the grammar derives the data in (59a–c), as illustrated in Tableaux 23 and 24.

Even though they satisfy MaxSN as well as MaxNeg, postverbal *sem*-expressions require the presence of a preverbal marker of negation *nem*, because of the high ranking of NegFirst. Tableau 23 illustrates that (59a) is the optimal candidate in this configuration. Preverbal *sem*-expressions satisfy MaxNeg, MaxSN and NegFirst. The insertion of *sem* incurs an unnecessary violation of *Neg in this configuration, as illustrated in Tableau 24. This rules out (59c) as a suboptimal candidate. Thus, (59b) (without *nem*) emerges as the optimal candidate. Unlike *sem*-expressions, *s*-pronouns do not satisfy MaxSN.

Under the assumption that *sem*-pronouns satisfy MaxSN, but *s*-pronouns do not, the insertion of preverbal *nem* is required with preverbal *s*-pronouns, as illustrated in Tableau 25 (cf. 57). As shown in Tableau 25, the high ranking of both MaxSN and NegFirst in Hungarian requires the insertion of *nem* in preverbal as well as postverbal position (57a, b).

The analysis relies on the insight that *sem*-expressions have a lexically incorporated marker of sentential negation. Whether or not a particular item satisfies

Tableau 23 Postverbal *sem*-expressions with preverbal *nem* (production)

Meaning ¬V∃x	Form	MaxSN	NegFirst	MaxNeg	*Neg
	V *senki sem*		*		*
	nem V somebody			*	*
☞	*nem* V *senki sem*				**

Tableau 24 Preverbal *sem*-expression without *nem* (production)

Meaning ¬∃xV	Form	MaxSN	NegFirst	MaxNeg	*Neg
☞	*senki sem* V				*
	nem V somebody			*	*
	senki sem nem V				**

Tableau 25 Preverbal *s*-pronoun with *nem* (production)

Meaning ¬∃xV	Form	MaxSN	NegFirst	MaxNeg	*Neg
	senki V	*			*
	nem V somebody			*	*
☞	*senki nem* V				**

the constraint MaxSN is a lexical matter, as argued in Section 6 with respect to the French particles *ne* and *pas* (and postverbal negation markers in other Romance languages). If *sem*-expressions satisfy MaxSN, but *s*-pronouns do not, the difference between the two classes of Neg-expressions is in the lexicon, but in a different way from Surányi (2006b).

Interestingly, the lexical semantics of *sem*-expressions offers new insights into their restricted use in the preverbal domain. If *sem*-expressions involve an appended scalar additive particle, it is quite conceivable that this limits the use of these expressions to the immediately preverbal position, in which focus markers naturally appear in Hungarian. The view that *sem*-expressions reside in the focus/operator position of Hungarian is supported by the fact that the verbal prefix (*el* in the examples in 58a, 59b, d, 60a, c) appears in postverbal, rather than preverbal position.

Farkas and de Swart (2003: Chapter 4) use the postverbal position of the verbal prefix in incorporation constructions as an argument that the preverbal bare nominal lives in a special syntactic position, which is used as a diagnostics for an incorporation construction. In (62a), the prefix *fel* is in the preverbal position, but in the presence of an incorporated nominal in (62b), it is in postverbal position.

(62) a. Mari fel-olvasott Petinek.
Mari up-read.PAST Peti.DAT
'Mari read aloud to Peti.'
b. Mari kínai lampionat szerelt fel a plafonra.
Mari Chinese lantern.ACC set up the ceiling.on
'Mari set up a Chinese lantern on the ceiling.'

The postverbal position of *el* in the examples in (58a), (59b, d), (60a, c) suggests that the preverbal *sem*-expressions in these sentences live in the same special preverbal position that the incorporated nominal occurs in (62b). If other preverbal positions in Hungarian do not allow focus operators, this could very well block the multiplication of preverbal *sem*-expressions in (60b). It would also force them to be the last one in a sequence of Neg-expressions (60c). Many languages block special indefinites in the scope of negation from appearing in the preverbal domain for focus reasons, according to Haspelmath (1997: 316–317). A broader embedding of the theory of negation in a theory of focus and word order is called for in order to account for the precise restrictions that Hungarian is subject to, but the general lines are clear (cf. Surányi 2006a for further discussion).

In the postverbal domain, there is no special focus or operator position, but focus operators are not blocked from the postverbal domain. Surányi (2006b) provides the following example of multiple focus with a preverbal *sem*-expression, where focus is indicated with capitals.

(63) SENKI SEM SZAVAZOTT végül CSAK JÁNOSRA.
nobody.NOM SEM vote.PAST.3SG finally only János-for
'Finally, nobody voted for only Janós.'

If focus sensitive operators like *csak* 'only' can occur in the postverbal domain in Hungarian, this would explain why *sem*-expression involving an appended additive scalar particle may also occur there (60a).

If the preverbal/postverbal asymmetry involving *sem*-expressions indeed involves the special preverbal focus/operator/incorporation position in Hungarian, no special assumptions are needed for the postverbal domain. All that is needed is a language-specific constraint restricting the preverbal occurrence of *sem*-expressions to the special position immediately preceding the verb. Under the assumption that this constraint is sensitive to the enriched lexical semantics of *sem*-expressions involving scalarity, additivity, and focus, Hungarian emphasizes that a more indepth analysis of the interaction between focus and negation is required. This conclusion is likely to extend to other languages in which word order is sensitive to information structure. Kaiser (2006) offers a study of negation and information structure in Finnish.

The double negation reading of a sequence of *s*-pronouns illustrated in (58b) requires an extension of the OT analysis with a stochastic component, as will be outlined in Chapter 6 (Section 4). The double negation reading of a combination of a preverbal *sem*-expression and a marker of sentential negation (59d) cannot be derived in the strong bidirectional OT grammar of a negative concord language, but it follows under the weak bidirectional OT analysis I develop in Chapter 6 (Section 5).

Cross-linguistic evidence supporting the approach adopted here comes from similar data in Greek. The expression *oute* ('not even'), discussed in Giannakidou (2007) allows dropping of the negation marker *dhen* in constructions like (64a), even though Greek is a strict negative concord language and *dhen* obligatorily co-occurs with emphatic n-words like KANENAN ('nobody') (cf. Section 4):

(64) a. Oute o Janis (dhen) irthe. [Greek]
not.even the John (SN) arrived
'Not even/neither John arrived.'
b. O pritanis *(dhen) proskalese oute ti Maria.
the dean *(SN) invited.3SG even the Maria
'The dean didn't invite even Maria.'
c. Ki oute KANENAS me gnorize.
and not.even nobody me knew.3SG
'And nobody knew me.'

Dhen is obligatory when *oute* occurs in postverbal position (64b). N-words such as KANENAS can occur in the presence of OUTE (64c), and don't need the support of *dhen* in that case. In contrast to emphatic Greek n-words like KANENAS, *oute* is morphologically marked as negative: *ou* is the marker of sentential negation in Ancient Greek, and is used here as a bound morpheme. Given this analysis, and given the incorporation into the scalar focus particle, *oute* might very well compare to Hungarian *sem*-expressions, and accordingly be taken to satisfy MAXSN. Under a high ranking of NEGFIRST, the contrast between (64a) and (64b) are then explained along similar lines as the contrast between preverbal and postverbal *sem*-expressions in Hungarian (cf. Tableaux 23 and 24). If n-words like KANENAN do not satisfy

MAXSN, but *oute* does, *dhen* is not needed in (64c).[6] If the marker of sentential negation *dhen* is indeed optional when *oute* appears in preverbal position, this could imply that the internal composition of *oute* is not entirely transparent in modern Greek, and its use is therefore not stable. It is possible that we are witnessing a lexical change in progress.

5.9 Flemish: a Puzzle

Most Flemish dialects have a preverbal and a postverbal negation marker, as well as n-words. As outlined by Haegeman (1995, 1997) and Haegeman and Zanuttini (1996), the preverbal enclitic *en* by itself is unable to express negation (65a), and it needs to be doubled with the postverbal marker *nie* (65b), which constitutes the bearer of semantic negation. The preverbal enclitic *en* is optional; (65c) can also be used.

(65) a. *Valère en-eet dienen boek. [Flemish]
Valère SN has that book
b. Valère en-eet dienen boek nie.
Valere SN has that book SN
'Valère doesn't have that book.'
c. Valère eet dienen boek nie.
Valere has that book SN
'Valère doesn't have that book.'

The patterns in (65) confirm that Flemish is in an intermediate stage between discontinuous negation and postverbal negation (cf. Chapter 3, Section 4). N-words can bear semantic negation by themselves (66a), or spread over multiple Neg-expressions (66b). Both *en* (66c) and *nie* (66d) are compatible with n-words, but the presence of neither particle is obligatory (66a, b).

(66) a. da Valère niemand kent. [Flemish]
that Valère nobody knows
'that Valère doesn't know anybody.'
b. K'een an niemand niets gezeid.
I have to nobody nothing said.
'I didn't say anything to anyone.'
c. K'en-een an niemand niets gezeid.
I SN have to nobody nothing said.
'I didn't say anything to anyone.'
d. K'een t niemand nie gevroagd.
I have it nobody SN asked
'I didn't ask anyone.'

[6] Thanks to an anonymous reviewer for drawing my attention to the example of *oute* in relation to the Hungarian *sem*-expressions.

The examples in (66a–d) are easily accounted for in the OT analysis. If *niemand* and *niets* are analyzed as Neg-expressions, and Flemish has the ranking MAXNEG >> *NEG >> INTNEG of a negative concord language, the Neg-expressions contribute semantic negation (66a), but combine to express resumptive negation (66b) (negative spread). An overlapping range between MAXSN and *NEG in the OT syntax allows for the optional presence of a marker of sentential negation besides the n-word. In Flemish, both the enclitic *en* (66c), and the postverbal marker *nie* (66d) can fulfill this role. Flemish thus emphasizes the independence of MAXSN and NEGFIRST found in Afrikaans (cf. Section 4).

Interestingly, the discontinuous negation in Flemish follows a diachronic path different from that of standard French, as discussed in Section 6. The postverbal marker of negation *pas* does not participate in negative concord in standard French. I argued that *pas* does not satisfy MAXSN, and is therefore blocked by the economy constraint *NEG. However, this is a specific lexical restriction on *pas*. Other Romance languages (Occitan, Valdovain, Brazilian Portuguese) allow postverbal markers of negation to participate in negative concord. So there is nothing unusual per se about the fact that West Flemish postverbal *nie* satisfies MAXSN, and participates in negative concord, as in (66d).

Note that the only instance of postverbal *nie* that satisfies MAXSN is the *nie* that follows the Neg-expressions (66d). According to Haegeman and Zanittini (1996), the scope of negative concord is to the left of *nie*. If *nie* precedes the Neg-expression, a double negation reading ensues, as in (67a) and (67b).

(67) a. Valère (en) eet nie niets kuopt. [DN]
Valère (SN) has SN nothing bought.
'Valère has never bought nothing.'
b. da Valère nie niets kuopt. [DN]
that Valère SN nothing buys
'that Valère doesn't buy nothing.'

The double negation reading of (67a) and (67b) is accounted for under the weak bidirectional OT analysis that will be developed in Chapter 6 (Section 5). The fact that the postverbal *nie* is tied to a particular position in (66d) supports the view that its presence is licensed by the syntactic constraint MAXSN. The markers generated by MAXSN function as scope markers that are often tied to a particular position.

Examples like (68a) and (68b), in which a Neg-expression combines with discontinuous negation, constitute a puzzle for the OT analyis.

(68) a. K'en-een t niemand nie gevroagd. [Flemish]
I SN have it nobody SN asked
'I didn't ask anyone.'
b. K'en-een niets nie gezien.
I SN have nothing SN seen
'I have't seen anything.'
c. Er en wil niemand niet dansen.
There SN wants nobody SN dance.
'Nobody wants to dance.'

The examples in (68a, b) are from Haegeman and Zanuttini (1996); the example in (68c) is from Van der Auwera and De Vogelaer (2008). According to Van der Auwera and Neuckermans (2004) and Van der Auwera and De Vogelaer (2008), the pattern in (68) is common in southern Oost-Vlaanderen, in regions where both the preverbal enclitic *en* and the postverbal *nie* are commonly used. In these regions, dialects in which *en* is used meet dialects in which *nie(t)* is used. If the only reason why the marker of sentential negation is present in a sentence containing a Neg-expression is to satisfy MaxSN, one marker (either *en* as in 66c, or *nie(t)* as in 66d) would be expected to do the job. The postverbal marker *nie(t)* follows the Neg-expression in (68), so it is in the 'standard' position in which it satisfies MaxSN, and is not expected to give rise to the double negation reading found in (67). However, the presence of multiple markers satifying MaxSN should be blocked by the economy constraint *Neg. Even in negative concord languages, in which the expression of negation is multiplied in the sentence, negative expressions are not generated randomly, let alone redundantly. I conclude that the system developed so far does not account for the combination of a Neg-expression and discontinuous negation in (68).

The pattern in (68) is also found outside of the Germanic languages. Besides Québécois examples of the type (42b) in Section 6, Martineau and Déprez (2004) provide the example in (69a). Similarly, besides examples like (42d), there are French internet examples of the type in (69b). In both cases, the combination *ne* plus *pas* and an n-word gives rise to a single negation reading.

(69) a. Personne (n') a (pas) pu rien nous dire. [Québécois]
No one (SN) has (SN) could nothing us tell
'No one could tell us anything.'
b. Je ne suis pas d'accord avec personne. [French]
I SN am SN of agreement with nobody
'I don't agree with anybody.'

The enclitic *ne* is frequently dropped in Québécois as well as in the informal registers found on the Internet. However, the examples in (68) indicate that the two can co-occur with an n-word.

Wheeler et al. (1999: 482) provide similar examples in Catalan (70).

(70) a. No he vist pas ningú. [Catalan]
'I haven't seen anyone.'
b. No pensem anar-hi pas mai.
'We don't intend to go there ever.'

As pointed out in Section 5, Catalan has a high ranking for NegFirst, which motivates the use of the preverbal marker *no* with the postverbal n-words *ningú* and *mai* in (70a) and (70b). According to Wheeler et al., the postverbal negator *pas* is optional and emphatic in discontinuous negation.

There are different approaches one can take to explain the data in (68)–(70). One could assume that the set of OT constraints used so far is not complete, and the two negation markers are licensed by two different faithfulness constraints above *Neg, rather than the one (MaxSN) appealed to so far. According to Breitbarth and

Haegeman (2008) and Haegeman and Lohnman (2008), *en* and *nie* do not have equal status, and the two do not enter into NC in the same way. In their view, *en* spells out an emphatic feature on a high polarity phrase (PolP). Under the assumption that Pol as such has an unvalued feature, it needs to be valued by a clause mate negative constituent. The observation that discontinuous negation in West Flemish is quite stable provides support for this analysis, for the standard development in the Jespersen cycle would expect this to be a relatively unstable intermediate period between preverbal and postverbal negation (cf. Chapter 3, Section 4).

A reinterpretation of the view advocated by Breitbarth and Haegeman and Haegeman and Lohndal in the OT framework developed here would require the introduction of additional constraints governing emphasis and polarity (cf. Chapter 4, Section 5). Such constraints might also be needed to deal with the Catalan cases in (70). Note however that Catalan is in an intermediate stage between preverbal and discontinuous negation, where it is the 'new' postverbal negation marker that is emphatic, whereas Flemish is in an intermediate stage between discontinuous and postverbal negation, for which Breitbarth and Haegeman claim that it is the 'old' marker developing a new status. So polarity related constraints would have to operate in different ways depending on the stage of the Jespersen cycle the language is in.

Another possibility, suggested by Van der Auwera (2008, personal communication), would be that in the Flemish dialects at hand, both *en* and *niet* can both be used to satisfy MAXSN. In regions in which the two markers overlap, they could freely be used individually as well as in combination with each other. Along these lines, Van der Auwera and Neuckermans (2004) explain the pattern in (68) without an appeal to emphasis or polarity marking, but as an intermediate stage in the Jespersen cycle. The old negator *en* is disappearing, and needs reinforcement by the new negator *niet*. Analogical pressure extends the use of the *niet* in predicate negation (*en..niet*) to the quantifier negotor (*niemand niet*), leading to the complex *en.. niemand niet*. This hypothesis could be extended to the combination of *ne* and *pas* in (69), which is also not clearly emphatic in nature. The double satisfaction of MAXSN would imply that the economy constraint *NEG is weakened in contact situations in which two distinct markers occurring in different syntactic positions are available to satisfy MAXSN.

The third option would be to investigate the application of NEGFIRST in the postverbal domain to these cases. Section 7e proposed a treatment of the combination of n-words with the postverbal marker *ddim* in informal Welsh in terms of variants of NEGFIRST. Along similar lines, the postverbal negators *pas* in French (69) and Catalan (70) might be motivated by the need to express negation 'higher' in the clause, in a position closer to the verb. This approach would not extend to the Flemish examples in (68), though.

These options will have to be worked out in more detail, and carefully evaluated for the language at hand before one can determine which of these avenues provides the best explanation of the data in (68)–(70). I leave this for further research.

5.10 Negative Doubling and Negative Spread

Chapter 3 (Section 3) showed that Afrikaans is a language that exemplifies discontinuous negation. The two markers are identical in form, and are both postverbal, as illustrated in (71).

(71) Ek het hom nie gesien nie.
I have him SN seen SN
'I haven't seen him.'

The two *nie*'s do not have the same status (Biberauer 2006 and references therein). Sentence-final *nie* has to be doubled by medial *nie* in order to convey propositional negation. The intuition is that sentence-medial *nie* is the bearer of semantic negation in sentences like (71), but sentence-final *nie* is independently licensed. If sentence-final *nie* is perceived as a marker that is present in all negative sentences, but is semantically non-negative, it can be treated similar to French *ne*. Accordingly, I take sentence-final *nie* to satisfy MAXSN, but not FNEG. This implies that examples such as (71) can be derived along similar lines of the French example (37a) in Tableau 17.

So far, this chapter has only studied the role of MAXSN in type I languages that have the ranking {MAXSN, MAXNEG} >> *NEG >> INTNEG. In such languages, all indefinite arguments under the scope of negation are realized as n-words (Greek, Slavic, Hungarian, etc.). Afrikaans was discussed in Section 4 as a language that uses a sentence-final marker to satisfy MaxSN, but it is not so clear that Afrikaans qualifies as a strict negative concord language.

As far as the combination of negation with n-words in Afrikaans is concerned, it is useful to go back to den Besten's (1986) distinction between negative doubling and negative spread. Negative spread is defined as the situation in which the negative feature of negative indefinites is distributed over any number of indefinite expressions. In terms of Haspelmath's classification, negative spread languages are type II negative concord languages, but negative doubling is not distinguished from strict negative concord as a separate category.

Negative doubling occurs when the negation marker is doubled by some other negative expression (either a negative indefinite, or a second negation marker). Under negative doubling, negation is not necessarily spread over all indefinites. den Besten hypothesizes that Afrikaans is a language which instantiates negative doubling, but not negative spread. This view is motivated by sentences such as (72a, b).

(72) a. Sy is nêrens ooit tevrede nie. [Afrikaans]
She is nowhere ever happy SN.
'She is never happy anywhere.'
b. Niemand is ooit tevrede nie.
Nobody is ever satisfied SN
'Nobody is ever satisfied.'

Only one of the indefinite arguments under the scope of negation is realized by means of an n-word, the other one is a regular indefinite. Negative indefinites are always doubled by the sentence-final marker *nie*. Sentence-medial *nie* does not occur in sentences with an n-word that convey a single negation.

Although the examples in (72) support the hypothesis that Afrikaans is a negative doubling language, this is clearly not the full story, for sentences like (73a, b) occur alongside those in (72a, b) (all quoted in van der Wouden 1994).

(73) a. Permissie het hy nog nooit van geniemand gevra nie.
permission has he yet never of nobody-at all asked SN
'He has never asked anyone's permission yet.'
b. Ek krij geen hulp van niemand nie.
I get no help from nobody SN
'I don't get any help from anybody.'

The situation is further complicated by the observation that examples like (73) are relatively infrequent, and sentences with multiple n-words also allow double negation readings (cf. Chapter 6, Section 3 and van Gass 2007). All in all, it is still possible that Afrikaans is more on the side of a negative doubling language (without multiplication of n-words), than on the side of a full-fledged strict negative concord language.

van der Wouden (1994) and Biberauer (2006) take the examples in (73) to be the exception rather than the rule, and propose a treatment of multiple Neg-expressions in terms of emphatic negation, rather than negative spread. Zeijlstra (2007) proposes a similar account of resumptive readings of n-words in Dutch (74).

(74) a. Zij heeft nergens geen zin in. [Dutch]
She has nowhere no lust in
'She doesn't feel like anything at all.'
b. Hij gaat nooit niet naar school.
He goes never SN to school
'He never goes to school.'

Standard Dutch is a double negation language, in which examples such as (74) should not be tolerated with a single negation reading (cf. Chapter 4, Sections 3 and 4). According to Zeijlstra (2007), emphatic negation is subject to adjacency requirements, requires stress on the first negative indefinite, and is idiosyncratic in nature.

In the polyadic quantifier semantics, there is no need to maintain a strict distinction between negative spread and emphatic negation. Both n-words and negative quantifiers are inherently negative expressions, so both phenomena instantiate resumption of negative quantifiers. This does not exclude the possibility that emphasis motivates the construction of a resumptive negative quantifier in standard double negation languages.

Suppose that examples like (72) instantiate the basic grammar of Afrikaans negation. Then Afrikaans qualifies as a negative doubling language with negative attraction, but without negative spread. The ranking {MAXSN, NEGATTRACT} >>

MaxNeg in the OT syntax generates a mixture of negative indefinites and regular indefinites, with an obligatory marker of sentential negation, as illustrated in Tableau 26. I added FocusLast as the constraint that realizes the obligatory marker in the sentence-final position. This is relevant for Afrikaans, but obviously, doubling languages could also come with a high ranking of NegFirst in their grammar, triggering a preverbal marker of sentential negation.

The preference for a Neg-expression over a marker of sentential negation in (72) is favored by the highly ranked constraint NegAttract. However, the low ranking of MaxNeg does not lead to the spread of negation over all indefinite arguments under the scope of negation. In the combination with n-words, the high ranking of MaxSN ensures negative doubling by sentence-final *nie*. Sentence-medial *nie* is redundant, because FNeg is satisfied by the Neg-expression. But Tableau 26 only covers the syntax of Afrikaans negation, what about the semantics?

The discontinuous negation used to convey propositional negation in (71) supports the view that sentence-final *nie* is inserted to satisfy MaxSN, but it does not satisfy FNeg. As far as examples (71) and (72) are concerned then, Afrikaans can be viewed as a double negation language. The ranking IntNeg >> *Neg in the semantics leads to the single negation reading of a sentence containing an n-word and a negation marker, as long as the negation marker is semantically non-negative, as illustrated in Tableau 27.

A purely existential interpretation is excluded by the high ranking of FNeg, for this candidate does not take into account the semantic contribution made by the negative indefinite. The non-negative nature of the sentence-final marker *nie* in Afrikaans implies that a single negation interpretation of the sentence does not violate IntNeg. The double negation reading incurs an additional, unmotivated violation of *Neg, and is therefore rejected as a suboptimal candidate.

Tableau 26 Negative doubling languages without negative spread (production)

Meaning ¬∃x∃y V(x,y)	Form	FNeg	MaxSN	NegAttr	*Neg	FocusLast	MaxNeg
	indef V indef	*	*	*		*	*
	neg V indef		*		*	*	*
	neg V neg		*		**		
	indef V indef SN			*	*		*
☞	neg V indef SN				**		*
	neg V SN indef SN				***		*
	neg V neg SN				***		

Tableau 27 Negative doubling languages (interpretation)

Form neg V indef $SN_{[-NEG]}$	Meaning	FNeg	IntNeg	*Neg
	∃x∃yV(x,y)	*	*	
☞	¬∃x∃yV(x,y)			*
	¬¬∃x∃yV(x,y)			**

The grammar developed so far, does not account for examples in which the n-word combines with sentence-medial *nie* as well as sentence-final *nie*. Biberauer and Zeijlstra (2009) provide the examples in (75):

(74) a. Hy het niks nie teen Hans gesê nie. [Afrikaans]
He has nothing SN to Hans said SN
'He didn't say anything to Hans.'
b. Hy is nooit nie tevrede nie.
He is never SN satisfied SN
'He is never satisfied.'

According to Biberauer and Zeijlstra (2009), the addition of sentence-medial *nie* signals emphasis ('anything at all', 'absolutely never'), which brings these examples in line with the cases of emphatic negation discussed in Section 9. However, the discussion in that section focuses on multiple satisfaction of MAXSN in negative concord languages, whereas the grammar so far treats Afrikaans as a double negation language.

The examples in (75) suggest that Afrikaans displays features that combine the system of double negation with that of negative concord. Further evidence of such a combination is provided by the strict negative concord examples in (73). The grammar in Tableau 27 does not derive the right interpretation of these examples, because they require the ranking *NEG >> INTNEG in the semantics. Van Gass (2009) develops an extension of the bidirectional OT analysis that reconciles the examples in (72) and (73) within the grammar of Afrikaans, but a discussion of that system is beyond the scope of this section.

What the observations in Sections 5 through 10 teach us is that there are a number of rankings that are theoretically possible, but that have not been fully explored in this book. They mostly involve more lexical/syntactic possibilities for variation, for the semantic variation across languages has been exhaustively investigated. However, a closer look at more unusual patterns might shed light on the syntax–semantics interface as it is explored in natural language. I leave this for further research.

5.11 Conclusion

In studies on negation, the position of the marker of sentential negation with respect to the verb and with respect to other negative indefinites in the sentence has always been an important object of study. Double negation languages like English and Dutch never combine negative indefinites with a marker of sentential negation in the expression of a single negation reading. This is a straightforward extension of the bidirectional grammar developed in Chapter 4.

Under the polyadic quantifier analysis, the marker of sentential negation is claimed to be semantically redundant in negative concord contexts. This makes it possible for negative concord languages to rely solely on Neg-expressions (Class II negative spread as in spoken French or Occitan), or to exploit the marker of negation as a scope marker. Nonstrict class III languages that display a preverbal/postverbal asymmetry use the marker of negation to satisfy NEGFIRST in contexts in which the n-word is not in preverbal position (Spanish, Italian, European Portuguese). Strict NC class I languages that always require the presence of a marker of sentential negation (Greek, Hungarian, Slavic, Afrikaans) use the negation marker to indicate clausal scope.

Several mixed cases arise from a stochastic ranking of the relevant constraints. Such stochastic rankings account for language change in progress (Catalan) and microvariation within a language (French, Welsh). Different sets of n-words in Hungarian underscore the importance of the lexicon. The discussion of Flemish and negative doubling in Afrikaans show that further variations on the constraint rankings in the OT syntax can be explored. The interaction of negation and n-words with the rest of the grammar (in particular word order and information structure) is also a relevant further development.

It would be interesting to investigate the interaction between the negation marker and negative indefinites in the Central African VO languages studied by Dryer (2007). The data in his paper are relevant for the position of the marker of sentential negation, and have been discussed in Chapter 3, Section 3. Unfortunately for me, Dryer does not discuss negative indefinites. Hopefully, future studies will be able to complement the typology.

In Chapters 4 and 5, I relied on a classification of languages as displaying either negative concord or double negation. However, there are certain cases of what looks like negative concord in double negation languages, and cases of double negation in negative concord languages. Such cases were expected under the polyadic quantifier analysis, but have not been spelled out in the OT analysis yet. Chapter 6 investigates these unusual combinations.

Chapter 6
Double Negation in Negative Concord Languages

Introduction and overview This chapter investigates double negation readings in negative concord languages. This may look like a contradiction in terms. After all, negative concord is a system in which multiple expressions of negation combine to convey a single negation reading. If so, then how is it possible to express double negation in such languages? I will discuss three cases in which this arises.

No resumptive negative quantifier is built in combinations of sentence negation with constituent negation (Section 1) and multiple clause configurations (Section 2). Resumptive negative quantification is defined over negative quantifiers, but not affixal negation. Furthermore, it is a clause-bound phenomenon, so Neg-expressions in different clauses lead to multiple negation readings in standard negative concord languages.

The ambiguities between double negation and negative concord readings of sequences of Neg-expressions play an important role in de Swart and Sag's (2002) analysis. A stochastic OT analysis is developed for such cases, in which overlapping constraints in the semantics allow for ambiguities (Section 3).

The third and most unusual case of double negation discussed in this chapter arises in the interaction of sentential negation and negative indefinites, as found in French, Spanish, Italian, Romanian, Afrikaans, West Flemish, Hungarian, and Welsh. Section 4 confirms that double negation readings are not expected under the strong bidirectional OT analysis developed in Chapters 4 and 5. A weak bidirectional OT analysis is developed to account for these cases (Section 5).

6.1 Affixal Negation

In negative concord languages, a sequence of Neg-expressions produces a single negation. Native speakers usually have quite strong intuitions that a double negation reading is blocked. Note that there is no conceptual ban on double negation readings for speakers of a negative concord language. In languages like English, the best examples of double negation involve the combination of a negation marker with an adjective carrying a negative prefix, as in (1) (from Horn 2001):

H. de Swart, *Expression and Interpretation of Negation: An OT Typology*,
Studies in Natural Language and Linguistic Theory,
DOI 10.1007/978-90-481-3162-4_6,

(1) a. It is not an impossible job.
b. She is happy or at least not unhappy.
c. Women's rights lawyers say such concerns are not unfounded.

Negative concord languages also exemplify double negation readings in combinations of negation with negative prefixes (Italian, Polish, and Greek data from S. Benazzo, 2006, personal communication, D. Klimek, 2007, personal communication, and an anonymous reviewer respectively):

(2) a. Il n'est pas incompétent. [written French]
he SN is SN incompetent.
'He is not incompetent.'
b. Il n'est pas impensable qu'elle puisse commettre un meurtre.
it SN is SN unthinkable that she can.SUBJ commit a murder
'It is not unthinkable that she might commit a murder.'
c. Non è n'impresa imposibile. [Italian]
SN is a enterprise impossible
'It is not an impossible enterprise.'
d. Ona jest szczęśliwa...a przynajmniej nie jest nieszczęśliwa. [Polish]
she is happy ...or at least SN is not-happy.
'She is happy or at least not unhappy.'
e. Dhen ine adinato afto. [Greek]
SN is impossible this
'This is not impossible.'

Morphological negation (as realized by prefixes) does not participate in negative concord, and a single negation reading is not available for sentences like (2) in any NC language that I am aware of. In terms of the analysis developed in this book, the observations indicate that affixal negation does not qualify as a Neg-expression. Accordingly, negative adjectives are not collected in the N-store, and do not participate in the construction of a resumptive polyadic quantifier.

Support for this analysis comes from the fact that affixal negation counts as constituent negation, rather than sentence negation, according to the criteria advanced in Chapter 1 (Section 1). In the languages under consideration, I can also advance another test. Recall that strict and nonstrict negative concord languages combine n-words with the marker of sentential negation in a certain configuration. Unlike Neg-expressions, the negative prefix never requires the presence of the marker of sentential negation. In (3), negation is solely expressed by the negative prefix, and all sentences convey a single negation reading.

(3) a. Il (*n')est incompétent. [written French]
he (*SN) is incompetent.
'He is incompetent.'
b. Il (*n')est impensable qu'elle puisse commettre un meurtre.
it (*SN) is unthinkable that she can.SUBJ commit a murder.
'It is unthinkable that she might commit a murder.'
c. È un'impresa impossibile. [Italian]
is a job impossible
'It is an impossible job.'

d. Ona jest nieszczęśliwa. [Polish]
she is not-happy.
'She is unhappy.'

If the prefix *un-* qualified as a Neg-expression, the preverbal clitic *ne* should be obligatorily present in written French, because this language displays strict negative concord (Chapter 5). The fact that the presence of *ne* is blocked in (3a, b) indicates that the adjectives *incompétent* and *impensable* do not qualify as Neg-expressions. The fact that a preverbal occurrence of *nie* is missing in the Polish example (3d) indicates that *nieszszesliwa* is not a Neg-expression either, for Polish is also a strict negative concord language (Chapter 5). Italian is a type III language, which requires negation to be expressed preverbally. A postverbal n-word thus requires the insertion of a preverbal marker of negation (Chapter 5). The absence of *non* in (3c) indicates that the postverbal adjective bearing the negative prefix does not have the status of an n-word in Italian.

The conclusion is that adjectives expressing affixal negation exemplify constituent negation in both double negation and negative concord languages, and constituent negation does not participate in resumptive polyadic quantification.[1] The observations made with respect to (2) and (3) support the view that double negation is not a conceptual problem in negative concord languages, but a grammatical issue involving sequences of Neg-expressions within a single clause. These data fit in with the analysis developed so far.

6.2 Multiple Clause Negation

Further support for the view that negative concord is a grammatical phenomenon and not a conceptual issue comes from the observation that negative concord is a clause-bound phenomenon. The clause boundedness of negative concord has been noted by Corblin (1996), Giannakidou (2000), de Swart and Sag (2002), Corblin and Tovena (2003), and Corblin et al. (2004). I address the general issues in Section 2.1, and focus on subordinate clauses introduced by *without* as a special case in Section 2.2. Exceptions to clause boundedness are discussed in Section 2.3.

6.2.1 Negation and Neg-expressions in Different Clauses

As Corblin and Tovena (2003) point out, n-words in independent sentences have their own domain of negation. This is clear in question–answer sequences like (4).[2]

[1] Affixal negation is to be distinguished from the instances of DP-internal negative concord signalled for Flemish by Haegeman (2002) and Haegeman and Lohndal (2008).

[2] Not all languages allow the type of question-answer sequences illustrated in (4); cf. Chapter 1 (Section 4) for discussion of blocking effects in Serbo-Croatian, discussed by Bošković (2008).

(4) a. Q: Chi non ha risposto? A: Nessuno. [Italian]
Q: who SN has answered A: nobody
Q: 'Who has not answered?' A: 'Nobody.'
b. Q: Qui n'a pas répondu? A: Personne. [French]
Q: who SN has SN answered A: nobody
Q: 'Who has not answered?' A: 'Nobody.'
c. Q: Kto nie odpowiedział? A: Nikt. [Polish]
Q: who SN answered A: nobody
Q: Who did not answer? A: 'Nobody'

Complex sentences, involving a main and a subordinate clause, also illustrate that negative concord is clause bound. The Italian sentence in (5a) gets a single negation reading, unlike its English counterpart in (5b), whereas the Italian sentence in (5c) gets a double negation reading, just like its English counterpart in (5d):

(5) a. Nessuno ha detto niente. [Italian]
nobody has said nothing.
'Nobody said anything.'
b. Nobody said nothing.
c. Nessuno ha detto che non era successo niente.
nobody has said that SN was happened nothing
'Nobody said that nothing had happened.'
d. Nobody said that nothing had happened.

Similar data are available for other negative concord languages. The examples in (6a) and (6c) are from Harries (1973), and are quoted by Horn (2001); (6b) is from an anonymous reviewer; (6d) is from D. Klimek (personal communication).

(6) a. Nem lehet nem nevet-ni. [Hungarian]
SN possible SN laugh-INF.
'It was not possible not to laugh.'
b. Dhen ipa na min erthi. [Greek]
SN said SUBJ SN come
'I didn't say that he should not come.'
c. Ja ne mog ne dat' emu nagrádu. [Russian]
I SN could SN give him reward
'I couldn't not reward him.'
d. Nikt nie powiedział, że nic się nie wydarzyło. [Polish]
nobody SN said that nothing REFL SN happened.
'Nobody said that nothing happened.'

Hungarian, Greek, Russian, and Polish are all strict NC languages in which a marker of sentential negation is obligatorily present in all negative clauses. The examples in (6) all imply two instances of the negation marker. When negation is present in both the main and the embedded clause, the complex sentence expresses double, rather than single negation.

Corblin (1996) and Corblin and Derzhanski (1997) point out that triple negation in natural language faces severe processing problems. Horn (2001) provides the example in (7), but notice that it involves multiple clauses.

(7) The Mets did not not re-sign Mike Hampton because they didn't want to pay him the money.
= It was not because they did not want to pay him that they did not re-sign Mike Hampton.

I will leave triple negation aside, and focus on double negation readings.

6.2.2 Subordinate Clauses Introduced by 'Without'

Sentences containing a subordinate clause introduced by *without* constitute an interesting case of multiple clause negation. The occurrence of n-words in *without*-clauses was first noticed by Zanuttini (1991). de Swart and Sag (2002) discuss the behavior of the French *sans* to illustrate the phenomenon.

The French *sans*, 'without', establishes negative concord relations with n-words embedded under the connective, so examples like (8a) have a single negation reading. However, an n-word in the main clause leads to double negation in combination with *sans* (8b).

(8) a. Il est parti sans rien dire à personne. [French]
he has left without nothing say to nobody.
'He left without saying anything to anyone.'
b. Personne n'est parti sans rien dire.
nobody SN has left without nothing say.
'Nobody left without saying anything.'

The sentences in (8) are constructed, but the following are attested examples from the Internet:

(9) a. Je n'y vois rien sans lunettes. [French]
I SN there see nothing without glasses
'I don't see anything without glasses.'
b. Ne faites rien pour nous sans nous!
SN do nothing for us without us
'Do nothing for us without us!'

de Swart and Sag (2002) take *sans* 'without' to be the negative counterpart of *en*, a conjunction that combines with a present participle to build a modifier of the VP:

(10) a. Anne est partie en chantant.
Anne is left CONJ singing
'Anne left singing.'
b. $\text{Leave}(a) \wedge \text{Sing}(a)$

The connective *en* builds an intersective modifier, so the conjunction of the VP and the participle in (10a) leads to the semantic representation seen in (10b). The treatment of *sans* as the negative counterpart of *en* implies that the connective in (11a) expresses the conjunction of the main verb and the negation of the property expressed by the infinitival complement (11b).

(11) a. Anne est partie sans pleurer.
Anne is left without cry
'Anne left without crying.'
b. Leave(a) $\wedge$ ¬Cry(a)

Syntactically, *sans* and the marker of sentential negation are quite different, but semantically they are very much alike. Both are expressions of propositional negation that are collected in the N-store. Given that they are not variable binding operators, their semantic contribution is absorbed in the resumptive negative quantifier.

The construction of a resumptive negative quantifier absorbing the contribution of *without* and a sequence of n-words in the infinitival complement of *sans* is illustrated in (8a), repeated in (12a). The resumptive interpretation strategy in (12b) leads to one negative quantifier binding the two variables contributed by the n-words *rien* and *personne*. The truth conditions of the sentence are spelled out in (12c).

(12) a. Anne est partie sans rien dire à personne. [French]
Anne is left without nothing say to nobody.
'Anne left without saying anything to anyone.'
b. Leave(a) $\wedge$ $\mathrm{NO}_{E2}^{\mathrm{INAN}\times\mathrm{HUM}}\ \mathrm{Say}_A$ [resumption]
c. Leave(a) $\wedge$ $\neg\exists x \exists y$ Say(a,x,y) [NC]

The resumptive strategy in (12) is possible because the connective *sans* takes scope over the infinitival complement, so *sans* and the n-words are all in the same N-store. N-words that are not in the infinitival complement of *sans* invariably create a double negation reading (9), because polyadic quantification is clause bound, and the construction of the resumptive negative quantifier cannot cross a clause boundary.

Counterparts of *without* in other negative concord languages display similar asymmetries between n-words embedded in the complement of *sans* and n-words in the main clause. Example (13a) is from Giannakidou (1997); (13b) is from E. Vlachou (2007, personal communication). Sentence (14a) is from Corblin and Tovena (2003); (14b) is from S. Benazzo (personal communication). Example (15a) is from Przepiórkowski and Kupść (1999); (15b) is from D. Klimek (personal communication), (16a, b) are from Isac (2004).

(13) a. O papus pethane [Greek]
the grandfather died.3SG
xoris na dhi KANENA apo ta egonia tu.
without SUBJ see.3SG none from the grandchildren his
'The grandfather died without seeing any of his grandchildren.'
b. Min kanis TIPOTA ja mas xoris emas
SN do nothing for us without us
'Don't do anything for us without us!'

(14) a. Ha riposto senza nessuna esitazione. [Italian]
has answered without no hesitation
'He/she answered without any hesitation.'
b. Non vedo niente senza i miei occhiali.
SN see nothing without the my glasses
'I don't see anything without my glasses.'

(15) a. Został bez niczego. [Polish]
Stayed-3SG without nothing
'He was left broke.'
b. Nic nie widzę bez moich okularów.
Nothing SN see-1.SG without my glasses
'I don't see anything without my glasses.'

(16) a. Nimic nam trimis fără să verific. [Romanian]
Nothing SN-have.1.SG sent without to check
'I have sent nothing without checking.'
b. Nada eu enviei sem corrigir [Eur. Portuguese]
Nothing have.1.SG sent without check
'I have sent nothing without checking.'

The behavior of *sans*, 'without', and its counterparts in other concord languages confirms the clause-bound character of resumption as an instance of polyadic quantification. Retrieval of the Neg-expressions at the clause boundary is built into the HPSG mechanism of de Swart and Sag (2002): the bottom–up interpretation process requires the N-store to be emptied before the next clause up is interpreted.

6.2.3 Exceptions to Clause Boundedness

Clause boundedness is conceived as a standard restriction on resumptive quantification (Corblin et al. 2004, and references therein), and languages generally do not establish negative concord across clause boundaries. Exceptions are the kind of cases illustrated in (17)–(20). Neg-raising as in (17) indicates that the clause boundary is permeable (Kayne 1984, Godard 2004), and the high vs. low construction of negation supports the view the negation behaves as a scope marker, as discussed for French in Chapter 5 (Section 6).

(17) a. Paul n'accepte de recevoir personne. [French]
Paul SN accepts to receive nobody
'Paul does not agree to see anybody.'
b. Personne n'accepte de rencontrer personne. [DN or NC]
Nobody SN accepts to see nobody
'Nobody agrees to see nobody/anybody.'

Permeability is favored by nonfiniteness (17) and (18) (from Przepiórkowski and Kupść 1999), as well as subjunctive mood (19).

(18) Jan nie chciał niczego kupować. [Polish]
Jan SN wanted nothing buy.INF
'Jan didn't want to buy anything.'

(19) Non pretendo che tu arresti nessuno. [Italian]
SN require that you arrest.SUBJ nobody
'I don't require that you arrest anybody.'

However, there is no one–one relation between mood and permeability. Although epistemic verbs in Greek subcategorize for indicative complements, Giannakidou (1997) shows that negative concord is allowed with neg-raising (20a). Tóth (1999) provides similar examples for Hungarian (20b).

(20) a. Dhe nomizo oti idhe KANENAN. [Greek]
SN believe.1SG that saw.IND.3SG no one.
'I don't believe that he saw anyone.'
b. Senkit sem hiszem hogy meghívtál. [Hungarian]
Nobody.ACC SN believe. 1SG that invite.2SG.IND
'I don't believe that you have invited anybody.'

The Hungarian *s*-expressions are normally not licensed in a subordinate clause when the negation marker is in the matrix clause (cf. Chapter 4, Section 5). Greek verbs that license long distance negative concord (Giannakidou 1997, 2000) also license inverse scope relations with universal quantifiers (Farkas and Giannakidou 1996). Inverse scope is arguably also a clause-bound phenomenon. In both Greek and Hungarian, the analysis of exceptional cases relies on an extension of the notion of predicate-argument structure, so that n-words in subordinate clauses count as arguments for a higher verb (cf. de Swart and Sag 2002). If resumption and inverse scope are subject to the same constraints on clause boundedness, and allow the same class of exceptions, these observations provide strong support for the polyadic approach.

If negative concord languages block double negation within the clause (or within a single predicate-argument structure), but double negation is not blocked conceptually in these languages, as argued in Sections 1 and 2, the question arises whether the borderline between NC and DN languages is absolutely sharp, or whether there are fuzzy cases in between. Chapter 3 (Section 4.3) brought up intermediate stages in the Jespersen cycle. These are found in languages that display a pattern in between preverbal and discontinuous negation, or in between discontinuous and postverbal negation. Such intermediate (stages of) languages required a stochastic extension of the OT analysis. I propose to use the same stochastic OT tools to account for intermediate cases involving Neg-expressions in Section 3.

6.3 Ambiguities with Multiple N-words

According to Chapter 4 (Section 3), an important argument in favor of the analysis developed by de Swart and Sag (2002) is that the HPSG grammar does not encode preferences for iteration or resumption. In principle, both interpretations of the polyadic quantifier (iteration and resumption) are available. Support in favor of this view comes from ambiguities arising with sequences of Neg-expressions, in double negation as well as negative concord languages. Section 3.1 reviews the empirical observations, and Section 3.2 proposes a stochastic extension of the OT analysis developed in Chapter 4.

6.3.1 Empirical Observations

Standard English and Dutch are double negation languages that occasionally allow resumptive readings. Examples (21) and (22) are repeated from Chapter 1.

(21) When *nobody* knows *nothing*, everybody is an expert. Nobody can seriously claim to be an expert on the collapse of the World Trade Center, simply because nobody had a chance to study the rubble. Everybody who has looked at the photographs and television news video knows as much about the collapse as the most knowledgeable scientists. Therefore, everybody who has viewed the photographs and video can claim to be an expert.

(22) Als *niemand* luistert naar *niemand* vallen er doden in plaats van woorden.
if nobody listens to nobody fall there deaths in stead of words
'When nobody listens to anybody, the conversation doesn't die, people do.'

Examples such as (21) and (22) are rare, but in so far as they are accepted by native speakers, they should be accounted for by the grammar.

Double negation readings with a sequence of n-words in negative concord languages constitute the mirror image of the resumptive readings in DN languages. Corblin (1996) was the first to bring French examples exemplifying double negation into the discussion. Corblin (1996), Corblin and Tovena (2003), and Corblin et al. (2004) argue that the French sentences in (23) are truly ambiguous:

(23) a. Personne n'a rien payé. [French]
Nobody SN has nothing paid..
= No one has paid anything [NC]
= Everyone has paid something. [DN]

b. Personne n'est le fils de personne
Nobody SN is the son of nobody.
= No one is the son of anyone. [NC]
= Everyone is the son of someone. [DN]

For (23a), the two readings are equally available. The DN reading of (23b) conforms to our world knowledge in ways that the NC reading of this sentence does not.[3]

[3] The NC reading is viable in models where quantification is restricted to a contextually relevant subset of people. For instance, if one thinks of a group composed of ten 20-year-old men, and ten 45-year olds, where they are all related one way or the other, it could be true that no one is the son of anyone. This would of course correspond to the NC reading of (23b).

Thus the sentences are ambiguous, but pragmatic factors bring out the DN reading, and make the NC reading of examples like (23a, b) less likely (Corblin 1996).

As outlined in Chapter 1 (Section 5), Corblin brought out the theoretical implications of the French double negation readings for the analysis of negative concord. However, the literature reports similar ambiguities in other Romance languages. According to Zanuttini (1991: 144, 145), (24) exemplifies double negation in Italian, and Herburger (2001) makes the same claim for the Spanish sentence (25). Falaus (2007a, b) offers ambiguous examples of Romanian sequences of n-words in (26) and (27).

(24) Nessuno è rimasto con niente in mano. [Italian]
no one is left with nothing in hand
= No one was left with nothing. [DN]

(25) Nadie nunca volvió a Cuba. [Spanish]
nobody never returned to Cuba
= Nobody ever returned to Cuba [NC]
= Nobody never returned to Cuba [DN]

(26) Nimeni nu vine de nicaieri. [Romanian]
Nobody SN comes from nowhere
= Nobody comes from anywhere [NC]
= Nobody comes from nowhere [DN]

(27) Nimeni nu moare niciodata [Romanian]
Nobody SN dies never
= Nobody ever dies [NC]
= Nobody never dies [DN]

The double negation reading of (26) is marked, and requires a special intonation, just like double negation readings in DN languages like English (cf. Chapter 1, Section 1.1). It is typically available in contexts involving the denial of a negative presupposition or assertion. The double negation reading of (27) is preferred for pragmatic reasons.

Outside the family of Romance languages, double negation readings have been reported for Bulgarian (Corblin and Derzhansky 1997), colloquial Welsh (Borsley and Jones 2005), Hungarian (Puskás 2002, 2006), West Flemish (Haegeman and Zanuttini 1996), and Afrikaans (R. Botha, 2006, personal communication). Many of these languages instantiate strict negative concord. The Bulgarian example from Corblin and Derzhansky (1997) is repeated in (28a); the Polish example in (28b) is from Anna Młynarczyk (2009, personal communication).

(28) a. Nikoj ne običa nikogo [Bulgarian]
nobody.NOM SN loves nobody.ACC
= No one loves anyone. [NC]
= Everyone loves someone. [DN]

b. Nikt nie jest synem nikogo [Polish]
nobody SN is son nobody.ACC
= No one is the son of anyone [NC]
= No one is the son of no one. [DN]

Borsley and Jones (2005: 75) take the ambiguity of examples (29) and (30) to provide support for their claim that negative dependents in colloquial Welsh have negative force:

(29) Does neb yn caru neb. [colloquial Welsh]
SN.be.PRES.3SG no one PROG love no one.
= No one loves anyone [NC]
= No one loves no one [DN]
(30) Does neb yn dweud dim byd. [colloquial Welsh]
SN.be.PRES.3SG no one PROG say SN world.
= No one is saying anything [NC]
= No one is saying nothing [DN]

In both (29) and (30) the form *does* is glossed as negative, because it is the form of the verb *to be* that occurs in negative sentences. However, it cannot provide sentential negation on its own (cf. Chapter 5, Section 7).

According to Puskás (2002, 2006), the examples in (31) can have a negative concord interpretation, or a double negation reading, depending on intonation.

(31) Senkivel nem beszélt semmiröl [Hungarian]
nobody.INSTR SN speak.PAST.3SG nothing.DELAT
= He didn't speak about anything with anybody. [NC]
= He didn't speak with anybody about nothing. [DN]

The double negation reading requires a fall-rise intonation on *semmiröl.* The word order also plays a role. Hungarian is well-known to have a relatively free word order, and permits two n-words in the preverbal domain (cf. Chapter 5, Section 8). The order in (32a) favors a negative concord interpretation, whereas the order in (32b) favors the double negation reading.

(32) a. Senkivel semmiröl nem beszélt. [Hungarian]
nobody.INSTR nothing.DELAT SN speak.PAST.3SG
'He didn't speak about anything with anybody.' [NC]
b. Semmiröl senkivel nem beszélt.
nothing.DELAT nobody.INSTR SN speak.PAST.3SG
'He didn't speak with anybody about nothing.' [DN]

Haegeman and Zanuttini (1996) offer the West Flemish example (33), which allows either a negative concord or a double negation reading.

(33) K'en een an niemand niets gegeven. [West Flemish]
I SN have to nobody nothing given.
= I didn't give anything to anybody. [NC]
= I gave nothing to nobody. [DN]

In order to obtain the double negation reading, special emphasis is needed on both negative elements, with a marked pause separating them. Haegeman and Zanuttini (1996: footnote 26) note that the double negation reading is often available as an alternative to the negative concord reading, but it is always the marked interpretation. Negation and negative concord in West Flemish is complex (Chapter 5, Section 9), but the possibility of a double negation reading favors the analysis of *niemand* and *niets* in (33) as inherently negative.

Afrikaans displays similar ambiguities. The context of (34) favors a negative concord interpretation; (35) brings out the double negation reading.[4]

(34) Ons ouers het dit niet breed gehad niet. Ma moes maar kos maak van wat toevallig in die huis was. En sy het ons *nooit niks* lekkers om te eet gegee *nie*.
'Our parents weren't very rich. Mother just had to make dinner with what happened to be in the house. And she has never given us anything nice to eat.' [NC]

(35) Ons ouers het dit nie breed gehad nie. Vir hos was daar nie veel geld nie. Maar Ma kon baie goed kook en sy het ons *nooit niks* lekker om te eet gege *nie*.
'Our parents weren't very rich. There was not much money for us.
But Mother was a very good cook, en she has never given us nothing nice to eat.[DN]

Resumptive readings in English and Dutch on the one hand, and double negation readings in several Romance languages, Bulgarian, Welsh, Hungarian, West-Flemish, and Afrikaans on the other hand provide strong support for de Swart and Sag's (2002) polyadic quantifier approach. In their analysis, Neg-expressions are inherently negative. All Neg-expressions are collected in the N-store, and interpreted upon retrieval. Upon retrieval, the sequence of Neg-expressions is interpreted in terms of iteration or resumption, leading to single and double negation respectively. If both interpretation strategies are freely available, an explanation of the ambiguities of the examples in (21) through (35) follows in a straightforward way.

However, most theories of negation in natural language treat such examples as marginal, and typically do not include them in their analyses. In line with the claim that languages are predominantly double negation languages, or predominantly negative concord languages, Chapters 4 and 5 developed separate OT grammars

[4] I am grateful to R. Botha (2006, personal communication) for help with the examples, and for providing the contexts in (34) and (35). Negative spread in Afrikaans is still a matter of debate. Compare Chapter 5, Section 10 for discussion and a proposal.

for double negation and negative concord languages. This raises the question whether de Swart and Sag's account of the ambiguities in (21) through (35) has been lost in the OT analysis. In this section I reconcile the two views by showing that the ambiguities are accounted for in a stochastic OT version of the semantics developed so far.

6.3.2 A Stochastic OT Semantics

According to ordinal OT, the ranking of constraint C_1 above C_2 means that C_1 is always stronger than C_2. But some linguistic phenomena cannot be accounted for in these terms. Chapter 3 (Section 4) shows that an ordinal ranking does not work for intermediate stages in the Jespersen cycle. Intermediate stages arise when a preverbal marker of negation is always required, but a postverbal marker is optional, or a postverbal marker of negation is always required, but a preverbal marker is optional. This optionality is accounted for by creating overlap between the two relevant constraints, according to the stochastic version of OT developed by Boersma (1998), and Boersma and Hayes (2001). If adjacent constraints have an overlapping range, their order can be reversed in certain outputs.

In Chapter 3, I proposed a stochastic evolutionary version of OT in order to ground the markedness of negation as compared to affirmation in frequency asymmetries. In Chapter 5, I referred to the possibility of a stochastic OT syntax to account for the optional presence of the marker of sentential negation in languages like Catalan and Brazilian Portuguese (Section 5). The cases under discussion in this section suggest an ordinal ranking in the syntax, but a stochastic OT semantics, so they constitute a new application of the stochastic approach.

Double negation languages like English adopt the ranking INTNEG >> *NEG in the semantics. Suppose that this is not an ordinal ranking but that there is some overlap between the range of the constraint INTNEG and the range of the constraint *NEG. If the overlap is small, this means that INTNEG usually dominates. Accordingly, a combination of two Neg-expressions is normally interpreted in terms of double negation (cf. Chapter 4). But in some contexts, *NEG might win, and a resumptive reading as in (20) or (21) constitutes the optimal interpretation.

Negative concord languages adopt the ranking *NEG >> INTNEG in the semantics. Suppose again that this is not an ordinal ranking, but there is overlap between the range of the constraint *NEG, and the range of the constraint INTNEG. If the overlap is small, *NEG will usually dominate. Accordingly, a combination of two Neg-expressions is usually interpreted in terms of negative concord (cf. Chapter 4). But in some cases, INTNEG might win, and a double negation reading as in (22)–(35) comes out as the optimal interpretation.

Syntactic structure (such as the form of the n-word, i.e. pronominal versus full DP, word order), intonation, and context play an important role in disambiguation in general (cf. de Hoop 2004 and references therein). They might affect the balance between the two adjacent constraints, and thereby play a role in the disambiguation of cases like (20)–(35). Not all languages permit the variability that arises under the stochastic analysis. Giannakidou (2000, 2006) claims that Greek n-words never allow double negation readings, so in this language the overlap between the two constraints *NEG and INTNEG would be (close to) zero.

The analysis in terms of stochastic OT leads to the following conclusions. First, the fact that both double negation and negative concord languages display ambiguities supports the view defended by de Swart and Sag (2002) that all Neg-expressions are inherently negative. Corblin (1996) argues that a view of n-words as negative polarity items is unable to derive the ambiguities, and so far, I have not seen an analysis which invalidates that claim. Second, the stochastic analysis maintains the core of the OT analysis developed in Chapter 4, but reconciles the OT typology with de Swart and Sag's (2002) claims about ambiguities by adding the possibility of a certain degree of overlap between the constraints *NEG and INTNEG in the semantic component. Third, the stochastic view suggests that languages may occasionally switch to a ranking that was qualified as 'unbalanced' in Chapter 4.

The fact that perturbation of the semantic constraints leads to an unbalanced relation between form and meaning might very well explain why the overlap between *NEG and INTNEG is small. However, such perturbations could also eventually lead to language change (Jäger and Rosenbach 2006). Double negation readings are a marginal phenomenon in negative concord languages, and resumptive readings are infrequent in double negation languages. Language users may appreciate the special effects created by these exceptions, but are unlikely to invert the ranking in the semantic component without re-ranking the syntactic constraint MAXNEG, because the balance between form and meaning lies at the heart of the bidirectional grammar.

6.4 DN and NC Languages in Strong Bidirectional OT

Aside from the cases discussed in Sections 1 through 3, the bidirectional grammar developed in Chapters 4 and 5 does not allow double negation readings in NC languages. Of course this observation is the starting point of most discussions of negative concord, on the basis of examples such as (36), repeated from Chapter 4.

(36) a. Nadie ha dicho nada. [Spanish]
Nobody has said nothing.
'Nobody said anything.'

b. Nessuno ha detto niente. [Italian]
Nobody has said nothing.
'Nobody said anything.'

c. Personne (n')a rien dit à personne. [French]
Nobody (SN) has nothing said to nobody.
'Nobody said anything to anyone.'

Tableau 1, repeated from Chapter 4, expresses this insight in the OT analysis. Tableau 1 shows that double negation readings of multiple Neg-expressions in single clauses are unintelligible in NC languages (cf. de Hoop 2004 and references therein). Similarly, it seems impossible to express double negation within a single predicate–argument structure. In OT terms, this is the problem of ineffability (cf. Legendre 2001 and references therein). Tableau 2 shows that if we try to express a double negation input with the constraints adopted so far, the optimal way to express this in an NC language would also be a sequence of Neg-expressions.

If both a single negation input and a double negation input lead to an optimal expression in terms of two Neg-expressions, the neg + neg combination might be expected to lead to ambiguities. But examples such as (36) show that is not the case. As illustrated in Tableau 1, the ranking *NEG >> INTNEG yields $\neg\exists x_1 \exists x_2$ rather than $\neg\exists x_1 \neg\exists x_2$ as the optimal meaning of the sequence neg + neg in NC languages. So although neg + neg is the best form to express the double negation meaning, the double negation meaning is not the best meaning for the neg + neg form in NC languages.

These results indicate that in a full theory of the syntax–semantics interface, optimization does not proceed unidirectionally. The speaker takes the hearer's perspective into account in the production process, and the hearer takes the speaker's perspective into account in the interpretation process. Several versions of bidirectional OT have been postulated that establish a tighter relation between forms and meanings than can be achieved by unidirectional optimization. Blutner (1998, 2000), Beaver and Lee (2004), and Beaver (2004) evaluate formalizations of bidirectional OT that are currently in use. In this section, I will show that the grammars developed in Chapter 4 reflect a strong version of bidirectional OT.

In Blutner's (1998, 2000) framework, a form–meaning pair is strongly optimal if there is no better (more optimal) meaning for the form at hand, and there is no better form for the meaning at hand. A pair <f,m> is strongly optimal if there is

Tableau 1 Negative concord interpretation of a sequence of two Neg-expressions

Form neg + neg	Meaning	FNEG	*NEG	INTNEG
	$\exists x_1 \exists x_2$	*		**
☞	$\neg\exists x_1 \exists x_2$		*	*
	$\neg\exists x_1 \neg\exists x_2$		**	

Tableau 2 Sequence of Neg-expressions to convey double negation in NC languages (production)

Meaning $\neg\exists x_1 \neg\exists x_2$	Form	FAITHNEG	MAXNEG	*NEG
	indef + indef.	*	*	
	neg + indef.		*	*
☞	neg + neg			**

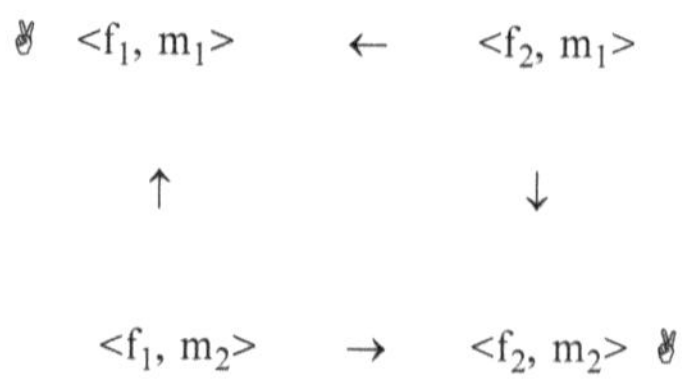

Figure 1 Strong bidirectional OT

no more optimal pair <f,m'> and there is no more optimal pair <f',m>. Strong optimality is illustrated in Figure 1 with two forms, f_1 and f_2, and two meanings, m_1 and m_2. The arrows indicate the preferences.

Figure 1 illustrates that $<f_1, m_1>$ and $<f_2, m_2>$ are strongly optimal form–meaning pairs, because both the horizontal and vertical arrows point at these pairs. For the interpretation of f_1, the meaning m_1 is preferred over m_2, and for the expression of m_1, the form f_1 is preferred over f_2. Similarly, for the interpretation of f_2, the meaning m_2 is preferred over m_1, and for the expression of m_2, the form f_2 is preferred over f_1. In strong bidirectional OT, optimization over forms and meanings converges. This convergence of forms and meanings constitutes the core of the bidirectional analysis developed in Chapters 3 and 4.

The expression and interpretation of propositional negation worked out in Chapter 3 is implemented in a straightforward way in strong bidirectional OT. Affirmative sentences usually do not contain any particular marking. Negative sentences always take a special form. The marker of sentential negation takes different forms in different languages, as illustrated by the examples in (37).

(37) a. John is not sick. [English]

b. Ou petetai Sokrates. [Ancient Greek]
SN flies Sokrates.
'Socrates doesn't fly'

c. On ne igraet. [Russian]
he SN plays.
'He doesn't play.'

d. Nid oedd Sioned yn gweithio. [formal Welsh]
SN be.IMPF.3SG Sioned PROG work
'Sioned was not working.'

e. János nem dohányz-ik. [Hungarian]
János SN smoke.3SG
'János doesn't smoke.'

Tableau 3 Propositional negation in strong bidirectional OT

Input [f,m] f_1: S; f_2: not S m_1: p; m_2: ¬p		FNEG	*NEG
[S, p]	✌		
[S, ¬p]		*	*
[not S, p]		*	*
[not S, ¬p]	✌		**

The bidirectional analysis is spelled out in Tableau 3, repeated from Chapter 3.

In the bidirectional Tableau 3, the input consists of two different forms (S and *not* S) and two different meanings (p and ¬p). In the optimization process over form–meaning pairs, we need to consider both whether there is a better form for a given meaning and whether there is a better meaning for a given form. Violations of constraints that arise in either direction of optimization are added up. FNEG and *NEG are both 'double-edged' constraints that are active in OT syntax as well as OT semantics. In the second candidate, FNEG is violated because the form does not reflect the nonaffirmative meaning. In the third candidate, FNEG is violated because the semantics does not reflect the nonaffirmative form. In the fourth candidate, *NEG is violated twice, because there is a negative form as well as a negative meaning. The comparison between forms and meanings leads to two strongly optimal pairs, indicated by the victory sign (✌).

The ranking FNEG >> *NEG is standard in the grammar of all natural languages, and reflects the syntactic and semantic markedness of negation with respect to affirmation (cf. Chapter 3). Accordingly, Tableau 3 treats S as the unmarked form, and *not* S as the marked form. Similarly, p is the unmarked meaning, and ¬p the marked meaning. The combination of an unmarked form with an unmarked meaning [S,p] comes out as a strongly optimal pair, and so does the combination of a marked form with a marked meaning [not S, ¬p] (examples 37).

The other two form-meaning pairs are not strongly optimal. The pairing up of an unmarked form with a marked meaning [S, ¬p] loses against [S, p], which is a better meaning for the same form. It also loses against [not S, ¬p], which is a better form for the same meaning. The combination of a marked form with an unmarked meaning [not S, p] loses against [not S, ¬p], which is a better meaning for the same form. It also loses against [S, p], which is a better form for the same meaning.

The relation between form and meaning in sentences expressing propositional negation is then a straightforward application of strong bidirectional OT in the way set up by Blutner (1998, 2000). This is a natural result in view of the markedness theory developed in Chapter 3.

The analysis can be extended to sequences of Neg-expressions in double negation languages such as (standard) English and Dutch, which are also accounted for

Tableau 4 Double negation languages in strong bidirectional OT

Input [f,m] f_1: neg + indef; f_2: neg + neg m_1: $\neg\exists x_1 \exists x_2$; m_2: $\neg\exists x_1 \neg\exists x_2$		FNEG	INTNEG	*NEG	MAXNEG
[neg + indef, $\neg\exists x_1 \exists x_2$]	✌			**	*
[neg + indef, $\neg\exists x_1 \neg\exists x_2$]				***	*
[neg + neg, $\neg\exists x_1 \exists x_2$]			*	***	
[neg + neg, $\neg\exists x_1 \neg\exists x_2$]	✌			****	

under a strong bidirectional OT analysis. The bidirectional analysis of examples such as (38), repeated from Chapter 4, is represented in Tableau 4.

(38) a. Nobody has nothing to hide.
b. Het is een stad uit een opgewekt verhaal of film, [Dutch]
waar (...) *niemand niets* te doen heeft, maar waar iedereen genoeg tijd over houdt.
It is a city from a beat-up story or movie, where (...) nobody has nothing to do, but where everyone has enough time left.

Markedness is a relative notion. Negation is marked with respect to affirmation, but single negation is unmarked in relation to double negation. The tableau is slightly simplified in the sense that it includes only candidates that respect FNEG. The pair [neg + indef, $\neg\exists x_1 \exists x_2$] comes out as a strongly optimal pair, because it is the best form–meaning pair under the constraint ranking INTNEG >> *NEG >> MAXNEG.

As far as the violation pattern in Tableau 4 is concerned, it would seem that neg + indef implies a better form for the double negation reading than neg + neg. However, I take polyadic quantifier theory to act as a filter on INT, the function that specifies the set of possible candidate meanings for a particular structure (cf. Chapter 2, Section 3). This theory does not offer an interpretation procedure for the pair neg + indef that assigns this form the interpretation $\neg\exists x_1 \neg\exists x_2$. In order to reflect that the pair [neg + indef, $\neg\exists x_1 \neg\exists x_2$] is not in fact a candidate under INT, this candidate has been shaded in Tableau 4.

The filtering role of polyadic quantifier theory leaves the candidate [neg + neg, $\neg\exists x_1 \neg\exists x_2$] as a strongly optimal pair: neg + neg is the best form to convey the double negation meaning, and $\neg\exists x_1 \neg\exists x_2$ is the optimal interpretation for a sequence of two negative indefinites. Under the assumption that polyadic quantifier theory filters the candidates generated by INT, the expression and interpretation of double negation in DN languages (examples 38a, b) is accounted for in strong bidirectional OT.

The ineffability and unintelligibility of double negation with sequences of Neg-expressions in negative concord languages constitutes the more interesting case. The combinations of forms and meanings in the examples in (36) lead to the patterns in Tableau 5.

Just like Tableau 4, Tableau 5 restricts the comparison to relevant candidates that satisfy FNEG. The candidate [neg + indef, $\neg\exists x_1 \neg\exists x_2$] has been shaded again, because it is filtered out as a possible form-meaning pair by polyadic quantifier theory. Negative concord languages adopt the ranking MAXNEG >> *NEG >> INTNEG as

Tableau 5 Negative concord languages in strong bidirectional OT

Input [f,m] f_1: neg + indef; f_2: neg + neg m_1: $\neg\exists x_1\exists x_2$; m_2: $\neg\exists x_1\neg\exists x_2$	FNEG	MAXNEG	*NEG	INTNEG
[neg + indef, $\neg\exists x_1\exists x_2$]		*	**	
[neg + indef, $\neg\exists x_1\neg\exists x_2$]		*	***	
[neg + neg, $\neg\exists x_1\exists x_2$] ✌			***	*
[neg + neg, $\neg\exists x_1\neg\exists x_2$]			****	

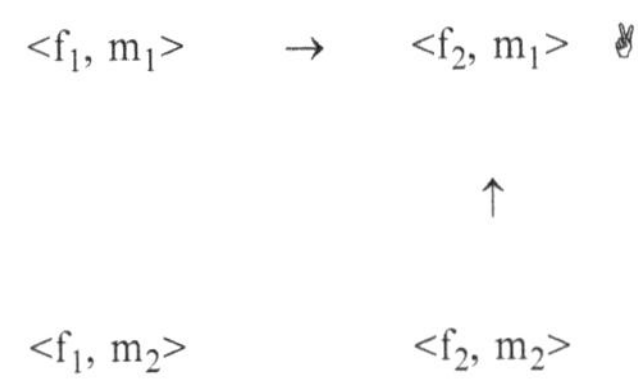

Figure 2 A single strong bidirectional pair in negative concord languages

established in Chapter 4. Tableau 5 shows that only one strongly optimal form–meaning pair arises under this ranking, namely [neg + neg, $\neg\exists x_1\exists x_2$]. This pair combines a marked form with an unmarked interpretation.

The pair [neg + neg, $\neg\exists x_1\neg\exists x_2$] is not a strongly optimal form–meaning pair, because the single negation meaning is preferred for this form, due to the ranking *NEG >> INTNEG. Because of the high ranking of MAXNEG in negative concord languages, the pair neg + indef is not strongly optimal for either the single or the double negation reading, because both meanings prefer to be expressed by the form neg + neg. This situation can also be represented in the arrow diagram in Figure 2.

The arrows indicate the preference relations. No arrows are pointing to or leaving from the form-meaning pair <f_1, m_2>, that is filtered out by polyadic quantifier theory. Arrow diagram 2 shows that under strong bidirectional optimality, double negation readings cannot be expressed by a combination of Neg-expressions and/or indefinites in negative concord languages, because these languages pair up the marked form (neg + neg) with the unmarked (single negation) meaning. The mismatch between syntactic and semantic markedness yields the effects of ineffability and unintelligibility that are familiar from the literature on negative concord.

The results obtained in this section confirm that the grammars NEGINT >> *NEG >> MAXNEG for double negation languages, and MAXNEG >> *NEG >> NEGINT for negative concord languages are instances of Blutner's strong bidirectional OT. However, besides strong bidirectionality, Blutner (1998, 2000, 2004) also defines a notion of weak bidirectional optimization or 'superoptimality'. As will appear in Section 5, weak bidirectional optimization comes into play in certain special combinations of n-words with a marker of sentential negation that give rise to double negation readings.

6.5 Double Negation in NC Languages in Weak Bidirectional OT

The conclusion that double negation cannot be expressed in NC languages is based on the behavior of sequences of Neg-expressions within a single clause, as analyzed in ordinal OT (for interpretations in stochastic OT, cf. Section 3). The results are expected on the basis of the bidirectional grammars developed in Chapter 4, and the semantics of polyadic quantifiers. In this section, I study unexpected cases that involve combinations of Neg-expressions with the marker of sentential negation. In special configurations, these combinations can lead to double negation readings, even in negative concord languages. Although the observations are stable, both within and across languages, the data have been mostly ignored in current analyses of negative concord. This section starts with a general reflection on the patterns in an OT perspective (Section 5.1), and then proceeds to develop a formal analysis in a weak bidirectional OT extension of the analysis of strict and nonstrict negative concord developed so far (Sections 5.2 and 5.3). Section 5.4 confirms that the mechanism stays within the cognitive boundaries that have been imposed upon recursive bidirectional optimization by Blutner et al. (2006).

6.5.1 Neg-expressions and the Marker of Sentential Negation

Chapters 1 and 5 establish three classes of negative concord languages as far as their use of the marker of sentential negation is concerned. The marker of sentential negation may be required for all negative sentences (Slavic, Greek, Afrikaans, etc.), due to the high ranking of the constraint MAXSN in the OT syntax. This leads to a type I language, exemplifying strict negative concord. The marker of sentential negation may be required for all negative sentences in which the n-word is post-verbal (Italian, Spanish, Portuguese), due to the high ranking of the constraint NEGFIRST in the OT syntax. This leads to a type III language, exemplifying nonstrict negative concord. Finally, the marker of sentential negation may be disallowed in sentences involving Neg-expressions, leading to negative spread (type II languages). This last situation arises in the grammar in case there are no syntactic constraints that require the negation marker to be present in sentences where a negative indefinite conveys semantic negation. Negative spread occurs in spoken French, varieties of Flemish, and certain Italian dialects, as discussed in Chapter 5 (Sections 6 and 9).

Empirical observations suggest that the circumstances under which double negation readings arise depend on the type of negative concord language at hand. Giannakidou (2006) states that double negation readings do not arise in strict negative concord languages. Her claim is too strong, though, and three sets of counterexamples present themselves. First, strict negative concord languages such as (written) French, Bulgarian, Polish, Romanian, Welsh, Hungarian, and Afrikaans allow double negation readings with a sequence of Neg-expressions, because of the possibility of a stochastic OT interpretation (Section 3), examples (23)–(35). The example in (23a) is repeated here as (39).

(39) Personne n'a rien payé. [French]
Nobody SN has nothing paid.
= No one has paid anything. [NC]
= Everyone has paid something. [DN]

Second, strict negative concord languages with discontinuous negation such as written French and Afrikaans allow double negation readings with one of the markers of the discontinuous negation, but not the other. For French, this is *pas* (but not *ne*) (40a). For Afrikaans this is the immediately postverbal *nie* (but not the sentence final *nie*) (40b). For Flemish, this is *nie* (but not *en*) (40c).

(40) a. Il n'est pas venu pour rien. [French]
He SN is SN come for nothing.
'He did not come for nothing.' [DN]
b. Hy kon nie niemand gesien het nie. [Afrikaans]
He could not nobody saw has SN
'He could not have seen nobody' [DN]
c. da Valère nie niets kuopt. [Flemish]
That Valère SN nothing buys
'that Valère does not buy anything.' [DN]

In Chapter 5, French *ne*, Flemish *en* and sentence-final Afrikaans *nie* have been characterized as co-negative markers, because they do not convey semantic negation (cf. Section 6 for French, Section 9 for Flemish and Section 10 for Afrikaans). The combination of the faithfulness constraint MAXSN and the economy constraint *NEG is responsible for these meaning effects, as will appear in Section 5.2.

Third, double negation readings arise in a strict negative concord language like Hungarian, where certain n-words (the *sem*-expressions) lexically satisfy MAXSN, and therefore do not occur with the negation marker in the expression of single negation. The example in (41) illustrates.

(41) Senki sem ma nem jött el. [Hungarian]
Nobody SEM today SN come.PAST.3SG PREF
'Nobody is such that it's today that he did not come along.' [DN]

However, Giannakidou is right that a double negation reading for the combination of a marker of sentential negation and a Neg-expression in a strict concord language not exemplifying discontinuous negation or Neg-expressions that lexically satisfy MAXSN is excluded. Falaus (2007a, b) confirms this for Romanian.

Double negation readings with a combination of negation and n-words are available in nonstrict negative concord languages, as illustrated for Italian in (42).

(42) Niente non giova mai. [Italian]
Nothing SN helps never
'Nothing is never useful.' [DN]

As argued by Zanuttini (1991), preverbal n-words do not need the support of the marker of sentential negation to convey a clausal scope of negation. To the degree that the negation marker is not ungrammatical in environments like (42), the sentence conveys double, rather than single negation. Nonstrict negative concord languages like Italian have a high ranking of NEGFIRST. However, NEGFIRST is satisfied by the preverbal n-word, so the negation marker is not needed to satisfy any syntactic faithfulness constraint. The relevance of NEGFIRST is confirmed by instances of double negation in negative concord languages that display NEGFIRST effects in the postverbal domain, such as Welsh (see examples in Section 5.3).

The core idea developed in this section builds on the assumption made by de Swart and Sag (2002) that the marker of sentential negation is semantically redundant in negative concord configurations. However, it can have a syntactic function as a scope marker. In Chapter 5, this insight was worked out by means of the two constraints MAXSN and NEGFIRST, which affect the marker of sentential negation in strict and nonstrict negative concord languages, respectively. The OT analysis supports the claim that double negation readings do not arise in negative concord systems if the marker of sentential negation is required to satisfy a faithfulness constraint (FNEG, MAXSN, NEGFIRST) in the OT syntax. If the negation marker is required in order to satisfy some faithfulness constraint ranked above *NEG, its presence is licensed by the syntax, and the ranking *NEG >> INTNEG guarantees the absorption of its semantic contribution in the resumptive polyadic quantifier.

However, if the presence of the negation marker is not required by any faithfulness constraint ranked above *NEG, economy considerations come into play, and the negation marker is not present in sentences expressing a single negation reading. By means of a process of weak bidirectional optimization, the presence of the negation marker in such sentences leads to an interpretation in terms of double negation, as will be shown in Sections 5.2 and 5.3.

The analysis makes the following predictions for type I, type II, and type III languages. In strict negative concord systems (type I languages) that do not have a discontinuous negation, double negation readings do not arise in the interaction of a Neg-expression with the negation marker, because the high ranking of MAXSN requires the presence of the marker of sentential negation for scope reasons. As far as nonstrict negative concord languages (type III) languages are concerned, double negation readings are expected to arise only if NEGFIRST is independently satisfied (e.g. by a preverbal n-word). In those circumstances, the marker of sentential negation is semantically redundant, and is blocked in sentences expressing a single negation for economy reasons. If it is inserted anyway, the sentence can only be grammatical under a double negation reading.

Type II languages exemplifying negative spread are predicted to convey double negation in all instances in which the negation marker combines with a Neg-expression. In the grammar of negative spread, there is no faithfulness constraint ranked above *NEG that requires the presence of the negation marker in configurations in which FNEG is independently satisfied by a sequence of Neg-expressions. Accordingly, the negation marker is blocked for economy's sake, and sentences involving the combination of a negation marker and a Neg-expression can only be grammatical under a double negation reading, licensed by a weak bidirectional optimization process.

Of course, double negation readings are hard to elicit, even in double negation languages, and they are subject to special context and intonation (cf. Chapter 1, Section 1), so the process of weak bidirectional optimization remains a fairly marginal phenomenon. Still, the fact that it works the way it works provides independent support for the analysis developed so far.

6.5.2 Double Negation Readings in Strict Negative Concord Languages

Double negation readings are well attested for (continental) French when it comes to combinations of an n-word and the negation marker *pas*, in both the written and spoken variety. A careful literature study reveals the existence of double negation readings in the combination of n-words and a negation marker in Welsh, Afrikaans, West Flemish, Hungarian, Spanish, and Italian as well. Given that the details of the French data are fairly well known, I will first analyze double negation readings in French, and then extend the analysis to other NC languages.

Written French is a clear instance of a type I language, exemplifying strict negative concord. Preverbal *ne* is mandatory in all negative sentences, whether or not there is a preverbal n-word, as illustrated in (32) in Chapter 5, repeated here as (43).

(43) a. Il ne vient pas. [written French]
He SN come SN
'He doesn't come.'

b. Il n'a rien vu.
He SN has nothing seen.
'He hasn't seen anything.'

c. Personne n'est venu.
Nobody SN is come.
'Nobody ever came.'

Chapter 5 viewed the obligatory presence of *ne* in written French as an instance of the high ranking of MAXSN in this variety of the language. The discontinuous negation pattern in (43a) illustrates a division of syntactic and semantic labor between the preverbal *ne* and the postverbal *pas*. The clitic *ne* is inserted to satisfy MAXSN, which cannot be satisfied by *pas*. *Pas* is inserted to convey semantic negation, because *ne* has lost its original semantic force (Bréal 1897, 1900, Godard 2004 and references therein). The way the syntactic role of *ne* and the semantic role of *pas* combine was shown in Chapter 5 (Tableau 17), and is reproduced here as Tableau 6.

Both *ne* and *pas* induce a violation of *NEG. Discontinuous negation emerges as the optimal candidate, because of the need to satisfy FNEG as well as MAXSN, at the expense of an additional violation of *NEG. NEGFIRST and FOCUSLAST take care of the position of one marker in the preverbal position, and the other in the postverbal position.

The claim that *pas* is inserted in sentences expressing plain propositional negation in order to satisfy FNEG predicts that *pas* will not be necessary in sentences that satisfy FNEG by independent means. Sentences that involve negation ranging over variables in argument or adjunct position exemplify that situation. Tableau 7, a reproduction of Tableau 13 in Chapter 5, illustrates the blocking of *pas* with a preverbal n-word (cf. 43c).

The ranking MAXNEG >> *NEG triggers the use of Neg-expressions rather than regular indefinites under the scope of negation. Tableau 7 is restricted to candidates that satisfy MAXNEG. Neg-expressions are inherently negative, so all the candidates in Tableau 7 satisfy FNEG. *NEG functions as an economy constraint that blocks candidates with more negations than necessary. Given that *pas* is not needed to satisfy any

Tableau 6 Discontinuous negation in written French (production)

Meaning ¬Come(x)	Form	FNEG	MAXSN	*NEG	NEGFIRST	FOCLAST
	Il vient	*	*		*	*
	Il *ne* vient	*		*		*
	Il vient *pas*		*	*	*	
☞	Il *ne* vient *pas*			**		
	Il *ne pas* vient			**		*
	Il vient *ne pas*			**	*	

Tableau 7 Generation of written French with preverbal n-word

Meaning ¬∃x Come(x)	Form	FNEG	MAXNEG	MAXSN	*NEG	NEG FIRST	FOCUS LAST
	Personne vient			*	*		*
☞	Personne *ne* vient				**		*
	Personne vient *pas*			*	**		
	Personne *ne* vient *pas*				***		

syntactic constraint (neither FNEG, nor MAXSN), it is not only semantically, but also syntactically redundant. Therefore, it is not used in written French to express a single negation in combination with a Neg-expression. The analysis supports the view that the proliferation of negative expressions in negative concord languages is not random, but governed by highly ranked faithfulness constraints. Tableau 7 illustrates the economy effect of the markedness constraint *NEG even in negative concord languages.

According to the ranking in Tableau 7, the combination of *pas* with n-words in modern French should lead to ungrammaticalities, for the candidate involving *pas* is suboptimal. In reality, the combination of *pas* with n-words is not ungrammatical, but leads to a double negation interpretation, both in written French (which preserves *ne*, 44a) and in spoken French (which drops *ne*, 44b).

(44) a. Il n'est pas venu pour rien. [DN]
he SN is SN come for nothing.
'He did not come for nothing.'
b. C'est pas rien. [DN]
it is SN nothing.
'It is not nothing.' (= It is quite something.)

According to the syntactic Tableau 7, the sentences in (44) should be ungrammatical. Furthermore, given the ranking *NEG >> INTNEG in the semantics of negative concord languages, all formal expressions of negation should be absorbed in the resumptive negative quantifier. So if the sentences in (44) were interpretable at all, they should have a single negation reading. The fact that these sentences are grammatical and exhibit a double negation reading is unexpected under a strong bidirectional OT analysis of negation in French. However, it can be explained as an effect of weak bidirectional optimality, also called 'superoptimality' in Blutner's (1998, 2000) approach.

Blutner (1998, 2000) defines a form–meaning pair <f,m> as a weakly optimal pair (also called 'superoptimal' pair) if and only if there is no other superoptimal pair <f',m> such that <f',m> is more optimal, and there is no other superoptimal pair <f,m'> such that <f,m'> is more optimal than <f,m>. It follows from this definition that strong pairs are superoptimal, but not all superoptimal pairs are strong. I illustrate the notion of superoptimality with the two forms, f_1 and f_2, and the two meanings, m_1 and m_2, in the arrow diagram in Figure 3. The arrows indicate preference relations, just like in Figure 1.

✌ <f_1, m_1> ← <f_2, m_1>

↑ ↑

<f_1, m_2> ← <f_2, m_2> ✌

Figure 3 Weak bidirectional OT

Tableau 8 French [*ne..pas* + neg, DN] in weak bidirectional OT

Input [f,m] f_1: *ne*..neg; f_2: *ne..pas* + neg m_1: $\neg\exists xV(x)$; m_2: $\neg\neg\exists xV(x)$	MAXSN	MAXNEG	*NEG	INTNEG
[*ne*..neg, $\neg\exists xV(x)$] ✌			***	
[*ne*..neg, $\neg\neg\exists xV(x)$]			****	
[*ne..pas* + neg, $\neg\exists xV(x)$]			****	*
[*ne..pas* + neg, $\neg\neg\exists xV(x)$] ✌			*****	

In Figure 3, form f_1 is always preferred over f_2, and meaning m_1 is always preferred over m_2. As a result, two arrows are pointing toward the pair <f_1, m_1>, and two arrows are pointing away from the pair <f_2, m_2>. The fact that two arrows are pointing toward the pair <f_1, m_1> indicates that this is a strongly optimal pair. The pairs <f_1, m_2> and <f_2, m_1> lose against this strongly optimal pair, because <f_1, m_1> has a better form for the same meaning, or a better meaning for the same form. As a result, <f_1, m_2> and <f_2, m_1> are neither strong nor superoptimal pairs.

Under strong bidirectional OT, the pair <f_2, m_2> is blocked, because there are better form-meaning pairs available, as the arrows indicate. Under weak bidirectional OT, the pair <f_2, m_2> arises as a weakly optimal (or 'superoptimal') form–meaning pair, because there is no superoptimal pair that has either a better form or a better meaning. In Figure 3, <f_2, m_2> is not in direct competition with the only other superoptimal pair <f_1, m_1>, because the two pairs differ in both their form and their meaning component. Weakly optimal pairs involve forms that are suboptimal in unidirectional generation, and meanings that are suboptimal in unidirectional interpretation. As a result, superoptimality pairs up marked forms and marked meanings that would not be available otherwise.

Blutner (1998, 2000, 2004) uses superoptimality to account for instances of partial blocking in the lexicon (cf. Chapter 2, Section 4). Here I view the unusual pattern of double negation readings with a negation marker and a Neg-expression as an instance of partial blocking in the grammar. Accordingly, the expression of single negation by means of an n-word, at the exclusion of *pas* constitutes a strongly optimal form–meaning pair in continental French (43b, c), in which *pas* is semantically ànd syntactically redundant. The insertion of *pas* creates a marked form that pairs up with the marked double negation meaning (44), as a weakly optimal pair. Tableau 8 spells out the bidirectional implementation.

The Neg-expression is unmarked compared to the combination of *pas* with a Neg-expression, because it is formally simpler and shorter. Just like in Tableaux 4 and 5, I take single negation to provide the unmarked meaning as compared to the marked double negation meaning. The pair [*ne*..neg, $\neg\exists xV(x)$] is strongly optimal, because there is no better interpretation for this form, and no better form to express this meaning. This unmarked form–meaning combination occurs in examples (43b, c). No other form–meaning pair is strongly optimal. [*ne*..neg, $\neg\neg\exists xV(x)$] is not a strongly optimal pair, because single negation is a better, more economical interpretation for this form. [*ne..pas* + neg, $\neg\exists xV(x)$] is not a strongly optimal pair, because *neg* is a better, more economical form to express this meaning.

In a secondary (not optimal but 'superoptimal') round of evaluation, the restriction of the comparison to superoptimal pairs eliminates both [*ne*..neg, $\neg\neg\exists xV(x)$] and [*ne*..*pas* + neg, $\neg\exists xV(x)$] from the set of competing form–meaning pairs. Both lose against the strongly optimal candidate [neg, $\neg\exists xV(x)$], and therefore cannot be superoptimal pairs themselves. The strong pair [*ne*..neg, $\neg\exists xV(x)$] does not directly compete with [*ne*..*pas* + neg, $\neg\neg\exists xV(x)$], because it has both a different form and a different meaning. Given that there is no better superoptimal alternative, [*pas* + neg, $\neg\neg\exists xV(x)$] itself qualifies as a weakly optimal or superoptimal form–meaning pair. The examples in (44) are instantiations of this marked form–meaning combination.

Obviously, this result cannot be obtained in a unidirectional system, because *ne..pas* + neg is itself a suboptimal form, and $\neg\neg\exists xp$ a suboptimal meaning. Weak bidirectional optimality thus allows the expression of double negation in negative concord languages, whereas this is impossible under unidirectional optimization or strong bidirectional optimality (cf. Section 4).

The same analysis can be applied to the double negation readings of certain Afrikaans examples. Afrikaans displays a pattern very similar to French, as illustrated by the data in (45) (from Waher 1978, van der Wouden 1994, K. van Gass, 2007 personal communication).

(45) a. Haar suster het nie haar verjaarsdag vergeet nie. [Afrikaans]
Her sister has SN her birthday forgotten SN
'Her sister didn't forget her birthday.'
b. Niemand het dit gesien nie.
Nobody has this seen SN
'Nobody has seen this.'
c. Hulle het nooit gesing nie.
They have never sung SN
'They have never sung.'
d. Hy kon nie niemand gesien het nie. [DN]
He could SN nobody saw has SN
'He could not have seen nobody'

Afrikaans is a doubling language with a discontinuous negation (cf. Chapter 5, Section 10). One occurrence of *nie* is in an immediately postverbal position (sentence-medial *nie*), and the other one appears in sentence-final position (sentence-final *nie*) (45a) (cf. Chapter 3, Section 3.4). Sentence-final *nie* co-occurs with the negative indefinites *niemand* and *nooit* in (45b, c) (cf. Chapter 5, Section 10). The insertion of sentence-medial *nie* in a sentence already containing a Neg expression leads to a double negation reading (45d). I explain this along the same lines as the co-occurrence restrictions between sentential negation and negative indefinites in French.

The constraint ranking in Afrikaans is FNeg >> {MaxSN, NegAttract} >> *Neg >> MaxNeg. Sentence-final *nie* satisfies MaxSN. Sentence-final *nie* is too weak to express propositional negation by itself, so negation is doubled in examples like (45a). The doubling *nie* is the bearer of semantic negation, and satisfies FNeg.

The negative indefinites express semantic negation, so the immediately postverbal marker *nie* is semantically redundant in (45b, c). As a result of the economy constraint *NEG, the doubling *nie* is left out.

As outlined in Chapter 5 (Section 10), the status of Afrikaans as a double negation or a negative concord language is under debate. If Afrikaans is perceived as a double negation language, the double negation interpretation of (45d) follows under the ranking INTNEG >> *NEG and the assumption that sentence-final *nie* is semantically non-negative. If Afrikaans is perceived as a negative concord language, the ranking *NEG >> INTNEG in the semantics should favor negative resumption, which blocks the double negation reading of (45d). The acceptability of (45d) with a double negation reading is then explained along the lines of weak bidirectionality: the example combines the marked form (n-word plus doubling *nie*) with the marked, double negation meaning, as illustrated in Tableau 9.

It is sufficient to replace *ne* and *pas* in Tableau 8 by sentence-final *nie* and sentence-medial *nie* in Tableau 9 to obtain the account of double negation readings in (the negative concord variety of) Afrikaans that mirrors the analysis of French.

The varieties of Flemish which display the discontinuous negation *en..nie* constitute another strict negative concord language that exemplifies double negation readings in certain configurations, as pointed out by Haegeman and Zanuttini (1996). They observe that (46a) has a single negation reading, but (46b) can only express double negation.

(46) a. da Valère niets (nie) kuopt. [NC][West Flemish]
that Valere nothing (SN) buys
'that Valère does not buy anything.'
b. da Valère nie niets kuopt. [DN]
That Valère SN nothing buys
'that Valère does not buy anything.'

According to Haegeman and Zanuttini (1996), the Neg-expression must precede the (optional) negator *nie* in order to establish a negative concord relation. They claim that Neg-expressions must scramble out of the VP in order to take sentential scope. Under the OT analysis, the fact that word order makes the difference between the single negation reading of (46a) and the double negation reading of (43b) suggests that the presence of *nie* in (46b) is not required for scope reasons, and should be blocked by economy. As a result, the sentence can only be grammatical under a

Tableau 9 Afrikaans [*nie..nie* + neg, DN] in weak bidirectional OT

Input [f,m] f_1: neg..*nie*; f_2: neg + *nie..nie* m_1: ¬∃xV(x); m_2: ¬¬∃xV(x)		MAXSN	NEGATTR	*NEG	INTNEG
[neg…*nie*, ¬∃xV(x)]	✌			***	
[neg…*nie*, ¬¬∃xV(x)]				****	
[*nie* + neg...*nie*, ¬∃xV(x)]				****	*
[*nie* + neg…*nie*, ¬¬∃xV(x)]	✌			*****	

double negation reading along the lines of the weak bidirectional optimization process outlined in this section. Similar constraints apply in Afrikaans, where the *nie* preceding the n-word leads to a double negation reading (40b), whereas Biberauer and Zeijlstra (2009) signal a double negation reading for examples in which the sentence-medial *nie* immediately follows the n-word (cf. Chapter 4, Section 10).

The Hungarian single and double negation examples discussed in Chapter 5 (Section 8), and repeated in (47) constitute a special case.

(47) a. Senki sem jött el sehova. [Hungarian]
nobody.NOM SEM come.PAST.3SG PREF nowhere-to
'Nobody came along anywhere.' [NC]

b. Senki sem ma nem jött el.
nobody SEM today SN come.PAST.3SG PREF
'Nobody is such that it's today that he did not come along.' [DN]

A preverbal *sem*-expression is sufficient to convey single negation, and enter into a negative concord relation with a postverbal n-word (47a). The insertion of the negation marker *nem* in a sentence with a preverbal *sem*-expression leads to a double negation reading (47b). At first sight, this constitutes a rather surprising observation, for Hungarian is a strict negative concord language, which ranks MAXSN above *NEG, and requires the presence of the marker of sentential negation in every negative clause. Chapter 5 (Section 8) accounted for the absence of *nem* in (47a) by the assumption that *sem* satisfies the constraint MAXSN, because the marker of SN is visible from its internal composition. The insertion of an extra marker of sentential negation in (47b) creates a marked form, which pairs up with a marked double negation meaning. The weak bidirectional Tableau 10 illustrates.

The form–meaning combination [*sem*, ¬∃xV(x)] constitutes a strongly optimal pair, under the bidirectional OT grammar {MAXNEG, MAXSN, NEGFIRST} >> *NEG >> INTNEG of Hungarian. Neither the pair [*sem*, ¬¬∃xV(x)] nor the pair [*sem* + *nem*, ¬∃xV(x)] constitutes a strongly optimal form–meaning pair. Given that the single negation reading is a better interpretation of the *sem* form, and the *sem* form is a better expression of the single negation reading, these pairs lose against the strongly optimal pair [*sem*, ¬∃xV(x)].

The form–meaning combination [*sem* + *nem*, ¬¬∃xV(x)] does not take part in this competition, because it involves both a different form and a different meaning. In a second round of optimization it can therefore emerge as a superoptimal pair.

Tableau 10 Hungarian [*sem* + *nem*, DN] in weak bidirectional OT

Input [f,m] f_1: *sem*; f_2: *sem* + *nem* m_1: ¬∃xV(x); m_2: ¬¬∃xV(x)		MAXNEG	MAXSN	NEGFIRST	*NEG	INTNEG
[*sem*, ¬∃xV(x)]	✌				**	
[*sem*, ¬¬∃xV(x)]					***	
[*sem* + *nem*, ¬∃xV(x)]					***	*
[*sem* + *nem*, ¬¬∃xV(x)]	✌				****	

The strict NC language Hungarian then aligns with French, Afrikaans, etc. as far as the treatment of double negation readings are concerned, even though it does not have a discontinuous propositional negation.

6.5.3 Double Negation Readings in Nonstrict Negative Concord Languages

The double negation readings in negative concord languages are not restricted to strict negative languages that display discontinuous negation, such as written French and Afrikaans, or that have n-words that incorporate negation and satisfy MAXSN, such as Hungarian. The fact that the meaning effects are the same in written (44a) and spoken French (44b) supports the view that negative spread is also governed by economy constraints.

As far as nonstrict negative concord languages are concerned, double negation readings occur in the combination of a preverbal n-word and preverbal sentential negation. Italian, Spanish, and European Portuguese are type III languages in which postverbal n-words require the presence of a preverbal marker of negation (48a, 49a), but preverbal n-words do not (48b, 49b).

(48) a. Mario non ha parlato di niente con nessuno. [Italian]
Mario SN has talked about nothing to nobody.
'Mario didn't talk to anybody about anything.'
b. Nessuno ha parlato con nessuno.
Nobody has talked with nobody.
'Nobody talked to anybody.'

(49) a. No he visto a nadie. [Spanish]
SN has seen nobody
'He hasn't seen anybody.'
b. Nadie ha dicho nada.
Nobody has said nothing
'Nobody said anything.'

Chapter 5 highlighted the role of NEGFIRST in nonstrict NC languages. NEGFIRST requires the expression of negation to be preverbal. In Italian and Spanish, the high ranking of this constraint triggers the insertion of a preverbal negation marker with postverbal n-words (48a, 49a). Both n-words and the marker of sentential negation can satisfy NEGFIRST. If preverbal n-words satisfy NEGFIRST, the marker of sentential negation is both syntactically and semantically redundant. The functioning of *NEG as an economy constraint penalizing unnecessary multiplications of negation predicts the insertion of a negation marker to be ungrammatical in contexts like (48b, 49b).

However, Zanuttini (1991: 151) observes marginal instances of double negation readings in such configurations, as illustrated by (50a). Corblin and Tovena (2003) make the same claim for (50b). They add that double negation readings are easier to obtain with determiners (as in 50b) than with pronominal n-words (as in 50c), which they give a question mark. A native speaker of Italian that I consulted rejected the distinction between determiners and pronouns, and accepted (50c) with the appropriate intonation contour in a context which supports the double negation interpretation.

(50) a. Nessuno degli studenti non è venuto. [Italian]
None of the students SN is come.
= None of the students hasn't come.
b. Nessun bambino non va scuola in questa società.
no child SN goes school in this society
'No child does not to go school in this society.'
c. Niente non giova mai.
Nothing SN helps never
'Nothing is never useful.'

Double negation readings with preverbal n-words in combination with the marker of sentential negation are also attested in Palestinian Arabic. Example (51) (from Hoyt 2006) is repeated from Chapter 5.

(51) a. ma-ʕind-hæ wɛla nɪtfɪt χağal. [Palestinian Arabic]
SN-at-her not.even bit shame
'She doesn't have the least bit of shame.'
b. wɛla ḥada fi:-hʊm šæ:f-ni.
not.even one.MS in-them saw.3MS-me
'Not even one of them saw me.'
c. wɛla yo:m ma-ʕağabni l-ɛkɪl. [DN]
not.even day SN-pleased.3ms-me the-food
'There wasn't one day the food didn't please me.'

(51a) shows that the postverbal n-word *wɛla nɪtfɪt* requires the support of the preverbal marker of sentential negation *ma*. In preverbal position, *wɛla ḥada* is not accompanied by the negation marker (51b). The insertion of the marker of sentential negation to the n-word in preverbal position leads to the double negation reading in (51c).

In nonstrict as well as strict negative concord languages, *NEG functions as an economy constraint, and the presence of the marker of sentential negation in negative concord constructions needs to be licensed by a faithfulness constraint. In type III languages, the role of NEGFIRST is crucial. Only if NEGFIRST is independently satisfied, a weak bidirectional optimization process leading to the expression of double negation can emerge.

Tableau 11 : Italian [*non* + neg, DN] with preverbal n-word in weak bidirectional OT

Input [f,m] f_1: neg V; f_2: neg + *non* V m_1: ¬∃xV(x); m_2: ¬¬∃xV(x)	NegFirst	*Neg	IntNeg
[neg V, ¬∃xV(x)] ✌		**	
[neg V, ¬¬∃xV(x)]		***	
[neg + *non* V, ¬∃xV(x)]		***	*
[neg + *non* V, ¬¬∃xV(x)] ✌		****	

Configurations of a negation marker with a preverbal n-word provide the ideal environment for nonstrict negative concord languages to allow double negation readings. The unmarked (single negation) reading is expressed by the unmarked form (with a preverbal n-word, without a negation marker) (48b, 49b), whereas the marked form (with a preverbal n-word and a negation marker) expresses the marked, double negation reading (50, 51c). The bidirectional OT grammar is {MaxNeg, NegFirst} >> *Neg >> IntNeg spelled out in Tableau 11 for Italian.

MaxNeg and NegFirst are satisfied by the preverbal n-word, so none of the candidates under consideration in Tableau 11 violates this constraint. The unmarked form-meaning combination [neg V, ¬∃xV(x)] constitutes a strongly optimal pair. The marked form–meaning combination [neg + *non* V, ¬¬∃xV(x)] emerges as a weakly optimal pair.

No double negation readings are predicted for the combination of a preverbal marker of sentential negation and a postverbal n-word, because in that case, the preverbal negation marker is required to satisfy NegFirst. Indeed, all the examples of double negation readings in type III languages that I have seen involve a preverbal n-word.

Nonstrict negative concord languages are not the only languages in which NegFirst plays an important role. Welsh constitutes a special case, because of the variety of NegFirst constraints governing the occurrence of negation in postverbal position in different dialects (cf. Chapter 5, Section 7). Recall that colloquial Welsh is a VSO language with an adverbial position immediately following the subject. Some Welsh dialects involve a modified (postverbal) version of NegFirst, which requires negation to precede all material following VS(A). Either the negation marker or an adverbial n-word in A can satisfy this constraint.

Interestingly, Borsley and Jones (2005: 133) report double negation readings for sentences in colloquial Welsh that combine two negative adverbs in postsubject position, the second of which is *ddim*:

(52) a. Dw i ddim ddim yn poeni. [informal Welsh]
be.PRES.1SG I SN SN PROG worry.
'I don't not worry.'

b. Dw i byth ddim yn poeni.
be.PRES.1SG I never SN PROG worry.
'I never don't worry.'

c. Dydy hi erioed ddim wedi helpu.
NEG.be3SG she never SN PERF help
'She has never not helped.'

In (52), the second negative adverb following the subject is the marker of sentential negation *ddim*. *Ddim* is not required to satisfy NEGFIRST, because there is an n-word (52b) or another instance of *ddim* (52a) in the adverbial position following the subject that takes care of that requirement. The additional presence of *ddim* means that the candidate incurs an extra violation of *NEG. Syntactically, it is then a less optimal form than its counterpart without the marker of sentential negation. Therefore, sentences like (52) might be expected to be ungrammatical, but they are not. However, they are only grammatical under a double negation reading. The felicity of this form–meaning combination arises as the result of the weak bidirectional optimization process spelled out in Tableau 12.

Tableau 12 is a variant of Tableau 11 with the appropriate postverbal version of NEGFIRST (here glossed as NEGFIRST (Welsh)). Other than the fact that a modified version of NEGFIRST is relevant to account for the data, the patterns are the same as those in nonstrict negative concord languages such as Italian.

6.5.4 *Concerns About Weak Bidirectional Optimization*

As pointed out in Chapter 2 (Section 8), a number of studies have argued against the recursive mechanism of weak bidirectional optimization (Zeevat 2000, Beaver and Lee 2004). According to Beaver and Lee (2004), full recursion implies the possibility of an infinite number of rounds of optimization. As a result, in weak bidirectional OT, every pair is a winner, as Beaver and Lee (2004: 126) phrase it. In certain nonlinguistic optimization processes, such as the matching partners in a

Tableau 12 Welsh [neg + *ddim*, DN] with immediately postverbal n-word in weak bidirectional OT

Input [f,m] f_1: V neg; f_2: V neg + *ddim* m_1: ¬∃xV(x); m_2: ¬¬∃xV(x)		NEGFIRST (Welsh)	*NEG	INTNEG
[V neg, ¬∃xV(x)]	✌		**	
[V neg, ¬¬∃xV(x)]			***	
[V neg + *ddim*, ¬∃xV(x)]			***	*
[V neg + *ddim*, ¬¬∃xV(x)]	✌		****	

dancing competition (Hendriks et al. 2009: Chapter 1, see also Chapter 2, Section 8), this is the desired outcome. In most linguistic applications, two rounds of optimization constitute a natural limit.

There are various ways to address the overgeneralizations made by weak bidirectional OT. Blutner et al. (2006: 149) impose general cognitive limitations on recursion. They constrain bidirectional optimization to at most two rounds, in agreement with the bounds observed for other cognitive tasks (such as the epistemic reasoning required for playing strategic games). As far as double negation is concerned, two rounds of optimization suffice to derive the patterns discussed in this section. The first round derives the single negation readings, the second round the double negation readings. I conclude that the analysis of double negation readings in negative concord languages is within the cognitive boundaries on recursive optimization.

6.6 Conclusion

This chapter focused on instances of double negation in negative concord languages. At first sight, such cases might seem problematic for the analysis developed so far, because the ranking *NEG >> INTNEG in the OT semantics should lead to a resumptive negative quantifier, and force a single negation reading throughout. In fact, the double negation readings that appear in negative concord languages provide independent evidence in favor of the OT analysis.

There is no conceptual ban on double negation in negative concord languages, but the OT grammar forces a resumptive interpretation for a series of Neg-expressions within a single predicate–argument structure. The double negation readings arising with morphological (affixal) negation and with negation in multiple clause constructions provide support for this view. The construction of a resumptive polyadic quantifier is restricted to Neg-expressions, and there is good reason to believe that affixal negation does not qualify as a Neg-expression. Polyadic quantification is clause bound, so the presence of Neg-expressions and/or a marker of sentential negation in different clauses should force a double negation reading. These two constructions are thus in line with the OT approach.

Other instances of double negation readings in negative concord languages require an extension of the OT analysis developed so far. Ambiguities with multiple n-words were an important motivation for the analysis developed by de Swart and Sag (2002). Their insights can be reconciled with the typology developed in Chapter 4 if there is overlap between the constraints *NEG and INTNEG in a stochastic extension of the OT semantics.

Finally, double negation readings arising in special combinations of the marker of sentential negation with an n-word are accounted for under a weak bidirectional extension of the OT analysis. Given that only one round of recursion is required to obtain the superoptimal double negation reading, this analysis remains within the cognitive limits that have been imposed upon weak bidirectionality.

The two phenomena of double negation readings with multiple n-words and double negation readings with an n-word and a negation marker have been marginally discussed in the literature, but are not properly explained in existing analyses of negation and negative concord. Obviously, the examples are highly marked, for they are dependent on particular intonation patterns, and are contextually restricted. However, these considerations apply to double negation in double negation languages like (standard) English and Chinese as well (cf. Horn 1989, 2001 and Chapter 1, Section 1.1). I conclude that the observations cannot easily be dismissed. The fact that these unusual patterns get a straightforward explanation in a weak bidirectional extension of the OT analysis developed in Chapters 4 and 5 constitutes an important new insight.

Chapter 7
Conclusion and Further Perspectives

Introduction and overview This chapter sums up the most important results of the study on the expression and interpretation of negation carried out in this book. I consider their relevance for linguistic theory (Section 1) and, more broadly, for a theory of human cognition (Section 2). Section 3 discusses the implications of my work for future research.

7.1 Summary of Research

Negation is a universal category of natural language (Dahl 1979, Chapter 3), and presumably of human cognition (de Swart 2009). However, systems of negation in natural language vary, and reflect highly complex grammaticalization processes, as witnessed by the wide range of data gathered in the literature, and as discussed in this book. It has proven extremely difficult to provide a unified view of negation, polarity, and concord across different languages, dialects, and diachronic stages of a language as part of a general theory of human language and cognition. In this monograph, I hope to have provided the main ingredients of just such an analysis. In this section, I sum up the most important results.

Language variation and language change are important topics in current linguistic research. The connections between optimality theory and game theory, with its applications in biology, economics, logic, and information retrieval make it possible to situate this particular research project in the broader context of research on human cognition. Bidirectional OT focuses on the communicative process between users of a language, and is embedded in a pragmatic theory of speaker–hearer interactions. These aspects make me believe that the project did not only lead to new results on negation in typology and language change, but benefits the interdisciplinary field of cognitive science that linguistics is a part of. The embedding in this broader perspective is the topic of Section 2.

Of course, there is always more to be done, and there are numerous possible extensions of the research reported on in this book. In Section 3, I sketch some potential lines of investigation.

H. de Swart, *Expression and Interpretation of Negation: An OT Typology*, Studies in Natural Language and Linguistic Theory,
DOI 10.1007/978-90-481-3162-4_7,

7.1.1 Negation as a Universal Category of Natural Language

The starting point of my study was the observation that all natural languages have ways to express negation, i.e. something that corresponds to the first-order logical connective ¬. In English, this would be *not*. It is a universal feature of human cognition that speakers are able to conceptualize the meaning ¬p as well as p, and want to express both affirmation and negation in their mother tongue. The analysis developed in Chapter 3 is rooted in the observation that there is an asymmetry between the expression of p and its negative counterpart ¬p in that negation is always overtly marked, but affirmation takes the 'zero' marker or a simpler expression. The asymmetry is illustrated for English in (1):

(1) a. It is raining.
    b. It is *not* raining.

The complexity in expression is related to the semantic markedness of negation. Under strong evolutionary pressure, languages form optimal systems of communication. A key feature of optimal communication is that natural languages respect Horn's (1984) division of pragmatic labor. According to this principle, unmarked meanings pair up with unmarked forms, and marked meanings with marked forms.

Zeevat and Jäger (2002), and Jäger (2003) and Mattausch (2005, 2007) develop an evolutionary model that derives Horn's (1984) division of pragmatic labor from frequency asymmetries. Under the assumption that negative meanings are relatively infrequent compared to affirmative meanings, an evolutionary stable system of communication arises in which the use of marked forms for unmarked meanings is avoided, because it violates the general economy constraint which avoids structure. Accordingly, zero or simple forms are used for affirmation, and more complex forms are used for negation. The translation of this model into the faithfulness and markedness constraints, familiar from optimality theory (Smolensky and Legendre 2006), leads to the formulation of the two constraints, FNEG and *NEG (repeated from Chapter 3).

- FNEG
  Be faithful to negation, i.e. reflect the nonaffirmative nature of the input in the output.
- *NEG
  Avoid negation in the output.

FNEG is a faithfulness constraint that establishes a correspondence between input and output. *NEG is a markedness constraint that avoids structure in the output. Under the assumption that human speakers want to distinguish between affirmative and negative statements in their language, evolutionary pressure leads to the universal ranking FNEG >> *NEG. In this way, the OT analysis accounts for Dahl's (1979) typologically based observation that negation is a universal category of natural language.

7.1.2 Integration of Negation in the Sentence Structure

The universal ranking FNEG >> *NEG imposes strong limits on the possible variation we find in natural language. This ranking requires negation to be overtly expressed in all languages. However, the ranking FNEG >> *NEG leaves a wide range of variation as to the means by which negation is expressed, and the integration of the negation marker in the grammar of a particular language. Chapter 3 focuses on variation in the position of the negation marker with respect to the verb. Building on Jespersen (1917, 1924, 1933), I distinguish three main possible constructions: preverbal negation, postverbal negation, and discontinuous negation.

The general markedness constraint *NEG and the faithfulness constraint FNEG are complemented with two faithfulness constraints governing the placement of negation in the sentence: NEGFIRST and FOCUSLAST (repeated from Chapter 3).

- NEGFIRST
  Negation precedes the finite verb.
- FOCUSLAST
  New information comes last in the sentence.

A high ranking of NEGFIRST leads to preverbal negation; a high ranking of FOCUSLAST leads to postverbal negation. If both NEGFIRST and FOCUSLAST outrank *NEG, discontinuous negation emerges (typically with one marker in the preverbal position, and the second in the postverbal position). A typology of languages arises as the result of re-ranking the three constraints with respect to each other (Table 1, repeated from Chapter 3).

The typology can be interpreted in a synchronic as well as a diachronic perspective. The three main phases of the Jespersen cycle are thus accounted for.

Overlapping ranges of constraints in the rankings corresponding with the three main phases of the Jespersen cycle account for intermediate systems in which a preverbal negation is always obligatory, but a postverbal marker is optional, or a postverbal negation is always obligatory, but a preverbal marker is optional. Such intermediate phases are diachronically unstable according to Haspelmath (1997), and require an extension toward stochastic optimality theory. It is likely that languages eventually stabilize on an ordinal ranking.

Table 1 Typology of placement of negation w.r.t. the verb

Preverbal negation	{NEGFIRST, *NEG} >> FOCUSLAST
Discontinuous negation	{NEGFIRST, FOCUSLAST} >> *NEG
Postverbal negation	{FOCUSLAST, *NEG} >> NEGFIRST

7.1.3 Neg-expressions, Double Negation, and Negative Concord

Dahl (1979), Horn (1989), and others claim that all languages have ways to express propositional negation. Many languages also have pronominal or adverbial expressions negating the existence of individuals having a certain property. In English, this would be *nobody, nothing, never, nowhere*. In this book they are called negative indefinites, or Neg-expressions. Their lexical semantics is subject to debate in the literature, as outlined in Chapter 1. Following Zanuttini (1991), Haegeman and Zanuttini (1996), de Swart and Sag (2002), Watanabe (2004), and others, I assign negative indefinites the lexical semantics $\neg\exists x$.

The syntactic and semantic properties of negative indefinites in double negation and negative concord languages are studied in Chapter 4. Under the assumption that knowledge of first-order logic is part of human cognition, negative indefinites might be expected to behave alike across languages. From empirical research carried out in typology and descriptive linguistics, we know that this is not the case. The key insight is that languages make use of the same underlying mechanisms, but exploit the relation between form and meaning in different ways.

Neg-expressions resemble the marker of sentential negation in their capacity to satisfy FNEG. Jespersen (1917) notices that negation is frequently attracted to an argument. The constraint NEGATTRACT captures this tendency.

- NEGATTRACT
  Realize (clausal) negation on an indefinite in argument or adjunct position

Whether or not a language realizes negation on an indefinite, rather than on the marker of sentential negation depends on the ranking of NEGATTRACT with respect to other constraints governing the expression of negation. Chapter 4 focuses on sequences of negative indefinites (Neg-expressions) across languages and distinguishes two main classes of languages: double negation languages and negative concord languages.

In languages like English, the combination of two negative indefinites gives rise to a double negation reading (2a). In negative concord languages such as Romance, Slavic, Greek, Hungarian, Afrikaans, and many others, the combination of two Neg-expressions leads to a single negation reading (2b).

(2) a. Nobody has said nothing. [English]
$\neg\exists x \neg\exists y$ Say(x,y)
b. Nessuno ha detto niente. [Italian]
Nobody has said nothing.
'Nobody has said anything.'
$\neg\exists x \exists y$ Say(x,y)
c. Nobody has said anything. [English]

The proper way to say in English what the Italian sentence (2b) expresses would be (2c), rather than (2a). The English *any* is not a Neg-expression, but an NPI, though (cf. Chapter 1, Sections 3 and 4 and Chapter 4, Section 5). Chapter 4 proposes a bidirectional OT analysis that captures the syntax-semantics interface of Neg-expressions in both classes of languages. In negative concord languages, the functional motivation that favors marking of 'negative variables' prevails. Double negation languages value first-order iteration. The bidirectional set-up is essential, for syntactic and semantic variation go hand in hand.

The analysis relies on de Swart and Sag's (2002) analysis of negative concord in terms of polyadic quantification. In this analysis, all Neg-expressions contribute semantic negation, and are stored in an N-store when the sentence is parsed. Interpretation takes place at the clausal level. A sequence of Neg-expressions can be interpreted in two different ways upon retrieval from the N-store. Iteration corresponds to function application, and leads to a double negation reading. Resumption involves the construction of an n-ary negative quantifier, binding all the variables of the Neg-expressions in the sequence.

The typology is built on four constraints. The two constraints FNEG and *NEG come into play in the expression and interpretation of sentential negation as well as Neg-expressions. The two new constraints are a syntactic faithfulness constraint, MAXNEG, and a semantic faithfulness constraint, INTNEG (repeated from Chapter 4).

- MAXNEG
  Mark 'negative variables' (i.e. mark indefinites in argument or adjunct position that are nterpreted in the scope of an anti-additive operator such as negation, as formally negative)

- INTNEG
  Force iteration (i.e. interpret every Neg-expression in the input form as contributing a semantic negation at the first-order level in the output)

FNEG is universally ranked higher than *NEG. MAXNEG, INTNEG, and *NEG can be re-ranked with respect to each other. The relevant bidirectional grammar for double negation and negative concord languages is spelled out in Table 2 (repeated from Chapter 4).

In the bidirectional grammars in Table 2, the choice between resumption and iteration in the semantics is explicitly related to the functional desirability of marking negative variables in the syntax. The functional desirability of marking 'negative variables in negative concord languages (MAXNEG >> *NEG in the syntax) is balanced by economy in the semantics (*NEG >> INTNEG favoring negative resumption). The preference for first-order interpretation in double negation languages (INTNEG >> *NEG in the semantics) is balanced by economy in the syntax (*NEG >> MAXNEG, producing regular indefinites or negative polarity items rather than n-words). Neither a purely syntactic nor a purely semantic analysis of negative concord, but a true syntax–semantics interface of negation and negative indefinites emerges from this analysis.

Table 2 Bidirectional grammar

Negative concord languages	FNEG >> MAXNEG >> *NEG >> INTNEG
Double negation languages	FNEG >> INTNEG >> *NEG >> MAXNEG

A factorial typology of the three constraints in Table 2 leads to six possible language classes. However, rankings other than the ones given in Table 2 are unstable, because the two directions of optimization (production and interpretation) are unbalanced. Languages vary in the ranking of these constraints synchronically as well as diachronically. Thus the OT analysis offers a theory of cross-linguistic variation as well as language change.

Once the markedness of negation, and the functional motivation for Neg-expressions is in place, the interaction of the marker of sentential negation and negative indefinites is studied. Chapter 4 maintains the view advanced by de Swart and Sag (2002) that negative concord is a semantic relation between negative indefinites. The marker of sentential negation is not driving negative concord (as it is in many existing analyses of negative concord). When it co-occurs with a Neg-expression, FNEG is satisfied by the Neg-expression, and the marker of sentential negation typically functions as a scope marker.

Three possible language classes are distinguished in Chapter 5: languages in which Neg-expressions always combine with the marker of sentential negation (strict negative concord), languages in which they never do (double negation or negative spread), and languages in which they sometimes do (nonstrict negative concord). In languages that block the co-occurrence of the negation marker and n-words, a high ranking of *NEG blocks the presence of the marker of sentential negation in sentences involving a negative indefinite. Neg-expressions contribute semantic negation, so FNEG is satisfied. Given that the basic role of a marker of sentential negation is to satisfy FNEG, it is redundant in sentences in which other expressions that have additional functions (such as Neg-expressions contributing a variable binding a particular argument position) assume that role. *NEG then works as an economy constraint, and rules out the use of a marker of sentential negation in combination with a Neg-expression as suboptimal. Depending on the position of MAXNEG and INTNEG with respect to *NEG, the negation system displays double negation or negative spread.

The presence of the marker of sentential negation is licensed in combination with a sequence of Neg-expressions in certain configurations if the negation marker is needed to satisfy other faithfulness constraints. In languages like Spanish, Italian, and European Portuguese, the constraint NEGFIRST plays a role. In these languages, preverbal Neg-expressions do not combine with a negation marker (2b, 3b), but postverbal Neg-expressions require the presence of a preverbal negation marker (3a).

(3) a. Mario *(non) ha parlato di niente con nessuno. [Italian]
Mario *(SN) has talked about nothing to nobody.
'Mario didn't talk about anything to anyone.'
b. Nessuno (*?non) ha parlato con nessuno.
nobody (*?SN) has talked with nobody.
'Nobody talked to anyone.'

A high ranking of the constraint NEGFIRST in Italian requires negation to be realized in a preverbal position, whether expressed by the marker of sentential negation (3a) or by the negative indefinite.

The third class of languages always requires the presence of a marker of sentential negation in a sentence expressing negation, independently of the presence of Neg-expressions. Strict negative concord is typologically very common, and found in Greek, Hungarian, Polish, Russian, Japanese, Romanian, and many other languages.

(4) a. Nimeni *(nu) a venit. [Romanian]
nobody *(SN) has come.
'Nobody came'
b. *(Nu) a venit nimeni.
*(SN) has come nobody.
'Nobody came'

None of the existing constraints takes care of this situation, so I define a new syntactic faithfulness constraint, MAXSN, in Chapter 5.

- MAXSN
  A negative clause (i.e. a clause that conveys a negative proposition) must bear a marker of sentential negation.

MAXSN is functionally motivated by the claim that the scope and focus of negation are marked by different expressions (negation and n-words, respectively). Whereas non-strict NC languages use either n-words or the marker of sentential negation to mark the clausal scope of negation, scope marking is the exclusive responsibility of the marker of sentential negation in strict NC languages. A high ranking of the constraint MAXSN is responsible for the obligatory presence of the negation marker *nu* in Romanian, independently of the preverbal (4a) or postverbal (4b) position of the Neg-expression.

The eight constraints, *NEG, FNEG, NEGFIRST, FOCUSLAST, NEGATTRACT, MAXNEG, MAXSN, and INTNEG, account for the full range of variation we find in the expression and interpretation of negation in a wide range of languages discussed in Chapters 3 through 5. Among these constraints, we find one core markedness constraint, namely, *NEG. *NEG aims at avoiding negation in either form (generation) or meaning (interpretation), so it is operative in the OT syntax as well as the OT semantics. The other seven constraints are faithfulness constraints.

FNEG operates on both forms and meanings, just like *NEG. INTNEG is the only semantic faithfulness constraint needed. The interaction between MAXNEG, *NEG, and INTNEG forms the core of our bidirectional grammar. The ranking of these three constraints determines whether a language exemplifies double negation or negative concord. NEGFIRST, FOCUSLAST, NEGATTRACT, MAXNEG, and MAXSN are all syntactic constraints, which relate a meaningful input to a particular output form.

The asymmetry between syntactic and semantic constraints indicates that a large amount of the cross-linguistic variation we find in the domain of negation is due to the way negation is expressed. However, the expression of negation cannot be separated from its interpretation. The study of the interaction of negative indefinites and the marker of sentential negation in Chapter 5 is crucially embedded in the bidirectional OT analysis developed in Chapter 4. The ranking of FNEG >> *NEG is universal, but all other constraints can be re-ranked with respect to each other, which gives rise to the typologies described in Chapters 3 through 5.

7.1.4 Double Negation Readings in Negative Concord Languages

Chapter 6 builds on the results obtained in Chapters 3 through 5, and treats double negation readings in negative concord languages. Affixal negation and multiple clause constructions are the easy cases. Affixes are not stored in the N-store, so they never participate in negative concord. Resumption, and polyadic quantification more, in general, operates within the boundaries of a single clause (de Swart and Sag 2002), so multiple clause constructions generally do not allow negative concord.

More complex cases involve the ambiguities that arise with a sequence of Neg-expressions as in (5).

(5) Personne n'a rien payé. [French]
nobody SN has nothing paid.
= No one has paid anything. [NC]
= Everyone has paid something. [DN]

Languages that display such ambiguities have a grammar that combines features from a double negation and a negative concord language. Negative concord languages adopt the ordinal ranking *NEG >> INTNEG in the semantics. A stochastic extension in which the syntactic ranking is that of a negative concord language, but the interpretive constraints have an overlapping range, accounts for ambiguities with a sequence of Neg-expressions in French, Romanian, Welsh, Afrikaans, etc. If there is a small overlap between the range of the constraint, *NEG, and the range of the constraint, INTNEG, *NEG will usually dominate. Accordingly, a combination of two Neg-expressions is usually interpreted in terms of negative concord. But in some cases, INTNEG might win, and a double negation reading as in (5) comes out as the optimal interpretation.

The most complex cases involve the combination of a Neg-expression and a marker of sentential negation in cases where this should be ruled out by the grammar, such as (6).

(6) a. Il n'est pas venu pour rien. [written French] [DN]
he SN is SN come for nothing
'He did not come for nothing.'
b. C'est pas rien. [spoken French] [DN]
it is SN nothing.
'It is not nothing.' (= It is quite something.)

Chapter 6 offers an account of such double negation readings in weak bidirectional OT. Written French has the grammar {FNEG, MAXSN} >> *NEG >> INTNEG of a strict negative concord language. The n-word is inherently negative, and satisfies FNEG. MAXSN is satisfied by the clitic *ne* in written French, which functions as a scope marker, but does not convey semantic negation. In spoken French MAXSN has been demoted, and the grammar FNEG >> *NEG >> INTNEG gives rise to negative spread. As a result, *pas* is not necessary in examples like (6a) and (6b) in order to express negative concord. The economy constraint *NEG thus rules out such sentences as suboptimal expressions of a single negation reading.

Under a strong bidirectional OT analysis, sentences like (6a) and (6b) are ruled out as ungrammatical. Furthermore, the ranking *NEG >> INTNEG in the semantics would predict a single negation interpretation, because all negative forms should be absorbed in the resumptive negative quantifier. In fact, the presence of *pas* in (6) does not lead to ungrammaticalities, but indicates a special interpretation as the superoptimal combination of a marked form (Neg-expression plus *pas*) and a marked interpretation (a double, rather than a single negation reading).

Such special double negation readings are rare, because double negation is highly marked in the semantics of all languages. However, these unusual form-meaning combinations have been attested in French, and are found in Afrikaans, Hungarian, West Flemish, Italian, and Welsh as well. Interestingly, existing accounts of negation and negative concord in the literature have difficulty accounting for the type of examples illustrated in (6), but they fall out naturally from the OT analysis. The appeal to violable constraints, stochastic orderings, and weak bidirectionality makes it possible to integrate such exceptional cases with the analysis of main patterns of double negation and negative concord.

7.2 Embedding into a Broader Theory of Cognition

The analysis developed in this book is primarily a linguistic account of the range and limits of the expression and interpretation of negation found in natural language. The use of optimality theory as the analytical framework implies that the proposals are explicitly embedded in a broader theory of human cognition. As pointed out in Chapter 2, OT is a linguistic theory rooted in a cognitive architecture based in neural computation. The view that well-formedness is grounded in optimization principles relies on the connectionist view of the brain that is common in cognitive neuroscience.

Chapter 3 shows that we can use an evolutionary version of optimality theory to derive universal features of negation, and ground the ranking FNEG >> *NEG we find in all natural languages in optimal communication. This ranking is viewed as an instantiation of general principles of optimal communication, as captured by Horn's division of pragmatic labor. Furthermore, the analysis developed in Chapter 4 is the first full analysis of double negation and negative concord that roots typological variation not exclusively in the syntax or in the semantics, but in the syntax–semantics interface.

The recent development of bidirectional OT as a model of speaker-hearer interactions allows the formulation of a bidirectional grammar of negation, which make forms and meanings mutually dependent on each other. The strong implication of this analysis is that the variation found in natural language is guided by principles of optimal communication. The extensions of this grammar in Chapters 5 and 6 show the power of the bidirectional OT approach in capturing typologies that have mostly been described to some degree in the literature, but have hardly been accounted for in full so far. At the same time, the approach highlights the economy principle encoded in *NEG as a core principle of language structure, language use, and human cognition.

7.3 Perspectives for Further Research

The main aim of the book was to explore the range and limits of cross-linguistic variation in the expression and interpretation of negation. A clear focus tightens the argumentation, and preserves coherence, but it also necessarily implies that certain research questions, by themselves interesting and relevant, are set aside. The main issue that was not fully explored in this book concerns the syntax of negation, and the relation between negation and other parts of the grammar (in particular word order, speech act theory, and information structure).

Given that the primary aim of this book was to describe the syntax-semantics interface of negation, a full analysis of the syntax of negation is outside the scope of this study. Even though the number of syntactic constraints in this book is higher than the number of semantic constraints (cf. Section 1), it is obvious that the syntactic analysis of negation developed so far needs to be worked out in more detail. As far as word order is concerned, the constraints NEGFIRST and FOCUSLAST capture the distinction between preverbal and postverbal negation (Chapter 3). These constraints focus on the position of the negation marker and the main verb, and leave the interaction of negation with other material (auxiliaries, the subject, the object, and adverbials) to be explored more thoroughly. A section on complications in the postverbal domain offers a refinement of NEGFIRST (Chapter 3, Section 3.4; Chapter 5, Section 7), and a section on subordinate and nonfinite clauses (Chapter 3, Section 5) indicates lines of development, but it is clear that the syntax of negation, and in particular of postverbal negation, requires more investigation, and the current formulation of NEGFIRST and FOCUSLAST is insufficient to capture the cross-linguistic variation that we find.

This book ignored all speech acts other than assertion. In particular, I did not study the use of negation in questions or in imperatives. Imperatives have played an important role in syntactic studies of negation (cf. Zanuttini 1996, Zeijlstra 2004), so it would be worth exploring an extension of the OT syntax in this direction. As far as the use of negation in questions is concerned, it is obvious that semantic as well as syntactic and intonational considerations play a role here (cf. Horn 1989, Krifka 2001, 2003, Ladusaw 2004 for discussion). The literature on the syntax of negation covers a wide range of languages, and a wealth of data is available to investigate synchronic and diachronic variation. This book only discusses the main patterns, but hopefully a refinement of the syntactic constraints will be possible, and will produce results strengthening the analysis developed so far.

A more precise analysis of negation and word order is not only relevant for syntactic considerations. In many languages, the position of the negation marker is flexible, and has implications for the focus of negation. In Germanic languages, the position of negation is particularly flexible in the so-called 'Mittelfeld' (Jacobs 1991 for German) (7).

(7) a. daß *nicht* viele Linguisten freiwillig jeden zweiten Tag duschen [German]
that SN many linguists voluntarily every second day shower
b. daß viele Linguisten *nicht* freiwillig jeden zweiten Tag duschen
c. daß viele Linguisten freiwillig *nicht* jeden zweiten Tag duschen
d. daß viele Linguisten freiwillig jeden zweiten Tag *nicht* duschen

Jacobs represents the meaning of the sentences in (7) as in (8):

(8) a. NOT (VIELE LINGUISTEN (FREIWILLIG (JEDEN ZWEITEN TAGE (DUSCHEN))))
b. VIELE LINGUISTEN (λX (NOT (FREIWILLIG (JEDEN ZWEITEN TAGE (X DUSCHEN))))
c. VIELE LINGUISTEN (λX (FREIWILLIG (NOT (JEDEN ZWEITEN TAGE (X DUSCHEN))))
d. VIELE LINGUISTEN (λX (FREIWILLIG (JEDEN ZWEITEN TAGE (NOT (X DUSCHEN))))

There is a strong correspondence between the linear order of constituents in (7) and the scope of negation in (8). According to Jacobs, when a constituent X occurs in the semantic scope of negation, the German grammar requires X to follow Y in the linear structure. According to Dahl (1979), the placement of negation is sensitive to focus in Russian as well.

In languages with less flexible word order, the focus of negation may be fixed with the help of intonation. English is a well-known example of a language in which emphatic stress can help to determine the focus of negation.

(9) a. Colyn did not buy the blue sweater.
b. Colyn did not buy the BLUE sweater (she bought the green one).
c. Colyn did not buy the blue SWEATER (she bought the blue skirt).
d. Colyn did not BUY the blue sweater (she stole it).
e. COLYN did not buy the blue sweater (Kevin did).

If we pronounce the sentence with neutral intonation, as in (9a), negation scopes over the entire proposition. With emphatic stress on specific constituents, the scope of negation can be narrowed down by that constituent. The rest of the sentence becomes background information, and a contrastive reading results in (9b-e). In the absence of syntactic or phonological indications of the focus of negation, the scope of this operator is usually fixed with the help of contextual information. The role of negation as a focus operator is made explicit in the discussion of Hungarian in Chapter 5 (Section 8), but a full analysis of the syntax-semantics interface of negation and focus in the OT framework developed here is outside the scope of this book.

7.4 Conclusion

The conclusion to the concluding chapter of this book need not be long. I enjoyed exploring the phenomenon of negation across a wide range of languages, finding patterns, and being surprised by unexpected and fascinating complications. I have long been puzzled by the cross-linguistic variation we find in the semantics of negation and negative indefinites, and I was glad to discover optimality theory as a tool that can account for such patterns.

One thing leads to another, so many extensions of the proposals outlined in this book can be conceived. The typological validity of my claims would benefit from more empirical work on negation and negative indefinites, especially outside the family of Indo-European languages. More research on diachronic patterns, and the relation between polarity and concord could improve the tentative description of the Jespersen cycle I gave in this book. An extension to positive polarity items would also be attractive.

All in all, I hope the reader will find some inspiration in the views and ideas developed here, and will carry on where this book ends.

References

Adger, Carolyn Temple, Walt Wolfram and Donna Christian (2007). *Dialects in schools and communities*, 2nd edition, London: Routledge.

Aissen, Judith (1999). Markedness and subject choice in optimality theory, *Natural Language and Linguistic Theory* 17, 673–711.

Aissen, Judith (2003). Differential object marking: iconicity vs. economy, *Natural Language and Linguistic Theory* 21, 435–483.

Ashby, William (1981). The loss of the negative particle *ne* in French: a change in progress. *Language* 57, 674–687.

Ashby, William (2001). Un nouveau regard sur la chute du *ne* en français parlé tourangeau: s'agit-il d'un changement en cours? *French Language Studies* 11, 1–22.

Atkinson, Jo, Ruth Campbell, Jane Marshall, Alice Thacker, and Bencie Woll (2004). Understanding 'not': neuropsychological dissociations between hand and head markers of negation in BSL, *Neuropsychologia* 42, 214–229.

Baker, Carl Leroy (1970). Double negatives, *Linguistic Inquiry* 1, 169–186.

Barbosa, Pilar, Danny Fox, Paul Hagstrom, Martha McGinnis and David Pesetsky (1998). *Is the best good enough? Optimality and competition in syntax*, Cambridge, MA: MIT Press.

Barwise, Jon and Robin Cooper (1981). Generalized quantifiers and natural language, *Linguistics and Philosophy* 4, 159–219.

Bayer, Joseph (1990). What Bavarian negative concord reveals about the syntactic structure of German, in: J. Mascaro and M. Nespor (eds.). *Grammar in progress*, Dordrecht: Foris, 13–24.

Beaver, David (2004). The optimization of discourse anaphora, *Linguistics and Philosophy* 27, 3–56.

Beaver, David and Hanjung Lee (2004). Input-output mismatches in OT, in: R. Blutner and H. Zeevat (eds.). *Optimality theory and pragmatics*, Houndmills, Basingstoke, Hampshire: Palgrave/MacMillan, 112–153.

Benmamoun, Elabbas (2000). *The feature structure of functional categories: a comparative study of Arabic dialects*, Oxford: Oxford University Press.

Benthem, Johan van (1989). Polyadic quantifiers, *Linguistics and Philosophy* 12, 437–464.

Bernini, Giuliano and Paolo Ramat (1996). *Negative sentences in the languages of Europe*, Berlin: Mouton de Gruyter.

Besten, Hans den (1986). Double negation and the genesis of Afrikaans, in: P. Muysken and N. Smith (eds.). *Substrata versus universals in Creole languages*, Amsterdam: John Benjamins, 185–230.

Biberauer, Theresa (2006). Doubling vs. omission: insights from Afrikaans, *Proceedings of the workshop on syntactic doubling*, Amsterdam: Meertens Instituut. Available at http://www.meertens.knaw.nl/projecten/edisyn/Papers_workshop-on-Doubling.html.

Biberauer, Theresa (2007). Afrikaans negation and polarity: a comparative investigation of the two *nies*, in: H. Zeijlstra and J.-P. Soehn (eds.). *Proceedings of the workshop on negation and polarity*, Tübingen: Collaborative Research Center 441, 1–7. Available at http://www.sfb441.uni-tuebingen.de/negpol/negpol07.pdf.

Blutner, Reinhard (1998). Lexical pragmatics, *Journal of Semantics* 15, 115–162.
Blutner, Reinhard (2000). Some aspects of optimality in natural language interpretation, *Journal of Semantics* 17, 189–216.
Blutner, Reinhard (2004). Pragmatics and the lexicon, in: L. Horn and G. Ward (eds.). *Handbook of pragmatics*, Oxford: Blackwell, 488–514.
Blutner, Reinhard and Henk Zeevat (eds.) (2003). *Optimality theory and pragmatics*. Houndmills, Basingstoke, Hampshire: Palgrave Macmillan.
Blutner, Reinhard, Petra Hendriks and Helen de Hoop (2003). A new hypothesis on compositionality, *Proceedings of ICCS 2003*, Sidney, 53–57.
Blutner, Reinhard, Petra Hendriks and Helen de Hoop (2006) *Optimal communication*, Stanford, CA: CSLI Publications.
Boersma Paul (1998). *Functional phonology*, Ph.D. dissertation, University of Amsterdam.
Boersma, Paul and Bruce Hayes (2001). Empirical tests of the gradual learning algorithm, *Linguistic Inquiry* 32, 45–86.
Borsley, Bob (1996). *Modern phrase structure grammar*, Oxford: Blackwell.
Borsley, Bob and Bob Morris Jones (2005). *Welsh negation and grammatical theory*, Cardiff: University of Wales Press.
Bošković, Željko (2008). Licensing negative constituents and negative concord, *Proceedings of NELS* 38.
Bosque, Ignacio (1980). *Sobre la negácion*, Madrid: Catedra.
Bréal, Michel (1897). *Essai de Sémantique*, Paris: Hachette.
Bréal, Michel (1900). *Semantics*, New York: Henry Holt & Co. Translation of Bréal (1897).
Breitbarth, Anne. A hybrid approach to Jespersen's cycle in West-Germanic, *The Journal of Comparative Germanic Linguistics* 12, 81–114.
Breitbarth, Anne and Liliane Haegeman (2008). Not continuity, but change: stable stage II in Jespersen's cycle, M.S. University of Cambridge and Université de Lille III.
Bresnan, Joan, Shipra Dingare and Chris Manning (2001). Soft constraints mirror hard constraints: voice and person in English and Lummi, *Proceedings of the LFG 2001 conference, Hong Kong*, Stanford, CA: CSLI Publications, 13–32.
Bresnan, Joan, Deo Ashwini and Dewani Sharma (2007a). Typology in variation: a probabilistic approach to *be* and *n't* in the Survey of English dialects, *English language and Linguistics* 11, 301–346.
Bresnan, Joan, Anna Cueni, Tatiana Nikitina and Harald Baayen (2007b). Predicting the dative alternation, in: G. Bouma, I. Krämer and J. Zwarts (eds.). *Cognitive foundations of interpretation*, Amsterdam: Royal Dutch Academy of Sciences, 69–94.
Bresnan and Hay (2008). Publication details: Gradient grammar: an effect of animacy on the syntax of *give* in New Zealand and American English, *Lingua* 118, 245–259.
Chierchia, Gennaro (2006). Broaden your views: implicatures of domain widening and the "logicality" of natural language, *Linguistic Inquiry* 37, 535–590.
Christensen, Susan (2003). Analysis of the standard French *ne*-drop phenomenon, M.S.. Swarthmore College.
Clark, Brady (2004). A stochastic optimality theory approach to syntactic change, Ph.D. dissertation, Stanford University.
Clarke, Charles C. (1904). The actual force of the French *ne*, *Modern Philology* 2, 279–287.
Corblin, Francis (1996). Multiple negation processing in natural language, *Theoria* 17, 214–259.
Corblin, Francis and Ivan Derzhanski (1997). Multiple negation, optional arguments and the reification of eventualities, in: F. Corblin, D. Godard, J.-M. Marandin (eds.). *Empirical issues in formal syntax and semantics. Selected papers from the Colloque de Syntaxe et de Sémantique de Paris*, Bern: Peter Lang, 219–243.
Corblin, Francis and Lucia Tovena (2003). L'expression de la négation dans les languages romanes, in: D. Godard (ed.). *Les langues romanes: 'problèmes de la phrase simple*, Paris: CNRS Editions, 279–342.
Corblin, Francis, Viviane Déprez, Henriëtte de Swart and Lucia Tovena (2004). Negative concord, in: F. Corblin and H. de Swart (eds.). *Handbook of French semantics*, Stanford, CA: CSLI Publications, 417–452.

Croft, Bill (1991). The evolution of negation, *Journal of Linguistics* 27, 1–27.
Cuervo, Cristina and Natalia Mazzaro (2008). Beyond negative concord: negative doubling in a variety of Spanish, paper presented at the 2008 meeting of the Canadian Linguistics Association.
Dahl, Östen (1979). Typology of sentence negation, *Linguistics* 17, 79–106.
Dalrymple, Mary, Makoto Kanazawa, Yookyung Kim, Sam Mchombo and Stanley Peters (1998). Reciprocal expressions and the concept of reciprocity, *Linguistics and Philosophy* 21, 159–210.
DeGraff, Michel (1997). Verb syntax in creolization (and beyond), in: L. Haegeman (ed.). *Elements of grammar*, Dordrecht: Kluwer, 64–94.
Dekker, Paul and Robert van Rooy (2000). Bi-directional optimality theory; an application of game theory, *Journal of Semantics* 17, 217–242.
Déprez, Viviane (1997a). Two Types of Negative Concord, *Probus* 9, 103–143.
Déprez, Viviane (1997b). A non-unified analysis of negative concord, in: D. Forget et al. (eds.). *Negation and polarity*, Amsterdam: John Benjamins, 53–75.
Déprez, Viviane (1999). The roots of negative concord in French and French based creoles, in: M. DeGraff (ed.). *Language creation and language change*, Cambridge, MA: MIT Press, 329–375.
Déprez, Viviane (2000). Parallel (A)symmetries and the internal structure of negative expressions, *Natural Language and Linguistic Theory* 18, 253–342.
Doetjes, Jenny (2005). The cameleontic nature of French *ni*: negative coordination in a negative concord language, *Proceedings of Sinn und Bedeutung* 9, 72–86.
Dryer, Matthew (1988). Universal of negative position, in: M. Hammond, E. Moravcsik and J. Wirth (eds.). *Studies in syntactic typology*, Amsterdam: John Benjamins, 93–124.
Dryer, Matthew (2000). Verb-object-negative order in Central Africa, in: N. Cyffer, E. Ebermann and G. Ziegelmeyer (eds.). *Negation patterns in West Africa*, Amsterdam: Benjamins (to appear).
Espinal, M. Teresa (1992). Expletive negation and logical absorption, *Linguistic Review* 9, 333–358.
Espinal, M. Teresa (1993). The interpretation of *no-pas* in Catalan, *Journal of Pragmatics* 19, 353–369.
Espinal, M. Teresa (2000). On the semantic status of n-words in Catalan and Spanish, *Lingua* 110, 557–580.
Falaus, Anamaria (2007a). Le paradoxe de la double négation dans une langue à concordance négative stricte, in: F. Floricic (ed.). *La négation dans les langues romanes*, Amsterdam: Benjamins, 75–97.
Falaus, Anamaria (2007b). Negative concord and double negation: the Romanian puzzle, in: J. Camacho et al. (eds.). *Romance linguistics* 2006, Amsterdam: John Benjamins, 135–148.
Falaus, Anamaria (2008). Romanian n-words as negative quantifiers, *University of Pennsylvania working papers in linguistics*, 14 (1), 119–134.
Farkas, Donka and Anastasia Giannakidou (1996). How clause-bound is the scope of universals? *Proceedings of SALT VI*, Cornell, Ithaca: CLC Publications, 35–52.
Farkas, Donka and Henriëtte de Swart (2003). *The semantics of incorporation. From argument structure to discourse transparency*, Stanford, CA: CSLI Publications.
Farkas, Donka and Henriëtte de Swart (2008). The semantics and pragmatics of plurals, M.S. UCSC/Utrecht University.
Fauconnier, Gilles (1975). Pragmatic scales and logical structures, *Linguistic Inquiry* 6, 353–375.
Fauconnier, Gilles (1979). Implication reversal in a natural language, in: F. Günther and S.J. Schmidt (eds.). *Formal semantics and pragmatics for natural languages*, Dordrecht: Reidel, 289–301.
Von Fintel, Kai (1999). NPI-licensing, strawson entailment and context-dependency, *Journal of Semantics* 16, 97–148.
Von Fintel, Kai and Lisa Matthewson (2008). Universals in semantics, *The Linguistic Review* 25, 139–201.
Franco, Jon and Alazne Landa 2006. Preverbal n-words and anti-agreement effects, in: Sagarra, N. and Toribio, A.J. (eds.). *Selected proceedings of the 9th Hispanic linguistics symposium*, Somerville, MA: Cascadilla Proceedings Project, 34–42.

Furukawa, Yukio (2007). Unembedded 'negative' quantifiers, in: T. Washio et al. (eds.). *New frontiers in artificial intelligence*, Berlin: Springer, 232–245.

Gaatone, David (1971). *Étude descriptive du système de la négation en français contemporain* (Publication romaines et françaises, CXIV), Genève: Librairie Droz.

Gass, Kate van (2007). Multiple n-words in Afrikaans, *SPIL Plus* 35, 167–186.

Gazdar, Gerald, Ewan H. Klein, Geoffrey K. Pullum and Ivan A. Sag (1985). *Generalized phrase structure grammar*. Oxford: Blackwell/Cambridge, MA: Harvard University Press.

Giannakidou, Anastasia (1997). *The landscape of polarity items*, unpublished Ph.D. dissertation, Groningen University.

Giannakidou, Anastasia (1998). *Polarity sensitivity as (non)veridical dependency*, Amsterdam: John Benjamins.

Giannakidou, Anastasia (1999). Affective dependencies, *Linguistics and Philosophy* 22, 367–421.

Giannakidou, Anastasia (2000). Negative...concord? *Natural Language and Linguistic Theory* 18, 457–523.

Giannakidou, Anastasia (2006). N-words and negative concord, in: H. van Riemsdijk et al. (eds.). *The syntax companion*, Oxford: Blackwell, 327–391.

Giannakidou, Anastasia (2007). The landscape of EVEN, *Natural Language and Linguistic Theory* 25, 39–81.

Giannakidou, Anastasia (2008). Negative and positive polarity items: licensing, compositionality and variation, in: C. Maienborn, K. von Heusinger and P. Portner (eds.). *Semantics: an international handbook of natural language*, Berlin: Mouton de Gruyter (to appear).

Givón, Talmy (1979). *On understanding grammar*, New York: Academic.

Godard, Danièle (2004). French negative dependency, in: F. Corblin and H. de Swart (eds.). *Handbook of French semantics*, Stanford, CA: CSLI Publications, 351–390.

Greenberg, Joseph (1966). *Language universals, with special reference to feature hierarchies*, The Hague: Mouton.

Grimshaw, Jane (1997). Projection, heads and optimality, *Linguistic Inquiry* 28, 373–422.

Groot, Casper de (1993). Hungarian, in: P. Kahrel and R. van den Berg (eds.). *Typological studies in negation*, Amsterdam: Benjamins, 143–162.

Haegeman, Liliane (1995). *The syntax of negation*, Cambridge: Cambridge University Press.

Haegeman, Liliane (1997). The syntax of n-word and the Neg-criterion, in: D. Forget, P. Hirschbühler, F. Martineau and M.L. Rivero (eds.). *Negation and polarity*, Amsterdam: Benjamins, 115–137.

Haegeman, Liliane (2001). Antisymmetry and verb-final order in West Flemish, *The journal of Comparative Germanic Linguistics* 3, 207–232.

Haegeman, Liliane (2002a). Some notion on DP-internal negative doubling, in: S. Barbiers (ed.). *Microvariation*, Amsterdam: Meertens Instituut.

Haegeman, Liliane (2002b). West Flemish negation and the derivation of SOV order in West Germanic, *Nordic Journal of Linguistics*, 25, 154–189.

Haegeman, Liliane (2003). Note on long adverbial fronting in English and the left periphery, *Linguistic Inquiry* 34, 640–649.

Haegeman, Liliane and Raffaella Zanuttini (1996). Negative concord in West Flemish, in: A. Belletti and L. Rizzi (eds.). *Parameters and functional heads*, Oxford: Oxford University Press, 117–180.

Haegeman, Liliane and Terje Lohndal (2008). Negative concord and (multiple) agree: a case study of West Flemish, M.S. Lille III/University of Maryland.

Haeseryn, W., K. Romijn, G. Geerts, R. de Rooij and M.C. van der Toorn (1997). *Algemene Nederlandse Spraakkunst*, Groningen: Martinus Nijhoff.

Harries, Helga (1973) A note on double negation and modals as main verbs, *Stanford working papers on language universals* 11, 211–217.

Haspelmath, Martin (1997). *Indefinite pronouns*, Oxford: Clarendon Press.

Haspelmath, Martin (2006). Against markedness (and what to replace it with), *Journal of Linguistics* 42, 25–70.

Heinämäki, Orvokki (1978). *Semantics of English temporal connectives*, unpublished Ph.D. dissertation, University of Texas, Austin.

Hendriks, Petra (2004a). Optimization in focus identification, in: R. Blutner and H. Zeevat (eds.). *Optimality theory and pragmatics*. Palgrave Macmillan, 42–62.

Hendriks, Petra (2004b). Coherence relations, ellipsis and contrastive topics, *Journal of Semantics*, Houndmills, Basingstoke, Hampshire, 21, 133–153.

Hendriks, Petra and Helen de Hoop (2001). Optimality theoretic semantics, *Linguistics and Philosophy* 24, 1–32.

Hendriks, Petra and Jennifer Spenader (2005). Why be silent? Functions of ellipsis. In *Proceedings of the ESSLLI 2005 workshop on cross-modular approaches to ellipsis*, Edinburgh, 29–36.

Hendriks, Petra and Jennifer Spenader (2006). When production precedes comprehension: an optimization approach to the acquisition of pronouns, *Language and Acquisition* 13, 319–348.

Hendriks, Petra, Helen de Hoop, Irene Krämer, Henriëtte de Swart, and Joost Zwarts (2009). *Conflicts in interpretation*, London: Equinox (in press).

Herburger, Elena (2001). The negative concord puzzle revisited, *Natural Language Semantics* 9, 289–333.

Hoeksema, Jack (1994). On the grammaticalization of negative polarity items, *Proceedings of Berkeley Linguistic Society* 20, 273–282.

Hoeksema, Jack (1997). Negation and negative concord in middle Dutch, in: D. Forget, P. Hirschbühler, F. Martineau and M.-L. Rivero (eds.). *Negation and polarity: syntax and semantics*, Amsterdam: John Benjamins, 139–158.

Hoeksema, Jack (2000). Negative polarity items: triggering, scope and c-command, in: L.R. Horn and Y. Kato (eds.). *Negation and polarity. Semantic and syntactic perspectives*, Oxford: Oxford University Press, 123–154.

Hoeksema, Jack (2002). Polarity-sensitive scalar particles in early modern and present-day Dutch, *Belgian Journal of Linguistics* 16, 53–64.

Hoeksema, Jack (2005). In days, weeks, months, years, ages: a class of temporal negative polarity items, in: D. Gilbers and P. Hendriks (eds.). *Rejected papers: Feestbundel voor Ron van Zonneveld*. Groningen: Rijksuniversiteit Groningen.

Hoop, Helen de (2004). The problem of unintelligibility, in: A. ter Meulen and W. Abraham (eds.). *The composition of meaning. From lexeme to discourse*. Amsterdam/Philadelphia: John Benjamins, 69–81.

Hoop, Helen de and Henriëtte de Swart (2000). Adjunct clauses in optimality theory, *Revista di Linguistica/Italian Journal of Linguistics* 12 (1), 107–127.

Horn, Laurence (1984). Towards a new taxonomy of pragmatic reference: Q-based and R-based implicature, in: D. Schiffrin (ed.). *Meaning, form, and use in context: linguistic applications*, Washington: Georgetown University Press, 11–42.

Horn, Laurence (1989). *A natural history of negation*, Chicago, IL: University of Chicago Press. Reprinted (2001) with a new introduction, Stanford, CA: CSLI Publications.

Horn, Laurence (2001). The logic of logical double negation, in Y. Kato (ed.). *Proceedings of the Sophia symposium on negation*, Tokio: Sophia University, 79–111.

Horst, Joop van der and Marijke van der Wal (1979). Negatieverschijnselen en woordvolgorde in de geschiedenis van het Nederlands, *Tijdschrift voor Nederlande Taal en Letterkunde* 95, 6–37.

Hoyt, Frederick (2006). Long distance negative concord and restructuring in Palestinian Arabic, in: P. Dekker and H. Zeijlstra (eds.). *Concord phenomena and the syntax-semantics interface*, Malaga: ESSLLI Publication, 27–32.

Isac, Daniela (2004). Focus on negative concord, in: R. Bok-Bennema et al. (eds.). *Selected papers from Romance languages and linguistic theory 2002*, Amsterdam: John Benjamins, 119–140.

Israel, Michael (1996). Polarity sensitivity as lexical semantics, *Linguistics and Philosophy* 19, 619–666.

Itô, Junko and Armin Mester (1995). Japanese phonology, in: J.A. Goldsmith (ed.). *Handbook of phonological theory*, Oxford: Blackwell, 817–838.

Itô, Junko and Armin Mester (1999). The phonological lexicon, in: N. Tsujimura (ed.). *Handbook of Japanese linguistics*, Oxford: Blackwell, 62–100.

Jacobs, Joachim (1991). Negation, in: A. von Stechow and D. Wunderlich (eds.). *Semantik. Ein internationales Handbuch der zeitgenössigen Forschung*, Berlin: Mouton, 560–596.
Jäger, Agnes (2005). Negation in Old High German, *Zeitschrift für Sprachwissenschaft* 24, 227–262.
Jäger, Agnes (2008). *History of German negation*, Amsterdam: John Benjamins.
Jäger, Gerhard (2003). Learning constraint sub-hierarchies, in: R. Blutner and H. Zeevat (eds.). *Pragmatics and optimality theory*, Houndmills: Palgrave MacMillan, 251–287.
Jäger, Gerhard (2007). Evolutionary game theory and typology. A case study, *Language* 83, 74–109.
Jäger, Gerhard and Annette Rosenbach (2006). The winner takes it all - almost. Cumulativity in grammatical variation, *Linguistics* 44, 937–971.
Jäger, Gerhard and Robert van Rooy (2007). Language structure: psychological and social constraints, *Synthese* 159, 99–130.
Jakobson, Roman (1932). Zur Struktur des russischen Verbums, in: *Charistera Guglielmo Mathesio*, Prague, 74–84 (also in *Selected Writings* II).
Jakobson, Roman (1939). Signe zéro, in: *Mélanges de linguistique offerts à Charles Bally*, Genève (also in *Selected writings* II).
Jakobson, Roman (1962). *Selected writings I. Phonological studies*, the Hague: Mouton.
Jakobson, Roman (1971). *Selected writings II. Word and language*, the Hague: Mouton.
Jespersen, Otto (1917). *Negation in English and other languages*, Copenhagen: A.F. Høst. Reprinted in: *Selected writings of Otto Jespersen* (1962), London: George Allen & Unwin, 3–151.
Jespersen, Otto (1924). *The philosophy of grammar*, London: George Allen & Urwin.
Jespersen, Otto (1933). *Essentials of English grammar*, reprinted 1964, Tuscaloosa, AL: University of Alabama Press.
Joan Bresnan and Hay, Jennifer (2008) Gradient grammer: an effect of animacy on the syntax of give in New Zealand and American English, *Lingua* 118, 245–259.
Kadmon, Nirit and Fred Landman (1993). Any, *Linguistics and Philosophy* 16, 353–422.
Kager, René (1999). *Optimality theory*, Cambridge: Cambridge University Press.
Kaiser, Elsi (2006). Negation and the left periphery in Finnish, *Lingua* 116, 314–350.
Kayne, Richard (1984). *Connectedness and binary branching*, Dordrecht: Foris.
Keenan, Ed (1987). Unreducible n-ary quantifiers in natural language, in: P. Gärdenfors (ed.). *Generalized quantifiers*, Dordrecht: Reidel, 109–151.
Keenan, Ed and Dag Westerståhl (1997). Generalized quantifiers in linguistics and logic, in: J. van Benthem and A. ter Meulen (eds.). *Handbook of logic and language*, Amsterdam: Elsevier, 837–893.
Kirby, Simon and Jim Hurford (1997). *The evolution of incremental learning: language, development and critical periods*, Technical report, Language Evolution and Computation Research Unit, University of Edinburgh.
Klima, Edward (1964). Negation in English, in: J. Fodor and J. Katz (eds.). *The structure of language*, Englewoord Cliffs, NJ: Prentice Hall, 246–323.
Koontz-Garboden, Andrew (2004). Language contact and Spanish aspectual expression: a formal analysis, *Lingua* 114, 1291–1330.
Krifka, Manfred (1995). The semantics and pragmatics of polarity items, *Linguistic Analysis* 25, 209–257.
Krifka, Manfred (2001). Quantifying into question acts, *Natural Language Semantics* 9, 1–40.
Krifka, Manfred (2003). Quantifiers in question, *Korean Journal of English Language and Linguistics* 3, 499–526.
Labov, William (1972). Negative attraction and negative concord in English grammar, *Language* 48, 773–818.
Ladusaw, William (1979). *Polarity sensitivity as inherent scope relations*, Ph.D. Dissertation. University of Texas, New York: Garland Press.
Ladusaw, William (1992). Expressing negation, *Proceedings of SALT* 2, Columbus, OH: Ohio State University, 237–259.
Ladusaw, William (1996). Negation and polarity items, in: S. Lappin (ed.). *The handbook of contemporary semantic theory*, Oxford: Blackwell, 321–341.

Ladusaw, William (2004). Framing the issues: the biasing effect of polarity items in questions, M.S. UCSC.

Lahiri, Utpal (1995). Negative polarity items in Hindi, *Proceedings of SALT V*, Ithaca, NY: Cornell University Press.

Lahiri, Utpal (1998). Focus and negative polarity in Hindi, *Natural Language Semantics* 6, 57–123.

Laka, Itziar (1990). *Negation in syntax: on the nature of functional categories and projections*, Ph.D. dissertation, Cambridge, MA: MIT Press.

Lascarides, Alex and Nicholas Asher (1993). Temporal interpretation, discourse relations and commonsense entailment, *Linguistics and Philosophy* 16, 437–493.

Legendre, Géraldine (2001). Masked second-position effects and the linearization of functional features, in: G. Legendre, J. Grimshaw and S. Vikner (eds.). *Optimality-theoretic syntax*, Cambridge, MA: MIT Press, 241–277.

Legendre, Géraldine, Paul Hagstrom, Anna Vainikka and Marina Todorova (2002). Partial constraint ordering in child French syntax, *Language Acquisition* 10, 189–227.

Levinson, Stephen (2000). *Presumptive meanings*. Cambridge, MA: MIT Press.

Linebarger, Marcia (1980). *The grammar of negative polarity*, Ph.D. thesis, Cambridge, MA: MIT Press.

Linebarger, Marcia (1987). Negative polarity and grammatical representation, *Linguistics and Philosophy* 10, 325–387.

Manning, Chris, Ivan Sag and Masayo Iida (1999). The lexical integrity of Japanese causatives, *Nordic Journal of Linguistics* 21, 107–144.

Martineau, France and Viviane Déprez (2004). Micro-parametric variation and negative concord, in: J. Auger, J.C. Clements and B. Vance (eds.). *Contemporary approaches to Romance linguistics*, Amsterdam: Benjamins, 139–158.

Martins, Ana Maria (2006). Emphatic affirmation and polarity: contrasting European Portuguese with Brazilian Portuguese, Spanish, Catalan and Galician, in: J. Doetjes and P. Gonzalez (eds.). *Romance languages and linguistic theory 2004. Selected papers from 'Going Romance'*, Leiden, 9–11 December 2004; Amsterdam: Benjamins, 197–223.

Mattausch, Jason (2005). On the optimization and grammaticalization of anaphora, Ph.D. dissertation, Humboldt University Berlin, published as *ZAS Papers in Linguistics* 38.

Mattausch, Jason (2007). Optimality, bidirectionality and the evolution of binding phenomena, *Research on Language and Computation* 5, 103–131.

May, Robert (1989). Interpreting logical form, *Linguistics and Philosophy* 12, 387–435.

Mazzon, Gabriella (2004). *A history of English negation*, Harlow: Pearson Longman.

McCarthy, John (2002). *Optimality theory in phonology*, Oxford: Blackwell.

McCarthy, John and Alan Prince (1994). Prosodic morphology, in: J. Goldsmith (ed.). *A handbook of phonological theory*, Oxford: Blackwell, 318–366.

Merchant, Jason (2001). *The syntax of silence: sluicing, islands and the theory of ellipsis*, Oxford: Oxford University Press.

Mitchell, Erika (2006). The morpho-syntax of negation and the positions of NegP in the Finno-Ugric languages, *Lingua* 116, 228–244.

Moltmann, Friederike (1995). Exception sentences and polyadic quantification, *Linguistics and Philosophy* 18, 223–280.

Morimoto, Yukiko (2001). Deriving the directionality parameter in OT-LFG, in: M. Butt and T. King (eds.). *Proceedings of the LFG01 conference*, Stanford, CA: CSLI Publications. Available at http://cslipublications.stanford.edu/LFG/5/lfg00.html.

Muller, Claude (1991). *La négation en français*, Genève: Droz.

Newson, Mark (1998). On the nature of inputs and outputs. A case study of negation, in: P. Barbosa et al. (eds.). *Is the best good enough? Optimality and competition in syntax*, Camrbidge, MA: Cambridge University Press, 315–336.

Ouali, Hamid (2003). Sentential negation in Berber: a comparative study, in: J. Mugany (ed.). *Linguistic description. Typology and representation of African languages*, Trenton: Africa World Press, 243–256.

Ouali, Hamid (2005). Negation and negative polarity items in Berber, *BLS* 13, Berkeley, CA: Berkeley Linguistic Society, 330–340.

Payne, John (1985). Negation, in: T. Shopen (ed.) *Language typology and syntactic description I: clause structure*, Cambridge: Cambridge University Press.

Penka, Doris (2006). A cross-linguistic perspective on n-words, in: *Proceedings of BIDE05, International Journal of Basque Linguistics and Philology*.

Penka, Doris (2007). *Negative indefinites*, Ph.D. dissertation, Karl-Eberhads Universität Tübingen.

Pollock, Jean-Yves (1989). Verb movement, universal grammar and the structure of IP, *Linguistic Inquiry* 20, 365–424.

Ponelis, Fritz (1985). *Afrikaans sintaksis*, 2nd edition, Pretoria: J.L. van Schaik.

Portner, Paul and Raffaella Zanuttini (2000). The force of negation in Wh exclamatives and interrogatives, in: L. Horn and Y. Kato (eds.). *Studies in negation and polarity*, Oxford: Oxford University Press, 193–231.

Portner, Paul and Raffaella Zanuttini (2003). Exclamative clauses: at the syntax-semantic interface, *Language* 79, 39–81.

Posner, Rebecca (1984). Double negatives, negative polarity and negative incorporation in Romance: a historical and comparative view, *Transactions of the Philological Society* 1984, 1–26.

Posner, Rebecca (1985). Post-verbal negation in non-standard French: a historical and comparative view, *Romance Philology* 39, 170–197.

Postal, Paul (2000). *The SQUAT lectures*, M.S. New York University.

Postal, Paul (2004). A remark on English double negatives, in: E. Laporte et al. (eds.). *Syntaxe, lexique, et lexique-grammaire*, Amsterdam: Benjamins, 497–508.

Postal, Paul (2005). Suppose (if only for an hour) that NPIs are negation-containing phrases, M.S. New York University.

Postma, Gert-Jan (2002). De enkelvoudige clitische negatie in het Middelnederlands en de Jespersen cyclus, *Nederlandse Taalkunde* 7, 44–87.

Prince, Alan and Paul Smolensky (1997). Optimality: from neural networks to universal grammar. *Science* 275, 1604–1610.

Prince, Alan and Paul Smolensky (1993/2004) *Optimality theory: constraint interaction in generative grammar*, appeared as Technical Report CU-CS-696-93, Department of Computer Science, University of Colorado at Boulder, and Technical Report TR-2, Rutgers Center for Cognitive Science, Rutgers University, New Brunswick, NJ, April 1993, published 2004 by Blackwell.

Progovac, Ljiljana (1993). Negative polarity: entailment and binding, *Linguistics and Philosophy* 16, 149–180.

Progovac, Ljiljana (1994). *Negative and positive polarity*, Cambridge: Cambridge University Press.

Przepiórkowski, Adam and Anna Kupść (1999). Eventuality negation and negative concord in Polish and Italian, in: B. Borsley and A. Przepiorkówski (eds.). *Slavic in HPSG*, Stanford, CA: CSLI Publications, 211–246.

Puskás, Genoveva (2002). On negative licensing contexts and the role of n-words, in: I. Kenesei et al. (eds.). *Approaches to Hungarian*, vol. 8, Budapest: Akadémiaio Kiadó, 81–106.

Puskás, Genoveva (2006). Negation in Finno-Ugric: an introduction, *Lingua* 116, 203–227.

Quirk, Randolph, Sidney Greenbaum, Geoffrey Leech and Jan Svartvik (1985). *A comprehensive grammar of the English language*, London: Longman.

Richter, Frank and Manfred Sailer (2004). Polish negation and lexical resource semantics, in: L.S. Moss and R.T. Oehrle (eds.). *Electronic notes in theoretical computer science*, vol. 53, Amsterdam: Elsevier, 309–321.

Richter, Frank and Manfred Sailer (2006). Modeling typological markedness in semantics, in S. Müller (ed.). *Proceedings of the 13th international conference on HPSG*, Stanford, CA: CSLI Publications, 305–325.

Rizzi, Luigi (1997). The fine structure of the left periphery, in: L. Haegeman (ed.). *Elements of grammar*, Dordrecht: Kluwer, 281–337.

Rooth, Math (1985). *Association with focus*, Ph.D. dissertation, University of Massachusetts at Amherst.

Rooy, Robert van (2004). Signalling games select Horn strategies, *Linguistics and Philosophy* 27, 493–527.

Rowlett, Paul (1998). *Sentential negation in French*, Oxford: Oxford University Press.

Rullmann, Hotze (2003). Additive particles and polarity, *Journal of Semantics* 20, 329–401.

Rullman, Hotze and Jack Hoeksema (2001). Scalarity and polarity: a study of scalar adverbs as polarity items, in: J. Hoeksema, H. Rullmann, V. Sanchez-Valencia and T. van der Wouden (eds.). *Perspectives on negation and polarity items*, Amsterdam: John Benjamins, 129–171.

Sag, Ivan, Thomas Wasow and Emily Bender (2003). *Syntactic theory: a formal introduction*, Stanford, CA: CSLI Publications.

Sankoff, Gillian and Dianne Vincent (1977). L'emploi productif du *ne* dans le français parlé à Montréal, *Le français moderne* 45, 243–256.

Schwenter, Scott (2005). The pragmatics of negation in Brazilian Portuguese, *Lingua* 115, 1427–1456.

Schwenter, Scott (2006). Fine-tuning Jespersen's cycle, in: B.J. Birner and G. Ward (eds.). *Drawing the boundaries of meaning: neo-Gricean studies in pragmatics and semantics in honor of Laurence R. Horn*, Amsterdam: Benjamins.

Sells, Peter (2000). Negation in Swedish: where it's not at, in: M. Butt and T.H. King (eds.). *Proceedings of the LFG-00 conference*, Stanford, CA: CSLI Publications. Available at http://cslipublications.stanford.edu/LFG/5/lfg00.html.

Sells, Peter (2001). *Structure, alignment and optimality in Swedish*, Stanford, CA: CSLI Publications.

Smolensky, Paul and Géraldine Legendre (2006). *The harmonic mind: from neutral computation to optimality-theoretic grammar. Vol 1: cognitive architecture*; *Vol 2: linguistic and philosophical implications*. Cambridge, MA: MIT Press.

Stump, Gregory (1989). Further remarks on Breton agreement, *Natural Language and Linguistic Theory* 7, 429–471.

Stvan, Laurel (1998). *The semantics and pragmatics of bare singular noun phrases*. Ph.D. dissertation, Northwestern University.

Surányi, Balázs (2006a). Predicates, negative quantifiers and focus: specificity and quantificationality of n-words, in: K. É Kiss (ed.). *Event structure and the left periphery*, Dordrecht: Springer.

Surányi, Balázs (2006b). Quantification and focus in negative concord, *Lingua* 116, 272–313.

Swart, Henriëtte de (1996). Meaning and use of not..until, *Journal of Semantics* 13, 221–263.

Swart, Henriëtte de (1998a). *Introduction to natural language semantics*, Stanford, CA: CSLI Publications.

Swart, Henriëtte de (1998b). Licensing of negative polarity items under inverse scope, *Lingua* 105, 175–200.

Swart, Henriëtte de (1999). Position and meaning: time adverbials in context, in: P. Bosch and R. van der Sandt (eds.). *Focus: linguistic, cognitive and computational perspectives*, Cambridge: Cambridge University Press, 336–361.

Swart, Henriëtte de (2000). Scope ambiguities with negative quantifiers, in: K. von Heusinger and U. Egli (eds.). *Reference and anaphoric relations*, Dordrecht: Kluwer, 118–142.

Swart, Henriëtte de (2001a). Weak readings of indefinites: type-shifting and closure, *The Linguistic Review* 18, 69–96.

Swart, Henriëtte de (2001b). Négation et coordination: la conjonction *ni*, in: R. Bok-Bennema, B. de Jonge, B. Kampers-Manhe and A. Molendijk (eds.). *Adverbial modification*, Amsterdam: Rodopi, 109–124.

Swart, Henriëtte de (2006). Marking and interpretation of negation: a bi-directional OT approach, in: R. Zanuttini, H. Campos, E. Herburger and P. Portner (eds.). *Negation, tense and clausal architecture: cross-linguistic investigations*, Washington: Georgetown University Press, 199–218.

Swart, Henriëtte de (2009). Negation in early L2: a window on language genesis, in: R. Botha and H. de Swart (eds.). *Language evolution: the view from restricted linguistic systems*, Utrecht: LOT Publications, 59–100.

Swart, Henriëtte de and Arie Molendijk (1999). Negation and the temporal structure of narrative discourse, *Journal of Semantics* 16, 1–42.

Swart, Henriëtte de and Ivan Sag (2002). Negation and negative concord in Romance, *Linguistics and Philosophy* 25, 373–417.

Swart, Henriëtte de and Joost Zwarts (2007). Less form, more meaning: why bare nominals are special, *Lingua*, 119, 280–295..

Szabolcsi, Anna (1997). Strategies for scope taking, in: A. Szabolcsi (ed.). *Ways of scope taking*, Dordrecht: Kluwer, 109–155.

Szabolcsi, Anna (2004). Positive polarity - negative polarity, *Natural Language and Linguistic Theory* 22, 409–452.

Szabolcsi, Anna and Bill Haddican (2004). Conjunction meets negation: a study in cross-linguistic variation, *Journal of Semantics* 21, 219–249.

Tanaka, Shin (1994). Die Mehrfachnegation: ein Sprachgut der Unraffinierten?, in: D. Halwachs and I. Stütz (eds.). *Sprache—Sprechen—Handeln: Akten des 28 Linguistischen Kolloquiums, Graz 1993*, Tübingen: Niemeyer, 191–198.

Tesar, Bruce and Paul Smolensky (1993). *The learnability of optimality theory: an algorithm and some basic complexity results*. Technical Report CU-CS-678-93, Computer Science Department, University of Colorado at Boulder.

Tesar, Bruce and Paul Smolensky (1998). Learnability in optimality theory, *Linguistic Inquiry* 29, 229–268.

Tesar, Bruce and Paul Smolensky (2000). *Learnability in optimality theory*, Cambridge, MA: MIT Press.

Tesnière, Lucien (1959). *Eléments de syntaxe structurale*, Paris: Klincksieck.

Thomson, Hanne-Ruth (2006). Negation patterns in Bengali, *Bulletin of SOAS* 69, 243–265.

Tóth, Ildiko (1999). Negative polarity item licensing in Hungarian, *Acta Linguistica Hungarica* 46, 119–142.

Tovena, Lucia (1998). *The fine structure of polarity*, New York: Garland.

Tovena, Lucia, Viviane Déprez and Jacques Jayez (2004). Polarity sensitive items, in: F. Corblin and H. de Swart (eds.). *Handbook of French semantics*, Stanford, CA: CSLI Publications, 391–415.

Trubetzkoy, Nikolaj (1931). Die phonologische Systeme, *Travaux du cercle linguistique de Prague* 4, 96–116.

Trubetzkoy, Nikolaj (1939). *Grundzüge der Phonologie*, Göttingen: Vandenhoeck and Ruprecht.

Tubau, Susagna (2008). *Negative concord in English and Romance: syntax-morphology interface conditions on the expression of negation*, Ph.D. dissertation, University of Amsterdam.

Uribe-Etxebarria, Miriam (1994). *Interface licensing conditions on negative polarity items: a theory of polarity and tense interactions*, Ph.D. thesis, University of Connecticut.

Vallduví, Enric (1994). Polarity items, n-words and minimizers in Catalan and Spanish, *Probus* 6, 263–294.

Van der Auwera, Johan and Annemarie Neuckermans (2004). Jespersen's cycle and the interaction of predicate and quantifier negation, in: B. Kortmann (ed.). *Dialectology meets typology. Dialect grammar from a cross-linguistic perspective*, Berlin: Mouton de Gruyter, 453–478.

Van der Auwera, Johan and Gunther de Vogelaer (2008). Negatie en kwantificatie, in: S. Barbiers, J.van der Auwera, E. Boef, H. Bennis, G. de Vogelaer and M. van der Ham (eds.). *Syntactische atlas van de Nederlandse dialecten*, dl. II, Amsterdam: Amsterdam University Press.

Vasishth, Shravan (2000). Word order, negation and negative polarity in Hindi, *OSU Working Papers in Linguistics* 53, 108–131.

Vasishth, Shravan (2004). Word order, negation and negative polarity in Hindi, *Research on Language and Computation* 2, 127–146.

Vincent, Nigel Vincent (1999). Competition and correspondence in syntactic change: null arguments in Latin and Romance, in: S. Pintzuk, G. Tsoulas and A. Warner (eds.). *Diachronic syntax: models and mechanisms*, Oxford: Oxford University Press, 25–50.

Vinet, Marie-Thérèse (1998). Contrastive focus, French n-words and variation, *Canadian Journal of Linguistics* 431, 121–143.

Vlachou, Evangelia (2007). *Freedom of choice in and out of context. The semantics of French, Greek and English free choice items*, Ph.D. dissertation, Utrecht University.

Waher, Hester (1978). *Die problem van die bereik van die ontkenning met spesiale verwysing na Afrikaans*, Ph.D. dissertation, Kaapstad.

Wallage, Phillip (2005). *Negation in Early English: parametrization and grammatical competition*, Ph.D. thesis, University of York.

Wallage, Phillip (2008). Jespersen's cycle in Middle English: parametric variation and grammatical competition, *Lingua* 118, 643–674.

Watanabe, Akira (2004). The genesis of negative concord: syntax and morphology of negative doubling, *Linguistic Inquiry* 35, 559–612.

Weiss, Helmut (2002). Three types of negation: a case study in Bavarian, in: S. Barbiers et al. (eds.). *Syntactic microvariation*, Amsterdam: Meertens Instituut, 305–332.

Wheeler, Max, Alan Yates, Nicolau Dols (1999). *Catalan: a comprehensive grammar*, London: Routledge.

Wilson, Colin (2001). Bidirectional optimization and the theory of anaphora, in: G. Legendre, J. Grimshaw and S. Vikner (eds.). *Optimality-theoretic syntax*, Cambridge, MA: MIT Press, 465–507.

Winter, Yoad and Joost Zwarts (2000). A semantic characterization of locative PPs. In *Proceedings of semantics and linguistic theory (SALT) 7*.

Wouden, Ton van der (1994). *Negative contexts*. Ph.D. dissertation, University of Groningen.

Wouden, Ton van der (1997). *Negative contexts: collocation, polarity and multiple negation*, London: Routledge.

Wouden, Ton van der and Frans Zwarts (1993). A semantic analysis of negative concord, *Proceedings of SALT*, III, 202–219.

Xiao, Richard and Anthony McEnery (2008). Negation in Chinese: a corpus-based study, *Journal of Chinese linguistics* 36: 274–330.

Zanuttini, Raffaella (1991). *Syntactic properties of sentential negation: a comparative study of Romance languages*, unpublished Ph.D. dissertation, University of Pennsylvania.

Zanuttini, Raffaella (1994). Reexamining negative clauses, in: G. Cinque et al. (eds.). *Paths towards universal grammar. Studies in honor of Richard S. Kayne*, Washington, Georgetown University Press, 427–451.

Zanuttini, Raffaella (1996). *Negation and clausal structure: a comparative study of Romance languages*, Oxford: Oxford University Press.

Zeevat, Henk (2000). The asymmetry of optimality theoretic syntax and semantics, *Journal of Semantics* 17, 243–262.

Zeevat, Henk (2006). Freezing and marking, *Linguistics* 44, 1095–1112.

Zeevat, Henk and Gerhard Jäger (2002). A reinterpretation of syntactic alignment, in: D. de Jongh, H. Zeevat and M. Nilsenova (eds.). *Proceedings of the 3rd and 4th international symposium on language, logic and computation*, Amsterdam: ILLC.

Zeijlstra, Hedde (2004). *Sentential Negation and Negative concord*, Ph.D. thesis, Utrecht: LOT Dissertation Series.

Zeijlstra, Hedde (2007). Emphatic multiple negative expressions in Dutch, Amsterdam: *ACLC working papers* and lingBuzz/000344.

Zwarts, Frans (1986). *Categoriale grammatica en algebraïsche semantiek: een studie naar negatie en polariteit in het Nederlands*, Ph.D. dissertation, Groningen University.

Zwarts, Frans (1995). Non-veridical contexts, *Linguistic Analysis* 25, 286–312.

Zwarts, Frans (1998). Three types of polarity, in: F. Hamm and E. Hinrichs (eds.). *Plurality and quantification*, Dordrecht: Kluwer, 177–238.

Zwarts, Joost (2003). Paths round a prototype, in: P. Saint-Dizier (ed.). *ACL-SIGSEM workshop: the linguistic dimensions of prepositions and their use in computational formalisms and applications*, Toulouse: IRIT, 228–238.

Zwarts, Joost (2004). Competition between word meanings: the polysemy of (a)round, in: C. Meier and M. Weisgerber (eds.). *Proceedings of SuB8*. Konstanz: University of Konstanz Linguistics Working Papers, 349–360.

Index

M